ETHOLOGY OF MAMMALS

LOGOS PRESS LIMITED
Scientific Publications

Ethology of Mammals

R. F. EWER

Department of Zoology
University of Ghana

Springer Science+Business Media, LLC
1968

Library of Congress Catalog Card Number 68–21946

ISBN 978-1-4899-4658-4 ISBN 978-1-4899-4656-0 (eBook)
DOI 10.1007/978-1-4899-4656-0

Originally published by Plenum Press in 1968.
Softcover reprint of the hardcover 1st edition 1968
Published in Great Britain by
LOGOS PRESS LIMITED
in association with
ELEK BOOKS LIMITED
2 All Saints Street, London N1

CONTENTS

DEDICATION

To the memory of
GOLLUM, GANDALF AND GIMLI,
the meerkats
who first introduced me to
the Fellowship of King Solomon's Ring

LIST OF PLATES

Between pages 50 and 51

PREFACE

"... *a unique pattern of behaviour not seen in the albino rat*"
(1967)

"*In mammals ... biological needs are satisfied by means of complex variable patterns of activity that reflect predominantly the influence of learning and experience. Indeed, in many, it is difficult to define any pattern of behaviour as wholly innate.*"
(Written, 1950; quoted with approval, 1966)

This book was written because I felt it was time, in the light of the recent great outburst of new work on mammalian behaviour, for a reappraisal of the validity of ethological concepts as applied to this group. I wrote it in the hope that it would be of interest to fellow workers, both in summarising factual information and for its theoretical discussions.

After it had been completed, I happened to meet with the passages quoted above; and (slightly shocked that such views should still be current) was very thankful that I had written in a way designed to be comprehensible on its own, without requiring very much previous knowledge of ethological theory or mammalian taxonomy—for, of course, the quotations are not from the writings of zoologists but come from the allied fields of physiology and psychology. If this book can but make clear that there is no such thing as a "typical mammal" and, if there were, it would not be the albino rat; if it can demonstrate that mammals do not live by learning alone; then, surely, a genuine step will have been taken towards the closer linking of neurophysiology with ethology which both sides would welcome but neither is very clear how to attain.

A few explanations and apologies are necessary. Firstly, the primates: they have received a rather meagre treatment. This is because work on them is now so extensive that *primatology* has taken its place beside ornithology, herpetology, ichthyology and the rest, so that primate ethology alone would fill a book and no small one at that. This limitation automatically precludes any extensive consideration of human behaviour and its relations to that of other primates. Those wishing for more information will find an excellent general summary of higher primate behaviour in Chapter 6 of Morris and Morris's *Men and Apes** and a more detailed treatment of selected topics in the volumes edited by D. Morris (1967) and by Altmann (1967). There is a wealth of information about field studies in the symposium edited by DeVore. The two volumes edited by Schrier, Harlow and Stollnitz deal with laboratory investigations. The bearing of ethological studies on human behaviour is treated in many of the articles in the Royal Society symposium on Ritualisation, published in Volume 251 (1966) of the *Philosophical Transactions*, Series B and also in Morris's book, *The Naked Ape*.

Secondly, it is necessary to say a little about attempts to deduce the course of behavioural evolution. There can be no argument about the ultimately unprovable nature of such theories. They cannot even be given the vastly increased plausibility which fossil evidence confers upon theories about the course of structural evolution. Most of us, however, would agree that a theory which accommodated satisfactorily every detail of the behaviour of the relevant extant species would have a high probability of being not very far wrong. Such is human desire to create order, even theories which can offer no more than this as their ultimate goal seem worth striving after. This, of course, is not the only, nor even the major reason for inventing possible phylogenies: it is done because one literally cannot help it. To study live animals is to be amazed again and again by the complexity and the detailed adaptiveness of what they do. This offers an intellectual challenge, for unless the mechanisms which produce the behaviour are explicable in terms of natural selection working in the orthodox manner, we will be forced to postulate special creation or some unknown mystical-magical process. To show that, despite apparent complexity, the observed end result could have been reached, in the course of evolution, by simple steps, involving none of Samuel Butler's "cheating", may not prove that this is how it *did* happen. It does, however, show that there is no need to postulate any unorthodox process to account for its existence. In just the same way, a

* See references for all the works cited here.

neurophysiologist will often embellish his paper with a circuit diagram, showing how the phenomenon he studies could be produced, quite undeterred by the absence of any positive evidence that the neural mechanism concerned in fact bears any resemblance to his diagram. What the latter really signifies is simply "see—no magic". Even molecular biologists —surely engaged in a highly respectable pursuit—have recently taken to speculating about the evolution of the genetic code. Apart from the purely intellectual satisfaction they afford, these theoretical exercises have their uses, for they frequently suggest new experiments or else they direct attention to missing pieces of information.

Thirdly, some explanation is required about the general approach adopted. In certain circles it is regarded as outmoded and old-fashioned to speak of innate (or, as I prefer to say, endogenous) behaviour. This attitude reflects a failure to understand that there can be different ways of studying behaviour, that there are different sorts of question to be asked and that a concept may be more meaningful in one context than in another.

The study of the ontogeny of behaviour is an important one and is continually adding to our understanding of the nervous organisation underlying behaviour. From the point of view of such investigations, there may often be no great point in drawing a clear distinction between highly specific experimental effects and highly generalised ones. However, the situation is very different for those of us whose interest lies in studying the animal as a product of phylogeny as well as of ontogeny, and as a creature adapted to make its living in a certain way in a particular sort of environment. To understand the details of this adaptation it is required to make the distinction between specific experience and more generalised effects. A semi-fictitious example will make this clearer.

Wiesel and Hubel (1965) have found that in the cat, although much of the complex neural organisation of the visual system is present when the eyes first open, subsequent exposure to light is required for its maintenance. Deprivation of the visual stimulation which normally follows opening of the eyes can cause disturbance of this organisation. Since the patterning of the visual stimuli is not critical, this can be described as a generalised or non-specific effect. Whether there is any comparable effect of exposure to smells on the development of mammalian olfactory mechanisms is not known, but, for the sake of the present argument, let us assume that there is such an effect. It is known (see Chapter 6) that an adult golden hamster which has been reared without any experience of dogs or their smell will react defensively if exposed to the smell of a dog, whereas an agouti shows no such behaviour. It would be possible, by appropriate training procedures, to make the agouti react negatively to the smell of a dog.

From the viewpoint of morphogenesis, it may be important to stress the similarities between the early non-specific effects of olfactory stimuli and the elaboration of a later learnt response. In both cases the neurosensory system is being affected by the impulse traffic to which it is subjected, and it is important to determine the mechanisms and processes underlying this effect. On the other hand, the hamster and the agouti are different. The hamster, growing up in any natural environment, will respond negatively to the smell of dog the first time he encounters it. The agouti requires specific experience of a combination of this particular smell with an unpleasant experience before he will do so. This difference is important to the student of the evolution of adaptation. He will enquire about the functional significance of the difference, and will take it that the existence of the specific response in the hamster is the result of a neural mechanism which has been elaborated by natural selection during the animal's evolutionary past history. The difference between the two species he may describe by saying that the response of the hamster is innate, whereas the agouti must learn before it can behave in the same way. By this he does *not* mean that the hamster's response has no ontogeny nor that it is totally impervious to environmental influences, but he does mean that whether specific experience is or is not necessary for the elaboration of a response is a real difference and, from his particular viewpoint, an important one. Similarly an endogenous response to a parental alarm call and one that has to be learnt by association of the call with something dangerous have very different implications for individual survival, regardless of whether the ability to show auditory responses at all requires exposure to noises during earlier ontogeny or not.

Unfortunately it seems to be remarkably difficult to make clear to those who are not in the habit of regarding an animal as first and foremost an adapted product of natural selection the relevance of the distinction I have been attempting to explain. I do not therefore have any great hope of succeeding where others more knowledgeable and more gifted have failed. Fortunately, however, there are advocates more persuasive than any human voice can be. To those who feel that to speak of endogenous patterns does not greatly illuminate their own particular problems, and who have therefore concluded that the concept cannot be meaningful at all, I can only say this. Get a baby mammal—preferably a pair—of some wild species. Hand-rear them, give them the freedom of your house and live with them, watch what they do and how they do it and learn to speak their language. When you have mastered their vocabulary (and it usually takes several years), they will explain it all to you in a way that no member of your own species ever can.

Such, at least, has been my own experience. I literally did not understand Lorenz's writings until their meaning was patiently explained to me, over and over again, by the meerkats to whom, in gratitude, I have dedicated this book.

Lastly comes a point about the use of common names. Even in one's own language, these are not always familiar or memorable; in a foreign one, they are a nightmare. I still remember vividly how long it once took me to find out to what animal the name "ouistiti" referred. A French dictionary was at length found which acknowledged the existence of the word, but I was little advanced to learn that the creature was "the common wistit". Further research was required to penetrate this disguise and find beneath it what I call a marmoset. Because of this, I have included a glossary of common names, with the scientific equivalent and the order and family to which the animal belongs. When mentioning a species for the first time, I have given both scientific and common names; thereafter I have used one or the other, according to convenience rather than consistency. An exception is made for domestic animals and very familiar names, like lion and tiger.

Their scientific equivalents, however, are included in the glossary.

In conclusion, I would like to thank all those who have helped me to write this book. Directly or indirectly, this includes most of my scientific colleagues. Amongst them there are some from whose knowledge of mammals I have particularly benefited and to whom I owe not only information and the stimulus of discussion but hospitality and friendship too. For all these things, I thank Dr. C. K. Brain, Drs. G. K. and P. Crowcroft, Dr. L. S. B. Leakey and Dr. P. Leyhausen.

I should also like to thank a number of people for photographs: the Australian C.S.I.R.O. Wildlife Division for Plates IV and V, Dr. C. K. Brain for Plate VIII, H. F. Von Ketelhodt for Plate I and Dr. P. Leyhausen for Plate VII. I am indebted to a number of authors for allowing me to reproduce figures from their publications: to Dr. H. K. Buechner for Figure 13, Dr. I. Eibl-Eibesfeldt for Figure 8, Dr. P. Leyhausen for Figure 3, Dr. K. Lorenz for Figure 2, Dr. P. Pfeffer for Figure 11, Dr. E. Trumler for Figure 4 and Dr. F. Walther for Figures 6, 7, 9, 10 and 12. Messrs. Methuen and Co. have kindly allowed me to reproduce Figure 2; for Figure 11, I am indebted to the editors of *Mammalia*. All the others are by permission of the editors of the *Zeitschrift für Tierpsychologie*. They are taken from papers which appeared in Beiheft 2 (1956) and in Volumes 15, 16, 20, 22, 23 and 24.

Last but not least, I would like to thank my husband, not only for a critical reading of the manuscript, but also for having shared his home, at one time or another, with a variety of rather demanding and not always domestically convenient fellow mammals.

Legon
November 1967

R. F. EWER

CHAPTER 1

Some basic concepts

The principles on which the ethological approach to animal behaviour is based were worked out mainly from studies on vertebrates other than mammals. More recently, the same approach has been fruitfully adopted in the study of mammalian behaviour, particularly by German zoologists and a very considerable body of factual information has already been amassed. Although new information is still coming in at an ever increasing rate, it seems worth while to attempt some interim synthesis of what is already known; from this to see whether the principles deduced from studies on other groups require any modification, amplification or restriction when applied to mammals; and finally to try to clarify what are the particular features that characterise mammalian behaviour.

Before launching into a summary of what we now know about mammalian behaviour, it is desirable to give a brief outline of the basic concepts of ethology. This and the succeeding chapter are intended as no more than that—an outline of the sort of ideas that underlie the descriptions of behaviour that follow. In Chapter 12 some of these concepts will be discussed more critically.

In animals, structure, behaviour and mode of life form an integrated adaptive unity, a biological trinity, of which no one component is comprehensible except in relation to the other two. Geoffroy Saint-Hilaire, in 1854, gave official recognition to this fact by his use of the name *Ethology* to describe the study of the way in which these three are inter-related. For him, of course, the emphasis was mainly on structure, which he realised could be understood only in relation to habit and habitat and his ethology was, in effect, the study of structural adaptation. Modern ethology concentrates on behaviour rather than structure, but the inter-relations with the other two members of the trinity remain essential. In addition, modern ethology is evolutionary and the trio has become a quartette. Behaviour, structure and mode of life are only partly comprehended if we do not add to them *evolutionary history and relationships with other species.*

Many attempts to give a succinct definition of modern ethology have been made—perhaps as good as any is the semi-serious one; "ethology is behaviour studied by people who love their animals." In other words, an ethologist studies the behaviour of an animal for its own sake and because he wishes to understand that particular animal more fully; not because he believes that his work may throw light on human behavioural disorders, on the rules governing the process of learning or for some other extraneous or semi-extraneous reason. Konrad Lorenz, more than any other single person, has been responsible for the development of modern ethology. He has expressed this attitude very clearly (1950): "No man, . . . could physically bring himself to stare at fishes, birds or mammals as persistently as is necessary in order to take stock of the behavioural patterns of a species unless his eyes were bound to the object of his observation in that spell-bound gaze which is not motivated by any conscious effort to gain knowledge but by that mysterious charm that the beauty of living creatures works on some of us."

This, however, does not mean that ethology restricts itself to the proto-scientific level of observation and description. This leads to the formulation of hypotheses, and ways and means of testing these have to be found. Tinbergen (1963) defines ethology as "the biological study of behaviour". He explains that by this formulation he is defining (a) the object of study—the observable phenomena of behaviour, and (b) the method of study—the biological one, which involves asking and answering questions dealing with immediate physical-physiological causation, with function or survival value and with evolutionary and ontogenetic history. He then goes on to give a masterly summary of the way the ethologist poses these questions and the methods he has devised for finding the answers to them. This paper, together with his earlier one *The evolution of animal communication—a critical examination of methods* (*1962*) should be consulted by anyone who is inclined to feel that ethology can be written off as the semi-scientific amusement of semi-scientific animal-lovers and rather beneath the notice of a serious scientist.

There are, in fact, several fields on which the findings of ethology impinge very closely. The first of these is ecology, particularly those of its branches related to management and control. Since it is through its behaviour that an animal establishes its relations with its environment, including individuals of its own and other species, the relevance of ethology is obvious. Possibly less obvious are the relations between the ethologist and the experimental neurophysiologist. It might seem that this is a one-way traffic, in which the neurophysiologist provides information required by the ethologist to answer his questions about the immediate causes of behaviour, but the ethologist has nothing to give in return. This is not the case, for in physiological investigations of complex central nervous phenomena, particularly those involving brain stimulation by various methods, when a response is given, it is necessary to understand what that response really is. The animal may twitch its tail in a certain way; but is this a friendly greeting, a threat, a courtship gesture or is it maybe a purely artificial product whose exact counterpart is never normally seen? It is the business of the ethologist to provide the answer. The work of von Holst and von St. Paul (1963) is an example of how fruitful the uniting of the ethological and the physiological approaches in this type of investigation may be. One may also quote the way in which Leyhausen's (1956) descriptive studies have formed an indispensable basis for interpretation of the brain stimulation experiments of Brown and Hunsperger (1963).

The observational studies of the ethologist may be essential to the more analytical experimental worker in another way. If the animal's behaviour as a whole is not comprehended, it may be impossible to design a meaningful test or to know whether an experiment covers the whole or only a part of a behavioural phenomenon. For example, to place an animal in a strange environment and record what it does may provide some interesting information: to believe that one is thus investigating more than one aspect of exploratory behaviour is unjustified; to imagine that one can thus assess a rat's readiness to explore (Halliday, 1966a and b) is comparable with testing human eagerness to swim by pushing a number of people in at the deep end of a swimming bath. Similarly to expect that in a strange cage a rat will build a nest is, to any ethologist, so ludicrous that the actual carrying out of an experiment based on this assumption (Riess, 1954) is almost incredible.

Finally, we may find that in our own behaviour we have more in common with other mammals than we had previously suspected. Purely physiological investigations provide information about how various parts of the brain function and may suggest therapeutic methods of treating the central nervous system. An ethological approach, with its concentration on the animal as a whole, on the biological trinity of behaviour, structure and mode of life, may tend to direct attention in a different direction and to ask the question; does man necessarily have to be treated so as to make him fit the society in which he now lives? Should we not also ask whether perhaps the society should be adjusted to fit man?

The "central dogma" of the ethologist has been succinctly described by Tinbergen (1963): that an animal's behaviour is part and parcel of its adaptive equipment and, as such, can be studied in the same way as other aspects of its biology. Its immediate causation is a matter for physiological investigation; its survival value must be studied by systematic observation and controlled testing and its ontogeny and phylogeny dealt with like those of structural adaptations.

Implicit in this formulation is the idea that behaviour is something which an animal has got in the same way as it may have horns, teeth, claws or other structural features. It has not *merely* got a nervous system whose responses can be triggered off and moulded into shape by environmental stimuli. It is true, of course, that much of behaviour is modifiable by experience and that in the life of higher vertebrates, particularly mammals, learning plays an important part. A little reflection, however, shows that learning cannot account for everything the animal does. Learning has been defined as modification of behaviour (excluding the effects of damage) as a result of experience. While this definition may not be very satisfactory (see Chapter 12) it does serve to emphasise that the animal must show some behaviour before it has anything to modify. Moreover, if it did not possess a number of self-differentiating "built-in" patterns of activity, it would never survive long enough to learn anything. It would be perfectly possible to draw up the specification for the design of a relatively simple animal in which all of the behaviour was of this built-in type; indeed Dethier (1962) is of the opinion that in flies, learning ability is negligible; it would, however, be exceedingly difficult to design one possessed only of an ability to learn but devoid of any built-in patterns.

It is, then, one of the basic tenets of ethology that a very great deal of behaviour is, in fact, of this unlearnt, self-differentiating type—and in this, mammals are no exception. This type of behaviour comprises what used to be called instinct, but the term is so broad and general and has had so many meanings attached to it, that it has gradually passed more and more out of currency. One of its main disadvantages is that it tends to be applied to too large a unit; a whole major activity, such as nest building, for instance, being thought of as instinctive. Lorenz has stressed again and again that the inherited built-in elements are much smaller; the unit may be a very simple sequence of movements, or even a single movement. For these elements he has coined the term *Erbkoordination*—literally, hereditary coordination or pattern. The usual English equivalent is *fixed action pattern*, although *instinctive pattern* and *innate behaviour pattern* are also used. In some ways the term pattern is unfortunate. The German *Erbkoordination* can be used as legitimately for a single movement as for a sequence, but pattern would usually be assumed to mean only the latter. In practice, however, fixed action pattern is often used in exactly the same way as its German counterpart to describe either, which may sometimes be confusing. In simple descriptive studies, this is rarely very serious, as the context usually makes clear what is implied and fixed action pattern will be used in this sense in much of what follows. When a distinction is required, however, individual actions, whether occurring in isolation or as parts of a sequence are commonly referred to as *instinctive movements*. In many ways, the term *endogenous movements*, proposed by Lorenz (1950) would be preferable, but it has not become general usage. This name emphasises the fact that during the normal ontogeny of the individual, even if reared in isolation, these movements will differentiate, exactly as structures do: without any need to be learnt, in the sense of being elaborated as a result of external reward or punishment experienced in the situation to which they are appropriate.

One of the characteristics of this type of behaviour is that although it is so clearly adaptive, each pattern normally being performed in the appropriate situation, nevertheless, it also shows a sort of independence of external stimulation which is at first sight somewhat surprising. Innumerable observations and investigations have shown that some type of internal motivation for such patterns exists and that this may build up progressively so that the animal becomes more and more ready to perform the appropriate actions and will do so in response to less and less external stimulation. In extreme cases the pattern may, in the end, be discharged in the complete absence of any of the normal stimuli, as a vacuum activity. This type of response is largely a captivity artefact, since in its natural environment the animal is rarely cut off from places and situations where the normal stimuli are most likely to be encountered. The theoretical importance of vacuum activities, however, is in no way lessened by this, for they provide the clearest direct evidence for the existence of the endogenous motivation which is a key point in ethological theory.

Such a system, in which not only do the patterns of movement develop endogenously but their motivation is also endogenously generated, has obvious survival value. Just as one could not design an animal devoid of built-in behaviour, so too it would be very difficult to design one which could hope to survive if its patterns were merely released by appropriate sign stimuli. The animal would sit where it happened to be, responding to the stimuli that happened to impinge upon it until it presently died. This, however is not what happens. If no opportunity offers for the appropriate discharge of an endogenous pattern then, as its internal motivation builds up, the animal does not necessarily remain passive. Instead, it becomes more active and leaves the place where the requisite stimulation has failed to appear. It may move at random or may direct its course in relation to specific features of the environment or in relation to its own past experience. Even if the animal is not consciously searching for the missing stimuli, the effect of this behaviour is to increase its chances of encountering them and it is therefore known as appetitive behaviour.

Because they have always stressed the importance of internal motivational factors, ethologists have preferred to use the terms *releaser* and *released* rather than stimulus and stimulated when speaking of endogenous movements. The difference is not merely terminological; it signifies a completely different picture of the central nervous organisation underlying the behaviour. To speak of stimulus and response implies an essentially passive central nervous basis, from which no action will emerge in the absence of an external *vis a tergo*; to speak of a pattern being released implies that the basis of the activity is there, inside the central nervous system, waiting, like a car held up at the traffic lights, for the signal to go ahead.

This, in turn, implies a certain organisation of the perceptual apparatus. If each pattern has its own particular sign stimulus or stimuli which serve to release it, then there must exist a mechanism capable of performing upon the multiplicity of incoming sensory data some type of sorting or filtering operation, so that when a particular releaser appears, the information is relayed to the appropriate point for the release of the corresponding pattern. In the ethologist's terminology, each built-in motor element has an accompanying innate releasing mechanism (IRM). Investigation of IRM's has shown that the releasing stimuli normally comprise but a fraction of the total available information and are usually restricted to a few of the major features characteristic of the appropriate situation. Thus, in his classical work on the robin, Lack (1943) found that threat behaviour in a territorial male, the normal response to the sight of a rival, can be released simply by a bunch of red feathers. Similarly I have found that the response of its littermates to the "I'm alone" call of a young meerkat (*Suricata suricatta*) can be released by a cry resembling the natural one only in that it is high pitched and repetitive. This is very much what one would expect of a built-in system—the minimum of complexity that is compatible with ensuring that the response can be relied upon to occur in the appropriate situation with only a remote chance of its occurring elsewhere.

Of course, endogenous patterns do not comprise the whole of behaviour, but ethologists lay particular stress on them for two reasons. Firstly, by virtue of the fact that they are endogenous, they constitute the animal's basic behavioural equipment. They underlie everything else; they provide the basis on which learning can occur and thus they are the elements out of which, during its ontogeny, the animal's total behaviour must be constructed. Secondly, because they, like other self-differentiating characteristics, must be genetically determined, their phylogeny can be studied by the same sort of comparative methods as are used in corresponding anatomical studies and so, by adding an evolutionary dimension, our understanding of behaviour becomes vastly enriched.

This approach has proved extremely fruitful. One of its earliest products was a recognition of the importance of the patterns that belong to what may be termed social behaviour. Many actions are concerned with the animal's social relations with its fellows and, in fact, a whole array of patterns exists whose principal function is to influence the behaviour of a conspecific. A little reflection will make it clear that no form of social organisation could exist if there were not a communication system of some sort between the members of the society; without communication there could be nothing more than an aggregation without social structure. Social behaviour, acting as a means of communication, may thus be extremely important and indeed may constitute a significant proportion of an animal's total behavioural repertoire.

Like other forms of behaviour, social behaviour is a product of evolution. Since its function is to produce specific responses in a conspecific, there will have been selective pressure in the past for making the signals which one individual transmits and another receives as clear and unambiguous as possible. Such communication behaviour should therefore be relatively simple to study; it should be possible for us to eavesdrop, reading the behavioural messages and watching for the answers. It is therefore not surprising that the field of social behaviour is one to which ethologists have devoted much attention and one in which some of their richest harvests have been reaped. Moreover, it is in this domain, if anywhere, that fully endogenous patterns are likely to be encountered. The communication system most effective in maintaining a social organisation, particularly if there is not a long period of protected childhood, will be one in which the significance, at least of the major signals used, does not have to be learnt; one in which the actions of one animal can constitute social releasers for the responses of another. This implies not only that the response must be innate, but that the signals must be standardised and not subject to the vagaries of individual learning. In this field it may therefore be looking at the matter from the wrong angle if we say "this pattern is stereotyped because it is a built-in piece of behaviour"; it may be truer to say that if the exigencies of the situation require that a pattern be stereotyped, then natural selection will have seen to it that it has a built-in basis.

As already noted, most of the work from which these concepts have been derived has dealt with vertebrates other than mammals, particularly with fishes and birds. There is a rather widespread reluctance to believe that they apply with equal force to mammals, a repugnance to the idea that much of mammalian behaviour is unlearnt and its object not necessarily comprehended by the animal itself. This stems partly from the fact that to grant this for our fellow mammals is at once to open the door to the possibility that our own behaviour may not always be as rational as we like to suppose. Naturally, even if this door is only just ajar, our impulse is to bang it shut without further ado. Partly, also, it arises because if a piece of behaviour is clearly adaptive and obviously the sensible thing to do in the circumstances, it is not easy to believe that the performer really does not appreciate the telos of his actions. Yet it is often not difficult to convince oneself that this is so. Often some slight alteration in the circumstances makes the normal pattern inappropriate; a very minor adjustment might suffice to put things right and yet the animal obstinately persists in the old pattern. Alternatively, the changed circumstances may make the whole action pointless, and yet it is still performed. Time and again I have seen what happens if a kitten leaves its faeces not properly buried in its earth box and the mother comes and finds them. She sniffs them, and then tidily scrapes a little earth over them with her paw. If the droppings happen to lie near the edge of the box, so that when she comes up and sniffs them, her fore paws are outside the area of the box a surprising thing happens: she simply makes the usual scraping movements with her paw on the floor alongside the box. All that is needed to make her action effective is a slight change of position—and yet she goes on scraping the floor; if she understood what she was trying to do, she would succeed at once. Even more striking is the fact that if the kitten has failed to use the earth box and defaecated somewhere on the floor instead, then the mother, on smelling the droppings (sometimes even merely sniffing the place where they were before they were cleaned up), will make the same scratching movements with her paw on the floor. She is quite familiar with the floor, she has had plenty of experience of the fact that it remains where it is when scratched—and yet she persists in a pointless action. Clearly her behaviour is a response to the smell of the faeces, not something she has thought out or learnt by experience.

That a pattern should be adaptive without being comprehended is not really surprising. Consider the workings of our own gut musculature or our pupillary reflex. These involve adaptive, coordinated muscular actions, but we know perfectly well that we do not have to say "my breakfast has been in my stomach quite long enough, it is time I opened by pyloric sphincter", nor yet "the light is getting brighter, I had better cut down my pupillary aperture". We accept that these relatively simple muscle coordinations are built-in, but find it more difficult to believe that this could be true also of the more complex actions that comprise what we classify as behaviour. And yet, is it any simplification to say that the animal must have learnt to do thus and thus? If a pattern is performed at all, then the neural basis for it is there, inside the animal. If the pattern is a learnt one, then the animal has found out, by trial and error, what actions are effective in bringing some form of satisfaction or reward; these it repeats, while those actions that were unsuccessful it abandons and does not repeat. Thus the animal selects the appropriate actions so that in the end a perfectly adapted sequence is performed, according to a pattern which must, in some form or other, be now represented in the central nervous system. If, on the other hand the pattern has not been learnt but is built-in, then once again, its basis is there inside the animal—but it got there by a different pathway. During a long evolution there must have been selection of those individuals who dealt with the relevant situation most effectively and most rapidly; the pattern has been perfected in the phylogenetic history of the species in a manner at least analogous to the way in which the learnt pattern is elaborated during the ontogeny of the individual.

It is my belief that the relationship between learnt and built-in patterns is often closer than mere analogy. I find it difficult to believe that the neural basis of a complex set of actions should be built up out of chance mutations when in fact there is an apparently simpler route, by way of the process which Waddington (1953) has called genetic assimilation. Selection does not have to wait for a neural pattern to turn up complete before it can start to operate; it can work on the speed and ease with which the solution to a commonly recurring problem is found. Let us take a specific example. In animals with hair, toilet behaviour is very important and most mammals care for their coats very assiduously, each in its own characteristic manner. A regular toilet is performed, whether the coat requires it at the moment or not—just as we are told that Queen Elizabeth I took a bath every month, whether she needed it or not. Mammalian toilet patterns are certainly not learnt; right from the start a little kitten washes its face in the typical cat manner, using the inside of one forearm and licking the cleaning surface after every few wipes. Even before its eyes are open and while its balance is still so imperfect that it often falls over, a baby giant rat (*Cricetomys gambianus*) washes its face with the palms of both hands simultaneously and again licks the cleaning surface between wipes. Neither animal has to go through a series of trials to find out how best to clean itself. How has this come about?

Reptiles do not perform elaborate toilets, but they will wipe off a piece of dirt adhering to the skin with hind or fore foot as appropriate. Presumably the early mammals could do as much. With the development of hair, as skin care becomes increasingly important, there will first be selection for increasing responsiveness, so that cleaning is not restricted to the removal of the immediate irritation but is extended to deal with other parts of the body. At the same time there will be selection for quickly finding the most effective ways of cleaning. The result of these selective pressures will be an increasing readiness to respond in a particular and relatively complex way to the relevant stimuli and step by step a whole pattern of toilet procedures will be elaborated, with trial and error in finding the best way to clean becoming more and more abbreviated. The final stage will be reached when the actions, now so readily and easily performed, cease to need an external trigger in the form of a definite skin stimulus and acquire a motivation of their own. We do not yet know enough about the physiological basis of motivation to say what changes are necessary to reach this final stage, but it is not difficult to believe that the genetic alterations required may be relatively small. Similarly, licking of the cleaning surface itself must have begun as a method of getting rid of a piece of dirt which had been transferred from face to paw and only later have become an essential part of the procedure.

Since the kitten and the baby giant rat wash their faces in different ways, we must be able to account for this difference too. The adaptive significance is simple. The cat's paw, with its close set pads and long claws (not yet retractile in a young kitten) are quite unfit for use as face cloths, while the much flatter palms of the rat are perfectly suitable. The cat therefore must needs use a different surface, not its paw but its forearm. Having made this change, one sided cleaning must follow; if both arms were used at once, the paws would foul each other and in addition, the shortened reach makes it desirable to turn the head towards the cleaning arm. The kitten's ancestors may have had to find this out the hard way, but once more, selection for increasing ease of learning this particular "right" way of doing it would gradually assemble the neural arrangements required for doing so with greater and greater facility, until finally it just "comes naturally", and no trials are required because no errors are made.

This, however, is pure hypothesis and the fact remains that for many mammalian patterns all we know is that they are characteristic of the species rather than the individual; no detailed investigations of their ontogeny or of the ways in which they may be affected by or interwoven with learning having yet been made. Since we lack the necessary knowledge to do otherwise, it will be necessary to begin merely by describing the sort of patterns that mammals do show and trying to understand their adaptive significance, without prejudice as to how they came by such behaviour. We will return later (Chapter 12) to this problem and discuss what is actually known about the interweaving of learning and built-in patterns in the few cases where detailed investigations have been made. In the meantime, the term *fixed action pattern* or simply *behaviour pattern* will, of necessity, be used to denote any action or action sequence which is performed in a well-nigh identical form by every member of a species (or at least by all of one sex), regardless of whether we do or do not know anything about its ontogeny.

So widespread, however, is resistance to the idea that mammalian behaviour, like that of other vertebrates, is based on endogenous movements, that it does seem necessary at this point to ask ourselves one question. As a working hypothesis, is it better to assume that mammalian behaviour is built out of the same elements as that of other vertebrates, or should we take it that their behaviour is constructed on a radically different plan? It seems to me more likely that the complexities of mammalian behavioural organisation will prove to be explicable in terms of the way the basic elements are built into the total fabric of behaviour, rather than that these complexities imply, within the vertebrates, a fundamental neurophysiological dichotomy, for which there is no concrete evidence and against which the whole of evolutionary theory cries out in protest.

CHAPTER 2

Expression and communication

Anyone who keeps a cat or dog will tell you that from its expression he knows when his pet is miserable or happy, frightened or angry, playful or just peacefully content. Darwin (1872) describes vividly the change in expression shown by his dog when the animal realised that his master was setting forth from the house not for a walk, but merely to visit the hothouse. "His look of dejection", Darwin wrote, "was known to every member of the family, and was called his *hot-house face*. This consisted in the head drooping much, the whole body sinking a little and remaining motionless; the ears and tail falling suddenly down, but the tail was by no means wagged. With the falling of his ears and of his great chaps, the eyes became much changed in appearance, and I fancied that they looked less bright. His aspect was that of piteous, hopeless dejection . . . every detail in his attitude was in complete opposition to his former joyful yet dignified bearing."*

Darwin called his book *The expression of the emotions in man and animals* and there can be very few people who have attentively watched the behaviour of even a single species of mammal who do not believe that the actions they observe are accompanied by vivid emotions. Most would concur with Heinroth's opinion that their animals are "exceedingly emotional people possessing very little intellect". However this may be, the study of behaviour is an objective science, concerned with the observable phenomena of what the animal does. As such, it cannot deal with emotions, whose existence can only be inferred and whose qualities must remain inaccessible to us. It is, however, an objectively determinable fact that the readiness of an animal to behave in particular ways varies from time to time and that these changes are reflected in visible movements such as alterations in the set of its ears

* Quoted from the "Popular Edition" of 1904.

and tail, degree of erection of the hair and so on. If we use the the term *mood* to describe these changes in responsiveness we can define what we mean in a perfectly objective manner. If we have found out from experience that an animal showing a certain type of behaviour is in a condition where it is very likely to attack we are justified in describing it as being in an aggressive mood. Conversely, if we know that the animal's present expression is usually followed by flight or other purely defensive actions, we may properly say that it is in a fleeing or defensive mood. The behavioural changes may then be said to reflect the animal's mood. The word *tendency* is often used in a rather similar manner. If, for instance, the animal behaves in a manner which experience has shown signifies that it will now be very easily put to flight, we may say that it is showing flight tendency or that its flight tendency is activated. Its expression at any moment may thus also be described as mirroring the tendency (or tendencies) which are at present activated. In practice, however, the consistent usage of such objectivistic terminology often becomes exceedingly cumbersome; indeed, the effort to avoid the use of words like anxiety, fear, rage and the like can result in formulations which become frankly ludicrous. Such words are in fact justifiable, so long as our use of them is based on what the animal actually does; frightened, for example, is a word that describes an animal ready to take to flight and anxious denotes a lower intensity of the same condition. We must, however, remember that our own feelings in a comparable situation are not necessarily exactly the same as those of the animals we study. Fear, for instance, may seem to us to be a single emotion but Lorenz (personal communication) believes that since a hen uses totally different vocalisations in the face of ground predators and birds of prey it must experience two qualitatively different sorts of fear in the presence of the two sorts of enemy.

Mammals have structural characteristics which permit them to show a greater range of expression than any other animals. Their ears are mobile; the shape of the mouth, eyes and nostrils can be altered by muscular action; the hair can be erected or sleeked down; the angle at which head and neck are held can be varied; the articulation of the limbs is such as to permit the carriage to change from stiffly erect to a low crouch (see Figure 1); the tail is usually not directly involved in locomotion and can be moved in all directions (indeed, in a number of species its only function is to act as an organ of expression). The changes in demeanour which can thus be made are not

Figure 1. Ground squirrel (*Xerus erythropus*). (a) Alert: body carried high with fore limbs extended; tail over back but with hair only moderately erected. (b) Intense anxiety: crouched close to the ground; tail over back with maximal piloerection.

irrelevant incidentals to more important activities; many of them can be shown to constitute an important part of social communication behaviour. It is no mere coincidence that dogs, whose wild relatives have a rich and complex social life, should have a wide range of expressive movements, while bears, with a very meagre social existence should have very little facial expression, apart from a warning glint of the whites of the eyes when attack is imminent. This is why bears have the reputation of being "treacherous", but it is not very logical to blame them for it; since they are not social, they do not possess an extensive expressive vocabulary and have no need to preface their attack by obvious indications of hostile mood. The overtone of moral indignation implied in our use of the word "treacherous" is itself of some interest.

It no doubt arises partly from our being so accustomed to the clear warning signals given by dogs that we take it for granted that some such preliminary notification of intention to attack *ought* to be given.

Another factor which greatly increases the range of expression shown by mammals is the fact that their moods are not always simple; more than one tendency may be simultaneously activated. Introspectively we know that this is true of ourselves—we may be angry and at the same time afraid; at once attracted and repelled; and irritation and affection frequently war within us. In this we do not seem to be unique; other mammals are subject to the same sort of complexity of motivation and they are capable of expressing this complexity in their behaviour. There may be a set of expressions that denote readiness to attack and another set that signify readiness for flight, both showing a graded series from just perceptible to maximal. The two sets can be combined, usually in the form of a compromise, occasionally by rapid alternation, and so a whole new series of expressions denoting varying combinations of aggression and fear is produced.

Lorenz (1963) has analysed canine facial expressions which combine attacking and fleeing tendencies in this manner. Aggression is signified by baring the canine teeth and drawing the nose up so that the snout is thrown into wrinkles; the corners of the mouth are not drawn back and the ears are directed forwards. The opposite submissive or cowed tendency is shown by pulling the corners of the mouth back and laying the ears back, the eyes at the same time becoming somewhat narrowed. Lorenz's figure shows these two tendencies combined at various intensities: the resulting expressions (see Figure 2) will be familiar to anyone who has kept a dog.

Leyhausen (1956) treated the attitudes and expressions of combined aggressive and defensive tendencies in the cat in the same manner (see Figure 3). The familiar arched-back threat is a classical example of mixed motivation and expresses simultaneous maximal activation of attacking and defensive tendencies. The erected hair, dilated pupils, open mouth and laid back ears reflect defence, the stiff legs and raised tail signify attack. In this posture the expressions of the two tendencies are not equally distributed over the animal's body;

its head, closer to the enemy, expresses almost pure defence; its hindquarters, further from the danger, show more aggression. As a result the animal tends to withdraw its forequarters and at the same time advance its rear, so that its back is thrown up into the characteristic arch and the whole body is turned broadside on to the enemy.

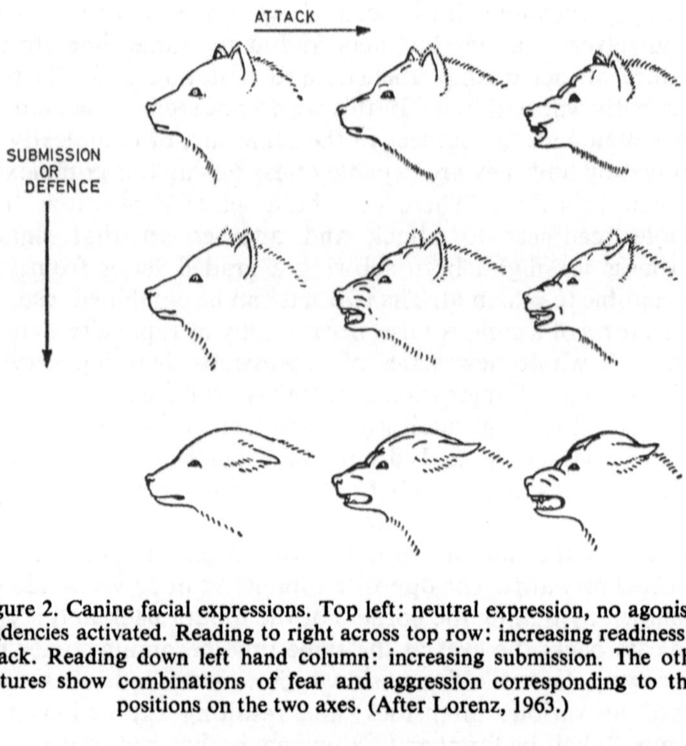

Figure 2. Canine facial expressions. Top left: neutral expression, no agonistic tendencies activated. Reading to right across top row: increasing readiness to attack. Reading down left hand column: increasing submission. The other pictures show combinations of fear and aggression corresponding to their positions on the two axes. (After Lorenz, 1963.)

It should be noted that the two tendencies activated in the cat and the dog are not quite identical. The cat's expressions are a combination of attacking and defensive attitudes; those of the dog are better described as combining aggression and submission. These terms may need some explanation. "Defensive" implies a condition where although no hostilities will be initiated, any attack that is made by the opponent will be countered; "submissive" implies the absence of any tendency to fight, even if the other party initiates an attack.

The difference relates to the modes of life of the wild relatives of the two species. The dogs are highly social animals, which implies that after a hostile encounter the defeated animal must still remain associated with the victor. This entails his being able to indicate his inferiority without actually fleeing while the victor must accept the surrender and refrain from

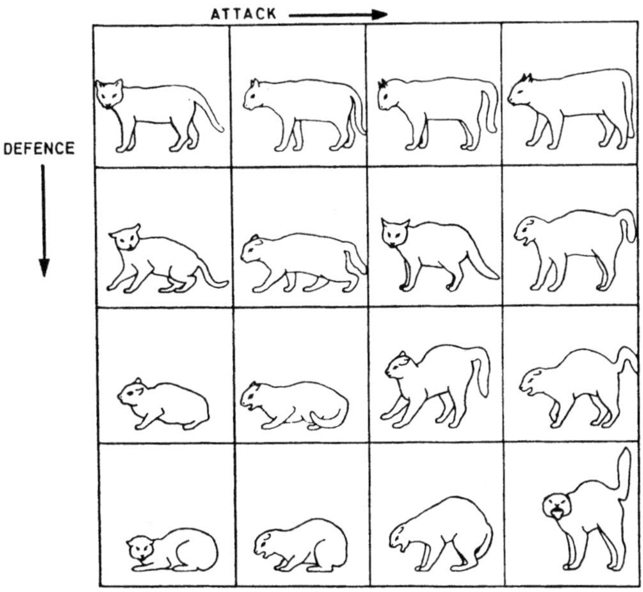

Figure 3. Moods expressed by body position in the cat. Conventions as in Figure 2: increasing attacking and defensive tendencies shown along horizontal and vertical axes. (After Leyhausen, 1956.)

further attack. The cat, largely solitary, has no such need. The vanquished contestant's aim is only to hold his attacker at bay until an opportunity for escape offers and so his expressions are defensive rather than submissive.

An expressive movement may originate in various ways, but if it once comes to act as a signal which is responded to by other members of the species, then a new selective force comes into play. If it is advantageous for a quick and correct response to be made, then there will be selection for making the signal more obvious and unmistakable. The changes which occur as a result of this selective pressure constitute what is known as ritualisation. Ritualisation of the movements involved may include making them more obvious by increasing their amplitude or by reiterating them. Concurrently other evolutionary changes may occur. On the one hand there may be evolution of structural characteristics which serve to draw attention to the signal by making it more visible or by adding an audible or an olfactory component. On the other, the central nervous basis of the ritualised signal may change; the motivation bringing it into play may alter and it may become a pattern in its own right with its own independent motivation, a process which Tinbergen (1952) has called emancipation. If this happens, then the original meaning may cease to have more than a symbolic value, or even be so obscured that only by the study of a series of related species can it be deduced.

A few examples will make these processes clear. In baboons (and other monkeys and apes), presenting the hindquarters as though inviting copulation, does not necessarily have a sexual connotation. This gesture serves symbolically as an indication of readiness for social contact, of friendliness; a response by a brief "token" mounting is a reply signifying "I accept your offer of friendship" (Heinroth-Berger, 1959). In deer of the genus *Rusa*, male threat includes lifting the lip in a manner which would be incomprehensible if we did not know that primitive deer, like the muntjac (*Muntiacus*), have long canine teeth and use them in combat and that they threaten by baring these weapons (Eibl-Eibesfeldt, 1957). The horse shows an intermediate condition (Trumler, 1959). Here only the male has canines, but both sexes have a threat expression which includes opening the mouth widely and pulling the corners up and back, so as to display the "weapons", whether these are really there or not.

Equine expressions also exemplify the principle of superposition of opposing tendencies. In threat, not only is there the ritual display of "weapons", but the ears are laid back, protecting them from a possible counter-attack. Horses, according to Trumler (1959), also have a distinctive "greeting face". Here the mouth assumes exactly the same menacing shape, but the threat is contradicted by the ears, which are turned forward towards the partner, indicating readiness for social contact (see Figure 4). The expression as a whole reflects the fact that in any encounter with a fellow there is an element of danger and the message conveyed is a double one which is the equivalent of saying: "I approach you with an offer of

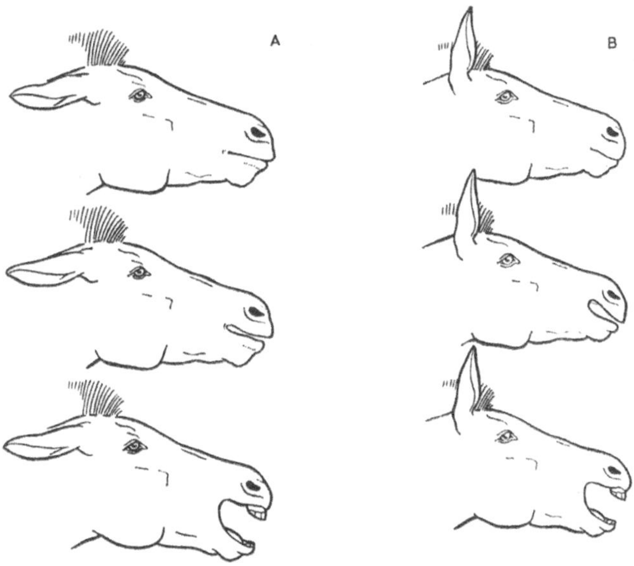

Figure 4. Facial expressions in the horse. (a) Threat: three stages of intensity. (b) Three successive stages as the greeting expression is assumed. (After Trumler, 1959.)

friendship, but do not interpret this as a sign of weakness on my part; I am perfectly able and ready to defend myself." In all Equidae that have been studied, with the exception of the horse, there is also a special "mating face", an expression assumed by the female when she permits the male to mount. This is identical with the threat face, but the facial expression is in this case contradicted by the attitude of the hindquarters, which are raised to invite copulation and the tail is turned aside. Trumler is of the opinion that the carriage of head and neck is also less elevated than in genuine threat, but the information available is not sufficient for this to be quite certain. The total message conveyed here is thus: "If it were not for the fact that I am ready to copulate, I would attack you."

In a number of rodents, a state of high general excitement is accompanied by small shivering movements of the tail, which do not appear to have any particular significance. In the mouse similar movements occur, but they are of larger amplitude and as the tail moves against the substratum, can produce a rustling sound which constitutes part of threat against a rival; the same is true of *Meriones persicus*, the Persian gerbil (Eibl-Eibesfeldt, 1951a). In the giant porcupine, *Hystrix cristata*, and in the smaller brush-tailed *Atherurus africanus*, tail shaking threat is much more impressive. In both genera the signal is made obvious by the development of special spines on the tip of the tail, which produce a rattling sound and which are white. The signal is audible anywhere quite independent of the type of substratum, moreover location of its source in the dusk is facilitated. The sound-producing spines have a thin base and at the tip a dilated hollow region. The sound is produced by these hollow tips rattling against each other. The detailed construction of the sound-producing dilations is totally different in the two genera, but the shaking movement is the same—a rapid small amplitude vibration, carried out with the tail held aloft. This all suggests that the movement acquired its characteristic form before the evolution of the specialised spines and that these represent independent solutions to the problem of how to change an ordinary defensive spine into a more effective sound-producing apparatus. In these porcupines no great excitement is necessary to produce tail rattling; it is the form of threat with the lowest threshold and may be used at low intensity to indicate no more than slight suspicion or uneasiness; the message which it then conveys is no more than "Watch out; remember, I am armed." The motivation of the pattern has thus changed; some emancipation has occurred and the high state of excitement in which the movement originated is no longer required to produce it. Porcupine tail rattling is thus an example of a ritualised pattern showing both emancipation and the evolution of structural adaptations increasing the effectiveness of the signal. Moreover, the response is probably innate at least in *Hystrix*, since it is shown by animals only a few days old, before the quills have developed sufficiently for any sound to be produced (Roth, 1964).

In *Thryonomys swinderianus*, the cane rat or grass cutter, and in *Myoprocta pratti*, the green acouchi (D. Morris, 1962), tail lashing also occurs, but here its significance is not threat but exactly the opposite; it is used by the male when courting the female and indicates that his approach is not aggressive but sexual. In the acouchi the tail is very short, but is embellished with a terminal tuft of white hairs and is held aloft when signalling. In the cane rat the tail is longer and naked; it is not held up, but is directed towards the female as it is lashed with a movement that is mainly from side to side. Although its significance is similar in these two species, the differences in the details of the movements suggest that they represent unrelated parallel evolution. In *Thryonomys* there is a further complication. I have seen youngsters tail wag when sniffed at, or sometimes when merely approached from the rear by an adult and also when taking food from directly under an adult's nose. The signal thus seems, in this species, to have acquired a generalised significance as a friendly or appeasing gesture.

Although tail lashing may thus have two distinct meanings in different species of rodents, it is easy to see how both could have been derived from a movement made originally as an incidental accompaniment of a high level of excitement, since this occurs both in sexual and hostile encounters.*

The same origin may also underlie the different meanings carried by tail wagging in cats and dogs. An encounter with an unknown conspecific is always fraught with tension and is a situation where incidental tail movements might be expected to occur. In the non-social cats, such encounters are more likely to be antagonistic than friendly and the tail signal has in them come to denote hostility; in the social dogs, the reverse has occurred and tail wagging has been ritualised into a friendly signal.

The most complex signalling systems are intra-specific, since it is with conspecifics that relationships are most varied

* Since this was written I have seen vibration of the tail occur in both types of situation in the black rat, *Rattus rattus*.

and complex; but inter-specific signals also exist. The commonest message that has to be conveyed to a member of another species is a socially negative one, designed to prevent a closer, probably dangerous approach; in other words, a defensive threat "Don't you dare touch me."

The response to such a warning could be based on learning if the species delivering the threat were the stronger, so that effective punishment followed failure to react by avoidance. Many warnings, however, are effective against a stronger species and many are made by species without effective weapons and are thus essentially bluff. In any case, a warning to which there is a built-in response would always be more advantageous than one whose significance had to be learnt. It is therefore not surprising that such inter-specifically valid signals are usually based on a few simple innate responses which appear to be of very general occurrence, not only in mammals, but within the vertebrates as a whole. The warnings based on these responses constitute a sort of universal code, an "esperanto of expression", as Leyhausen (unpublished) has called it.

One type of stimulation which evokes an innate response of this type is a sudden increase in the apparent size of an object. Presumably the basis of this lies in the fact that without very good binocular vision a sudden increase in size has the same optical effect as a very rapid approach. According to Leyhausen, we ourselves, despite our binocular vision, have this response; it is possible to produce a startle reaction in a human subject by getting him to watch the image of a dark spot on a light screen and then suddenly increasing its size. This principle is widely used in mammalian threat; sudden erection of the hair gives an abrupt increase in apparent size and many species possess especially long hair on certain regions of the body with the function of intensifying this effect. *Proteles cristatus*, the Aardwolf, for instance, whose threat is largely bluff as its dentition is very reduced, has very long hair on the tail and along the back and shoulders and this is erected in intimidatory threat (see Plate I). The effect of hair erection is often enhanced by the animal's turning broadside on so as to display its maximal dimensions to its enemy.

Two types of auditory threat appear to rest on a similar innate response basis. Any sudden explosive noise is likely to evoke an avoiding reaction, as witness our own tendency to start back if a gas fire gives a loud pop when we light it, or air in the pipe makes the water emerge in a sudden spurt when we turn on a tap. Sudden explosive spitting is a very common defensive warning in carnivores, especially *Felidae* and *Viverridae*, and it can be extremely effective. I have seen a large dog draw back hastily when a very small kitten "exploded" in his face in this way and it is surprisingly difficult to inhibit a withdrawal response from even an obviously harmless baby carnivore when it greets one's approaching hand with the defence spit. Defence spitting is not restricted to carnivores; it also occurs in some rodents and even in a marsupial, the tree-kangaroo, *Dendrolagus leucogenys*. A sudden hissing noise is also used as a defence sound in a number of marsupials and in the platypus, *Ornithorhynchus anatinus* (Tembrock, 1963a).

High-pitched rustling noises also are very generally intimidatory. Young kittens, for instance, are thrown into a frenzy of defensive responses by the rustling of tissue paper; a rustling noise also evokes a defensive avoiding reaction in ground squirrels, *Xerus erythropus*, and I have found that a young kusimanse, *Crossarchus obscurus*, flees to cover at once at the same sort of sound. Porcupine tail-rattling is an example of threat used both inter- and intra-specifically which is based on this response.

One case of a special signal with the function of counteracting this response is known. Lorenz and Schleidt (quoted by Busnel, 1963 and personal communication) have found that in the bank vole, *Clethrionomys glareolus*, which lives colonially, commonly amongst dead leaves, the effect of a rustling sound is to send the animals dashing to the safety of their holes. There is, however, no response to the rustling made by fellow members of the colony as they move about among the dry leaves. This was shown to be because every animal, before it moves, emits an ultra-sonic cry which inhibits the alarm response of its fellows and thus acts as a sign that the ensuing rustle is not dangerous. This cry is therefore functionally comparable with an aircraft recognition signal, carrying the message "Don't worry, it is one of us." This behaviour is also of interest in relation to the problem of the evolution of echolocation. A species which is sensitive to high frequencies, which reacts to the cries emitted by its fellows and itself emits an ultrasonic cry before it moves, has clearly got all the necessary equipment for echolocation: all that is now required is for it to attend also to the echoes of its own cries and begin to associate their characteristics with subsequent encountering of obstacles.

In view of the usefulness of signalling behaviour, it is easy to appreciate that if once signs of some sort are made, selection should act upon them to make them more obvious and more distinctive. Why any sort of signal should come to be made in the first place is, however, not so apparent. If the animal is about to attack, retreat, mate or whatever it may be, why does it not simply go ahead and do so? Tail rattling threat has been traced to an origin in a movement that was a mere incidental result of a high general level of excitement—but it is too much to ask that some such incidental should conveniently chance to turn up in every case and moreover, a different one for each eventuality. The answer to this problem lies in the fact that in the situations where signalling behaviour is highly developed, motivation is rarely simple. In many interactions between individuals there is simultaneous arousal of tendencies both to advance and to hold back or withdraw, so that there is some sort of conflict within the central nervous system between the mechanisms governing the various possible courses of action. A conflict situation may also arise if there is but a single motivation, but its expression is in some manner frustrated; maybe by the inaccessibility of the objective. In such circumstances animals commonly do one of two things. They may make as though to act and then stop; in fact, they may thus make alternating intention movements towards the various courses of action which might be appropriate to the situation. But they may also do something else. They may suddenly perform some action whose normal context is quite foreign to the situation—a displacement activity. Thus a cat, approaching a window to go out but finding it closed, will often give its face a few brief paw wipes, and I have often seen a *Cricetomys* do displacement face washing when she had climbed to a high ledge and was in difficulties about getting down again. Similarly, a ground squirrel, when she was cautionsly approaching an unfamiliar object and anxiety was warring with her need to investigate, made displacement burrow digging movements. These two types of behaviour—intention movements and displacement behaviour—so commonly seen in conflict situations—provide an ample source of raw material upon which selection can then operate and from which a more complex ritualised signal may be evolved.

The few examples which have been given will serve to illustrate the sort of results produced by the evolutionary process of ritualisation. In what follows there will be abundant evidence of the importance of expression and of ritualised signal movements in the social life of mammals; the different functions these signals serve and the different evolutionary origins from which they may stem will also become clearer.

CHAPTER 3

Food

1 FINDING FOOD AND FEEDING

The fact that mammalian teeth, jaws, jaw muscles and associated features of skull architecture reflect adaptation to particular feeding habits requires no stressing; it is a commonplace of elementary zoological teaching. What is not usually included in traditional textbook treatments is the fact that such structural adaptation presupposes appropriate behavioural adaptation. In the course of evolution, structural change must have been accompanied by the selection of corresponding behaviour patterns; indeed, grounds can be adduced for believing that behavioural evolution led the way (Ewer, 1960). It is therefore not surprising to find that most mammals show highly characteristic methods of feeding and of obtaining their food, as well as an ability to select the type of food normal for their species.

What is surprising is how little we actually know about mammalian feeding behaviour. In many cases all we know is what constitutes the normal food. In others we also know, in greater or less detail, the patterns involved in finding food and eating it, but very rarely has description been followed by analysis and experiment to elucidate the factors responsible for releasing and orienting the animal's responses. The same is true of the ontogeny of the patterns and the way in which learning and innate components are combined in the experienced adult. To take an example. Everyone knows that, unlike most herbivores, the elephant's neck is so short that it cannot reach up or down to gather food with the mouth and that this deficiency is made good by the possession of a long and mobile trunk, which is used to pluck food and carry it to the mouth. I do not, however, know of any detailed description of the movements used in dealing with food, nor of an analysis of the importance of tactile, chemical and visual stimuli in directing feeding activities, still less of any ontogenetic study or any assessment of the role played by learning in the development of skilled trunk movements.

Food choice is highly specific only in a few very specialised forms, such as the koala, *Phascolarctos cinereus*, which feeds exclusively on the leaves of various species of *Eucalyptus*. Most mammals are less demanding and may be found eating a greater variety of plant species or killing a variety of animal species as prey; moreover food may vary in different areas or at different seasons. Goodall, for instance, reports that the chimpanzees, *Pan troglodytes*, she studied in the Gombe Stream Reserve refused to eat paw-paw, with which they were not familiar. Kortlandt, on the other hand, found that in a different locality the chimpanzees regularly raided paw-paw plantations (verbal communications 9th International Ethological Congress). Bartlett and Bartlett (1961) found that black rhinoceros, *Diceros bicornis*, normally browsers, were able to subsist as grazers in the Ngorongoro crater, where their normal food was not available. Even ant-eaters do not live soley on ants; Krieg and Rahm (1956) record beetles and millipedes in the stomach contents of wild *Myrmecophaga* and a tame one ate fruit and meat. *Tamandua*, whose staple food is termites, will also eat fruit while the pangolin, *Manis tricuspis*, in captivity takes readily to a diet of scrambled egg flavoured with formic acid (Sikes, 1962). Indeed, most mammals can learn to accept foods which they would not encounter in natural conditions. Early experience may be important here, since a young animal is usually more easily induced to accept strange foods than is an adult, whose preferences are, to some extent at least, already determined. Dog keepers will know that if a puppy is brought up with a proportion of vegetable foods in its diet, these are readily accepted, whereas a grown dog reared on meat alone may refuse them even when very hungry. Similarly, even the much more strictly carnivorous cat can become surprisingly omnivorous if introduced to a variety of foods when young. This does not, of course, imply that adult mammals are incapable of changing their feeding habits; although they may be less flexible than the young and the process may require some time, most species show some degree of adaptability in this respect. For instance, I have found that in captivity adult cane rats gradually extended their choice of food over a period of several weeks and began to eat a variety of herbs and shrubs which were at first refused. When maize grains were first offered, they were accepted, but eaten last and without great enthusiasm. Within a week or two, however, this became the favourite food and was always taken first. Cattle also may be induced to change their feeding habits. In certain areas of the Eastern Cape Province of South Africa cattle are raised in areas where there is virtually no grass. Here they feed on the leaves of shrubs and bushes and have become browsers instead of grazers. If a cow is transferred to such an area from grassland, it at first becomes very thin and may take a few months before it adapts itself fully to the new circumstances, which entail a change not only in the food eaten, but in the manner of cropping it.

It is, of course, their adaptability in this respect that makes mammals important as agricultural pests. If man cultivates a rich crop of some nutritious food stuff, it is only to be expected that some of his fellow mammals will presently take advantage of it. Rodents are the most notorious as is sometimes reflected in their common names: the cane rat is so called because of the damage it does to sugar cane plantations, and in West Africa the local ground squirrel is known as the ground-nut thief. A more surprising instance of a new food habit is the way the black-backed jackal, *Canis mesomelas*, has recently taken to raiding pineapple farms in South Africa.

Amongst ungulates whose limbs are adapted for running, patterns of manipulation of the food are necessarily very simple and relate mainly to the way tongue, lips, incisor teeth and neck muscles cooperate. The ability of reindeer to clear away the snow to expose the underlying vegetation is an unusual complication. Other herbivorous species with less cursorial specialisation may show greater complexity, since the fore limbs may be used to assist in feeding. Most rodents, for instance, show highly characteristic methods of manipulating their food, which are constant within the species. To gnaw small objects such as seeds, nuts or fruits requires that they be held firmly while the incisor teeth go to work upon them. The exact details of how this is done vary from species to species and are related to the degree of manual dexterity, which in turn is correlated with general mode of life. Species which climb or burrow have relatively mobile paws and a considerable degree of pronation and supination is possible. Those that are purely terrestrial have less mobility in the fore limb. It is therefore not surprising that the purely surface living guinea pig hardly uses its paws at all in feeding. In natural conditions the vegetation is nibbled as it stands, its own roots holding it firm so that pieces can be held in the lips and teeth and pulled off with a backward jerk of the head, but when dealing with pre-cut food in captivity, the guinea pig is a very clumsy feeder. The paca, *Cuniculus paca*, is also unable to use its paws in feeding (Kunkel and Kunkel, 1964) but this limitation is unusual amongst rodents. In other species the usual habit is to hold the food between the two paws, although the exact details of how this is done vary from group to group. The feeding posture of the squirrel, sitting up on its haunches and gripping its food between its paws, is well known (see Plate II). Many murids can adopt a similar posture but may also eat in a lying position, leaning on their elbows. The brushtailed porcupine, whose spines make sitting up on the haunches impossible, habitually rests its elbows and forearms on the ground and holds its food between its two paws. The cane rat, when it eats small objects such as maize grains, does not sit on its haunches, but holds the food in one paw as it gnaws and supports itself on the other three legs. To deal with larger objects which require two hands to hold, it does sit up, to a variable degree—sometimes with the fore limbs barely clear of the ground and never as erect as the squirrel. Although, like a guinea pig, this animal is predominantly a grass eater, its method of dealing with this type of food is completely different. The grass is first cut through close to the ground with the extremely powerful incisors. It is then picked up in the teeth, the animal rises on its haunches and grasps the grass on either side of the mouth with its paws and again slices it in two. As it is cut, the paws are brought together in front of the chest and the two cut ends may be fed into the mouth and rapidly chopped into pieces with the incisors or one piece discarded and the other held with one or both paws and eaten in the same way. After an inch or two has been taken into the mouth there is a pause while it is chewed with a sideways grinding of the cheek teeth and then swallowed. Then the incisors take over again and the cycle is repeated.

Although holding the food in the paws is the norm in rodents, only a few species actually use the paws to pick up their food; the majority use the lips and teeth. The species known to use the paws in picking up their food are as follows: the hare-mouse, *Lagidium peruanum;* the pacarana, *Dinomys branickii;* the coypu, *Myocastor coypus* (Kunkel and Kunkel, 1964); the pocket mouse, *Perognathus pacificus* (Bailey, 1939) and the kangaroo rats, *Dipodomys* (Shaw 1934; Culbertson, 1946).

It may be tempting to assume that these various patterns of food manipulation are simply the result of the animals learning what is most convenient and that it is merely uniformity of structure that produces uniformity of feeding methods within a species. Few systematic investigations on the ontogeny of rodent feeding patterns have been made, but enough is known to show that, in certain cases at least, there is more to it than that. The young guinea pig does not try to hold its food in its paws, find itself in difficulties and give up the attempt; it feeds in the adult manner right from the start. Young *Xerus*, on the other hand, when they first start to eat small pieces of solid food, pick them up and hold them in the typical squirrel manner. In fact, young which I hand reared from a stage when the eyes and ears were still closed, started to hold their food in this way at a stage when their motor coordination was still so imperfect that they often fell over and could have managed more easily by leaving the food on the ground and steadying it with one paw (as is, in fact, done with large pieces of food). A young *Cricetomys* starts to feed before the eyes are open and it does so holding the food in the paws in the typical rodent manner and, like the squirrel, while it still has difficulties in maintaining its balance. Similarly a young cane rat would try to manipulate grass in the characteristic "chew through the middle and then feed in both ends" pattern before it was able to make the cut through the centre effectively. In these cases it is clearly neither learning nor the direct effect of body build that determines the details of the motor pattern. No doubt the specific patterns have been evolved because, in relation to general structure, they are the most convenient and efficent, but the individual does not have to find out by trial and error how best to manipulate his food; his feeding pattern appears to be just as much a genetically controlled developmental phenomenon as is the structure with which it ultimately forms an integrated adaptive unity.

Many rodents open hard shelled nuts and seeds and eat the kernels. Their techniques for nut-opening also show specific differences. The squirrel's method is unique: once a hole has been made, the incisors are inserted and with a quick twist, the shell is split in two (Eibl-Eibesfeldt, 1963). Learning plays a part in perfecting the technique (see Chapter 12), but the characteristic splitting movement appears to be innate. The smaller mice and voles simply gnaw a hole and enlarge it until the kernel can be got at and scraped out piecemeal, but even here, there are specific differences. The field mice, *Apodemus sylvaticus* and *A. flavicollis*, both relatively long armed species, hold their nuts well away from the body and after the initial hole is made in the top, they insert the lower incisors and enlarge the hole, working at the far side of the opening and gnawing from inside to outside. The bank vole, *Clethrionomys glareolus*, much shorter in the arm, holds the nut close to its chest and inserts the upper incisors and so works at the near side of the hole, gnawing from outside to inside (Petersen, 1965). These differences are clearly correlated with the length of the arms, which limits how the nut can be held. What remains obscure is why a third species of *Apodemus*, *A. agrarius*, with arms capable of holding the nut like its congeners, in fact utilises the same technique as *Clethrionomys*. Petersen studied young individuals of all three species and although the animals learnt how to find the best place to make the hole, their basic techniques appeared to be innately determined and trial and error was not shown.

Amongst the Primates the correlation of feeding techniques with fore limb structure is even more clearly shown. Although the extremely mobile fore arm and grasping hand are primarily an adaptation to climbing, their potentialities are extensively exploited in feeding. The hands are used not only for holding food and picking it up, but also for plucking fruit, breaking off leaves etc. and the grasping foot may also be used to assist.

Goodall (1963) describes the feeding behaviour of wild chimpanzees and notes that here, in contrast to the rodents, individual feeding mannerisms are common, a reflection of the fact that learning plays a more important part in the lives of higher Primates than it does in other mammals. Furthermore, Goodall found that the chimpanzees actually use tools in food capture. At the start of the rains, when winged termites are preparing to leave the nests, the chimpanzees are able to extract them using a grass stalk or a small twig. The tool is broken to a suitable length and leaves or small side branches are trimmed off. One of the exit passages in the termite mound is then scraped open with the index finger and the stalk inserted, left for a few moments and then withdrawn with the termites clinging to it by their mandibles. It is then drawn sideways through the mouth and the termites picked off. On one occasion, when no suitable stalk was to hand beside the nest, a chimpanzee walked over to a clump of grass, carefully selected several stalks and returned with them to the termite nest, and twice a chimpanzee was seen to carry a tool for over half a mile, inspecting termite heaps as he went, apparently searching for one that was ready for working. Goodall (1964) also saw sticks used as tool for extracting ants from their nests. Beatty (1951) records seeing a chimpanzee using a stone to crack palm nuts against a flat piece of rock. The chimpanzee brought an armful of nuts to the flat rock, sat down and proceeded to crack them one by one by pounding them with a stone. Here, as in Goodall's case, a considerable amount of insight is involved, for the tool and the object on which it was used were not originally side by side, but had to be brought together by the animal; an operation which would seem to demand some degree of forethought.

Amongst carnivorous species behaviour concerned with food getting is naturally more complex than in vegetarians, since killing and some form of hunting are involved. Three phases may be distinguished, each of which has its own distinctive patterns: (i) approach to the prey, which may involve hunting, (ii) killing and (iii) eating. To deal comprehensively with all of these would require a volume in itself. In what follows a few examples will merely be taken, illustrating two general principles: firstly, that there are distinctive patterns involved in every aspect of catching and eating the prey; secondly, that the differences between different species are adaptively related to other aspects of the animal's mode of life.

The species about which we know most is the domestic cat. Leyhausen's monograph (1956) deals extensively with this species and compares it with other Felidae and also treats, to some extent, of other families of Carnivora. Much of what follows is drawn from this work, which may be consulted for a more detailed treatment than is possible here.

Cats stalk their prey, using a series of distinctive movements. When first alerted to the presence of prey at some distance, the cat crouches and then hurries towards it with the body flat to the ground in what Leyhausen graphically calls the slink-run. At a distance determined by the available cover she pauses and "ambushes", crouched low with the whole of the sole of the foot on the ground and the fore paws supporting the body directly under the shoulders, the whiskers spread and the ears turned forwards. For a few moments she watches the prey, her head turning as she follows its every movement, as though her eyes were tied to it by an invisible cord. Depending on distance and cover, a second slink-run and ambush may follow, or she may now stalk the prey, moving forward slowly and cautiously, to the last piece of available cover and here again she ambushes and prepares for the kill.* The heels are now raised from the ground, the hind legs shift back and may make alternating movements, while the tip of the outstretched tail twitches. From this posture the final attack is launched—not usually as a single leap but as a short run, flat to the ground; the final "spring" too is flat to the ground and is, in fact, a thrust rather than a jump. While the fore quarters are thrust forward to seize the prey, the hind feet do not leave the ground, but remain firmly planted, giving stability for a possible struggle to follow.

The individual actions of this hunting sequence mature independently of prey catching and may be seen in the play of kittens before they have yet encountered live prey. Moreover, they are so essentially part of prey catching behaviour

* Similar stalking in a kusimanse attempting to catch a small bird was graphically described by our Ghanaian cook: "he made himself small small and he was walking with his stomach."

that they will be performed even when prey is given on a bare floor and the ambushing and slink-run give no concealment. If a cat has no opportunity to hunt genuine prey, then the ambushing, slink-running and stalking patterns will be directed to substitute objects, a fly, a dead leaf, a ball of knitting wool. In extreme deprivation they will even be carried out as true vacuum activities.

A number of other felids show the slink-run, ambushing and stalking patterns: Leyhausen (1956) records them in the serval cat, *Felis serval.* This type of hunting is adapted to catching small rodents that live in burrows. With such prey it is essential for the predator to remain concealed long enough for the prey to get well away from the burrow; a premature attack will merely send it scuttling back to the safety of its hole. It is, however, not suited to the capture of small birds, for these do not usually remain on the ground very long and the prolonged ambushing gives them time to fly off again. Most cats are not very adept at catching birds, although a few do learn to modify their technique and become quite skilled at bird killing. A tom cat, for instance, learnt to catch small birds by concealing himself in long grass and remaining motionless until a bird came close enough to be captured with a single quick dash; in other words, he learnt to inhibit all but the last moves in the normal sequence. The ocelot, *Leopardus pardalis*, is a specialised catcher of small birds and does not ambush but attacks directly the instant the prey is visible. Leyhausen records that his ocelot never ambushed, even when given mammals as prey.

The most common killing technique is a bite delivered in the region of the nape of the neck. The canine teeth are driven in and the cervical spinal cord, or even the hind brain, is cut through and death is almost instantaneous. Leyhausen (1956) investigated this response, using as test objects rats which were either normal, headless, or with the head removed and stitched onto the posterior end. His results show that in cats the bite is aimed wherever there is a constriction, which in the normal prey is, of course, the place where the head joins the body. With long-necked prey, the first bite is made in the shoulder region, near the base of the neck and a second bite is then delivered just behind the head. This two-bite technique is used by a wide variety of felids when killing birds. Leyhausen records it in the fishing cat (*Felis viverrina*), ocelot, lynx (*Lynx lynx*), caracal (*Felis caracal*), serval, puma (*Felis concolor*) and cheetah (*Acinonyx jubatus*). The simple neck bite is not peculiar to the Felidae but is used also by mustelids and viverrids and by some canids, for instance the fennec, *Fennecus zerda*, (Gauthier-Pilters, 1962) when killing small mammals. It is clearly a very basic mammalian pattern, for it is not restricted to the Carnivora but occurs also in rodents and insectivores and I have seen it used by marsupials—the Tasmanian devil, *Sarcophilus harrisi*, and the little mulgara, *Dasycercus cristicauda*. Wild rats will kill a mouse with a neck bite (Eibl-Eibesfeldt, 1958) and Arvola et al. (1962) report a female lemming, *Lemmus lemmus*, killing foreign youngsters in this way. The little insectivorous grasshopper mouse, *Onychomys torridus*, when given a pocket mouse, killed it instantly with a bite at the base of the skull (Horner et al., 1964). The short-tailed shrew, *Blarina brevicauda* (Herter 1957) and the Turkestan desert shrew, *Diplomesodon pulchellum* (Heptner, 1939) also kill with a neck bite.

Amongst the Canidae, the usual killing techniques are slightly different. Species like foxes, that kill prey smaller than themselves, use a bite in the region of neck or shoulder combined with a violent side-to-side shake. The raccoon dog, *Nyctereutes procyonoides*, also uses this method. Like the neck bite, the death shake occurs in a number of different groups. It is common in viverrids and is also found in marsupials: both *Dasycercus* and *Sminthopsis crassicaudata*, the fat-tailed marsupial "mouse", will shake invertebrate prey. Amongst the Insectivora shaking of prey has been reported in a number of species: in the hedgehog *Erinaceus europaeus* (Lindemann, 1951), in *Blarina* and the tenrec, *Tenrec ecaudatus* (Herter, 1957), in the Turkestan desert shrew (Heptner, 1939) and in the otter shrew, *Mesopotamogale ruwenzorii* (Rahm, 1961).

Both the neck bite and the shake are thus widely distributed and they occur both in marsupials and in the viverrids, the least specialised of the Carnivora. Moreover, together they are effective without requiring a high degree of development of the canine teeth or a very powerful and accurately oriented bite. The selective advantage of the neck bite is obvious, but the origin of the death shake is less simple. Leyhausen (1965) says that according to his observations, the death shake does not usually break the neck of the prey but, by upsetting the labyrinthine reflexes, reduces its ability to fight back effectively. He considers that the death shake has been derived from a movement in which the prey is gripped, thrown to one side by a quick sideways jerk of the head and then bitten again before it has time to recover its balance. The throwing movement itself he regards as arising from a conflict of tendencies to seize and at the same time to avoid prolonged contact with the prey. A reduction of the latter tendency might then lead to retention of the grip and exaggeration of the throwing movement into the typical shake.

It is, of course, quite likely that the death shake has been evolved more than once and while Leyhausen's explanation may be valid for the Canidae, it does not appear exactly applicable to the smaller carnivores or to the marsupials. *Suricata* and *Crossarchus*, the two viverrids with which I am most familiar, kill a mouse (for them, relatively large prey) by a neck bite without associated shake. Small invertebrates dug out of the ground, however, are usually given a quick shake as they are picked up. The shake forms so essential a part of the pattern of dealing with small prey that it is performed even with dead food, such as pieces of meat or egg. Indeed, a tame *Suricata* always shook a piece of jelly, regardless of the fact that such treatment was not merely unnecessary but positively unsuitable for use with this type of food.

The function of the shake, as shown by these species, is not entirely clear. Its normal role might be to free the food from adhering soil particles or maybe to assist in dismembering it in the same way as Heptner (1939) describes a beetle being shaken to pieces by the Turkestan desert shrew. Some observations on *Dasycercus*, however, suggest another possibility. In this species, as in the viverrids, a mouse is killed by a neck bite, without a death shake. Small and harmless insects such as mealworms, little grasshoppers and crickets are grasped in one paw, pulled towards the mouth and seized in the jaws or sometimes taken directly in the mouth. On the other hand, larger and more dangerous creatures, such as scorpions and big centipedes, are attacked by a direct bite combined with vigorous shaking. The shake here prevents the prey, not easily immobilised by a single bite, from getting a grip on its attacker's snout with one of its numerous appendages. This I believe to be the primary role of the shake in dealing with invertebrates and it is not difficult to imagine it arising simply from the effort to dislodge the hold of appendages which have actually succeeded in clutching the face. In these circumstances, shaking the head would seem to be a very natural response and from this to shaking immediately before a grip can be established, is a minor change. In *Dasycercus*, shaking the prey is less fully built into the feeding pattern than it is in the two viverrids and it does not appear when dead food or quite innocuous prey is being dealt with: the same is true of *Sminthopsis*. The shake is, however, sometimes used by *Dasycercus* when killing a snake. Snakes are seized somewhere towards the anterior end, but the orientation is not precise. If the grip is not sufficiently far forward to immobilise the snake's head, then the bite is accompanied by vigorous shaking, which effectively prevents any striking back.

If this is indeed the origin of the death shake in smaller species, then it is not very different from what Leyhausen has postulated for the canid death shake. In both cases the origin lies in a movement originally used in dealing with prey which, although relatively small, may be capable of fighting back. Whether the centrifugal force generated acts directly or by affecting the semicircular canals of the prey is immaterial, as far as the predator is concerned: what counts is that it does prevent striking back. It is not difficult to believe that shaking of the prey might have originated with this function more than once. It has very probably been evolved in *Dasycercus* and *Sminthopsis* separately from the viverrids and may quite possibly have arisen independently in the Canidae also. I would, however, interpret the canid "throwing away" movement in the opposite way from Leyhausen, regarding it as an incomplete shake, rather than interpreting the shake as an elaboration based on the throwing away movement. The behaviour of *Dasycercus* towards a scorpion is possibly relevant here. If the initial bite and shake do not quickly immobilise the victim, then it is dropped and a second bite-and-shake administered. Although I have seen many scorpions despatched by *Dasycercus*, only once was a "throwing away" movement performed. This was exceptional and what actually happened was that the grip on the prey was released a little prematurely, coinciding with the end of the shake, instead of in the normal way, after its completion.

Shaking is, of course, also used in a slightly different way by dogs when eating, as distinct from killing. Combined with a sharp backward tug, a shaking movement assists in tearing off a lump of flesh from large prey, if a simple bite fails. A dog "worrying" a carcase is using the head shake in this way and the same movement is often used in play with inanimate objects. *Suricata* and *Crossarchus* as well as the Tasmanian devil, will also use this trick in fighting play with an object sufficiently large to offer resistance—a towel, a duster or a human playfellow's hand. Even fish will shake lumps of food to assist in getting a manageable piece loose, so this type of shaking is not peculiar to mammals. Whatever its origin, the death shake has been almost completely lost in the Felidae and this is related to the development of extremely effective canine teeth and a very accurately oriented killing bite. With this type of killing, the death shake is not merely unnecessary; it would be positively disadvantageous as it would interfere with the accurate insertion of the canine teeth.

It seems likely, on general grounds, that prey killing patterns were first evolved in relation to prey smaller than the predator. To kill animals larger than itself presupposes that the killer is already highly efficient. Amongst both the Felidae and the Canidae there are species that have evolved methods of dealing with prey larger than themselves, but this has happened quite independently in the two groups and the techniques are very different.

In the Felidae, with canines capable of delivering a single lethal bite, the problem in dealing with large prey is to get it into a position where the bite can be made effectively. There is no authoritative account of prey killing in the large felids such as lions and tigers but the following information is based on statements made by highly competent observers. All are agreed that there is considerable variation in technique. According to Dr. L. S. B. Leakey, a lion frequently kills by leaping up, grasping the shoulders of the prey with one paw and dragging its nose downwards with the other. The animal then falls headlong, and may break its neck in so doing, or may be finished off with a neck bite. Sometimes, however, the second paw grasps the neck or throat instead of the nose and the animal is pulled down and killed as before. Leyhausen (personal communication) emphasises that the prey is always pulled down, so that it falls towards the attacker; it is never knocked over by the impact of the lion's body hitting it. This, of course, recalls the fact that the domestic cat's final "spring" is not really a jump and the hind feet do not leave the ground.

A recent study made by Kruuk and Turner (1967) in the Serengeti area is in good agreement with these accounts. According to these authors, the lion, like the domestic cat, makes good use of cover in its preliminary stalking of the prey and the final rush is usually short. If the prey is not overtaken within 50–100 metres at most, the pursuit is generally abandoned. The prey is grasped from the side or rear and is dragged down, with the paws clutching it by chest and back, or back and flank. Death usually seemed to be by suffocation and the neck was not broken in any of the eight kills witnessed. No organised cooperation was seen and Kruuk and Turner were of the opinion that when lions hunt as a group the "cooperation" shown is no more than a matter of one animal taking advantage of game put up by another, just as may be done with prey startled by a motor car.

Eloff (1964) describes a modified technique used by lions in the Kalahari desert to deal with the long-horned gemsbok *Oryx gazella*. Two fresh kills examined had the back broken between the last lumbar and first sacral vertebrae and in an unsuccessful attempt he saw a lion attack by leaping at the buck's haunches. In both kills the break was described as being "upwards, and not downwards as one would perhaps expect it to be if it was caused by the impact of the lion's heavy body landing on the gemsbok's back". According to the Game Conservator, the Kalahari lions normally attack by leaping at the hind quarters; the lion then "digs its teeth deep into the haunches and with a jerking motion upwards breaks its victim's back at what appears to be a weak spot in its vertebral column". Dr. G. Schaller (personal communication) says that tigers show great variation in their techniques. The prey is brought down by leaping at any part of the body, from shoulders to hind-quarters, as opportunity offers. Once down, the prey is usually gripped by the throat and the head held down so that the animal is unable to rise. The tiger then simply waits and presently the prey dies, apparently largely as a result of strangulation. There is almost no movement of either attacker or victim, once the latter has been brought down and the whole process appears remarkably peaceful and quite unlike popular ideas of a vicious and bloody struggle.*

The cheetah, renowned as the fleetest of all mammals, uses a different technique, in which the pursuit is more prolonged and stalking plays no part. In its preliminary approach to its prey, the cheetah depends not on concealment but on the fact that most animals do not at once take flight on noticing the approach of something possibly dangerous. They may become alert but do not flee until the object of their attention comes within a certain minimal "flight distance". This distance depends on the species and on the animals' previous history of liability to, or immunity from attack. It also depends on the speed with which the object is moving: the more slowly it approaches, the closer it may come before evoking flight. The cheetah's hunting technique utilises this last characteristic. It usually makes no attempt at concealment but simply walks towards its prey, in full view. The moment the prey breaks and runs, the cheetah launches into its famous gallop. The chase may cover a few hundred metres and if successful, it ends with the prey being knocked down with a fore paw and bitten in the throat or muzzle.

Another characteristic of the cheetah's hunting method was very clearly shown in the Disney film *African Lion*. The cheetah does not necessarily pursue the nearest prey but the one that first takes to flight and, having once started a

* A detailed account of these observations has now been published, see Schaller (1967).

pursuit, it follows its chosen victim and is not deflected from it, even if other members of the herd run across its path. The importance of this ability to concentrate on one prey and refuse to be diverted from it is familar to anyone experienced in de-fleaing a dog or cat by hand: if you come on two fleas at once, your only chance of killing either is to follow the cheetah's example and concentrate on one.

Before leaving the Felidae, mention must be made of a curious, specialised technique which Dr. C. K. Brain (personal communication) has seen used by a serval cat in killing snakes. The snake's head was crushed by a downward slapping blow of the outspread fore paw, delivered with quite surprising violence. If, as sometimes happened, the first slap missed its mark, the blow was so hard that the serval sometimes hurt its paw. This method of dealing with venomous prey, keeping the face well away from the source of danger, is an interesting parallel with the secretary bird's technique of stamping a snake to death with its feet. Its use by the serval, however, is not restricted to snakes, for Leyhausen (1965) has seen exactly the same method used in dealing with rodents, particularly hamsters, whenever the prey showed fight.

In the Canidae, where the canines are less specialised and the neck bite less accurately oriented, a different method of dealing with large prey has been evolved. Typically, there is pursuit by a group of animals and the prey is brought down and killed by an accumulation of minor wounds. Wolves and the African hunting dog, *Lycaon pictus*, leap up and bite at the hindquarters or belly until the prey falls. No specific killing bite follows, but in *Lycaon* the point of attack, once the prey is down, is usually the belly (Kühme, 1965a). This piecemeal killing is, of course, particularly suited to group pursuit, where if one individual tires before the kill is made, the efforts of the next may finish the job. According to Kühme, a single *Lycaon* is usually unable to make a kill unaided since, by the time he has overtaken the prey he is too exhausted to leap up and deliver a sufficient number of bites. The normal hunting procedure is for each member of the group to start chasing its own prey. The individual pursuits are, however, abandoned as soon as it becomes apparent that one member is closer to making a kill than his fellows and all then join to pursue and bring down this one victim.

Kruuk and Turner (1967) have also studied *Lycaon* and their account is generally in agreement with Kühme's findings. The prey is usually first detected by sight, from distances of at least 2 kilometres. The preliminary approach is at a trot, with pauses now and then. At something under 500 metres there is a sudden change: the dogs now proceed at a slow slinking walk, with the head held low in line with the body and they usually adopt a single file formation. Exactly like the cheetah, they approach their prey slowly and in full view and break into a run only when the prey's flight distance is transgressed and the herd turns and flees. Kruuk and Turner agree with Kühme that once one dog is clearly nearest to success, the rest abandon their individual chases and join in pursuit of this one victim. There is no question of an organized relay system with the individual dogs taking the lead in turn and in the chases witnessed, the original lead dog almost always made the kill, even though the chases were sometimes extremely long. The record seen was a kill after a chase of 13 kilometres (just over 8 miles) but this was exceptional and the kill was usually made within half this distance. As the leading dog finally begins to close in, the prey often starts to dodge and may attempt to break back; the dogs behind the leader now play a role, heading it off and preventing any escape and they also assist in the final kill, once the leader has felled the prey by a bite in the flank.

Although hyaenas are highly adapted as scavengers, they are also capable of killing for themselves. Their nearest relatives are the viverrids, in which although the neck bite is commonly used, it is less highly oriented than in the Felidae. The techniques used in killing large prey are therefore of some interest. According to Dr. E. S. Higgs (personal communication) the striped hyaena, *Hyaena hyaena*, in North Africa will frequently kill donkeys, using a neck bite. Leakey confirms the use of the neck bite by this species but says that the spotted hyaena, *Crocuta crocuta*, kills more after the manner of the Canidae, leaping up at the prey and biting repeatedly, at the belly. Kruuk's studies (1966) confirm and extend this. According to him, crocutas may hunt alone, in small groups or in packs of up to thirty. Solitary hunts are mainly after small prey, are usually undertaken in daylight and are rarely successful. Kruuk notes that the death shake is often used in killing small prey but he does not mention the orientation of the bite. Pack hunting is a nocturnal occupation and prey as large as wildebeest are pursued, the technique used being much like that of *Lycaon*. Each crocuta chases individually, springing up and biting at flanks or legs. As soon as one animal is wounded enough to slow it down, it is surrounded by the whole group and killed by numerous bites, mainly directed at the hinder end. These observations are of considerable interest since they show that, within the family Hyaenidae, the neck bite, the death shake and group pursuit with biting at the hinder end are all used, depending on the circumstances and on the type of prey.

Kruuk worked in Serengeti and Ngorongoro. In these areas, the crocutas did little scavenging and were, in fact, one of the major primary predators. In over a thousand observations of feeding crocutas, 11% of the kills had been made by other species; 82% had been killed by the crocutas themselves and 7% were doubtful. Indeed, in Ngorongoro, where lack of cover makes it difficult for lions to make a kill, the usual order of things is reversed and the lions get most of their food from kills made by the crocutas.

In the Carnivora, in addition to patterns concerned with catching and killing the prey, there are others concerned with starting to eat it. The domestic cat always eats a mouse from the head down. Even a young cat, given prey which it has not itself killed will do this. Leyhausen (1956), in an extensive series of experiments, has investigated the factors responsible for this orientation. Half grown rats and "skin sausages", consisting of a rat's skin filled with chopped meat, were used in his tests. Using a series of combinations in which either the head, the body or both were skinned, the tail was either present or removed, and the head was either left in place, removed or attached to the posterior end, he demonstrated that the orientation is not visual but tactile and the directing factor is the lie of the hair on the prey. In dealing with small birds, the lie of the feathers plays the same role but here, working "against the grain" may lead either to the head, or to the base of the wing and, in fact, a cat may start to eat a bird at either the head or the wing base.

Other small carnivores, such as viverrids and mustelids, also eat the prey from the head down and Leyhausen (1956) records that a coati, *Nasua rufa*, did the same. The greater cats, however, are dealing with prey too large for this method of eating and their usual habit is to start at the belly or groin. According to Lindemann and Riek (1953) wildcats do the same with large prey. Leyhausen (1956) records that his tree-ocelot, *Leopardus wiedi*, when eating a rabbit almost as big as itself, made repeated efforts to start at the head and only bit elsewhere after finding it impossible to break the skull bones. In this connection the behaviour of a male *Crossarchus* in dealing with a mouse is of some interest. Starting at the head presents great difficulty, mainly because the unspecialised carnassial teeth do not readily cut through the skin of the neck. Repeated bites are made at the head and the skull is crushed, but still the skin remains intact. After each attempt the mongoose lets go of the head and makes a tentative bite at the belly. As it does so, it moves its nose along the body of the prey (a movement Leyhausen also describes in cats) and at once the mongoose's snout shifts forwards, travels up the body and a new attempt on the head is made. The whole performance suggests that the animal is trying to start eating at the soft abdomen but is prevented from doing so by its response to the lie of the hair, which directs it back again to the head. When presented with a skinned but otherwise intact mouse, this animal was completely disoriented. He first bit here and there in a totally random manner and finally started to eat from the belly, thus confirming the importance of the lie of the hair in orienting the eating pattern in this species.

I have also found the skin to be essential for the orientation to the head in two dasyurids, *Dasycercus* and the Tasmanian devil. Both ate normal prey (mice and rats respectively), whether they had killed it themselves or had been given it already dead, from the head down, but began at the belly or groin of skinned prey. The devil was given a rat's skin stuffed with minced meat, with the head and tail removed and the two ends stitched up again. He smelt it, his nose at once travelled forward and there was then an abrupt check as he reached the place where the head should have been. He returned to the belly and repeated the move forwards. Finally, although clearly somewhat disturbed by the curious consistency of the skin sausage, he ate it from the front. This need not represent independent evolution, since the habit of working against the lie of the epidermal structures in finding the place to start eating is also shown by reptiles. It was probably originally evolved in relation to scales and, with the evolution of the mammals, was simply used in the same way in response to the lie of the hair. Placentals and marsupials may well have independently made the translation from scales to hair.

Since the response to the lie of the hair is so widespread, it would be extremely interesting to know whether it has been lost in the large cats or whether at first, like Leyhausen's ocelot, they try to start at the head and have to "unlearn" this response before they are able to show the adult pattern of starting at the groin or abdomen.

Although the smaller viverrids are able to kill prey of moderate size, as already described, many of them live principally on insects and other small invertebrates dug out of the ground and have appropriate food finding behaviour. In most Herpestinae the digging is done entirely with the claws; although the animal sniffs as it digs, this is only concerned with the detection of prey. Amongst the Viverrinae, *Genetta* and *Viverricula*, when they remove insects from the soil, do so with the snout, not the claws. *Crossarchus obscurus* is intermediate: it digs mainly with the claws, but also uses the snout to assist in moving the soil away. These differences all relate to mode of life. The more arboreal Viverrinae have claws which are adapted for climbing, not for digging, but the terrestrial mongooses are poor climbers and their fore limbs are primarily adapted for digging. *Crossarchus* again is intermediate: it lives in forested areas and is a good climber, although it does most of its food finding on the ground. Of all the Herpestinae *Suricata* is the most adapted to an open semi-desert habitat and almost all its food is found by digging. In captive animals this pattern, lacking its normal outlet, is discharged by quite amazingly persistent scratching at every available hole, crevice or cranny.

A number of these small viverrids also have special behaviour patterns for dealing with hard shelled objects such as molluscs and eggs. The marsh mongoose, *Atilax paludinosus*, grasps them in the fore paws, rears up and throws them vertically down and repeats the process until they break. *Herpestes pulverulentus*, the Cape grey mongoose, hurls them backwards between the hind legs, while *Mungos mungo*, the banded mongoose (Dücker, 1957), may use either type of throwing. The crab-eating mongoose, *Herpestes urva*, also uses the throwing technique to break hard shelled objects. All these species will perform their shell breaking patterns with substitute objects when the normal ones are not available. This is recorded for *Mungos mungo* by Kinloch (1964) and for *Herpestes urva* by Brownlow (1940). A pair of *H. pulverulentus* which I once kept would throw small stones and I know of a pet *Atilax* that would pick up any electric light plug left out of its socket and throw it on the floor with great persistance until it succeeded in smashing it. The force with which the throwing is done is quite surprising; in captivity both an *Atilax* (Steinbacher, 1951) and a *Mungos* (Davis, 1966) succeeded in breaking open nuts in this manner. An interesting parallel to these viverrid patterns is found amongst the Mustelidae in the spotted skunk, *Spilogale putorius*. The technique described by van Gelder (1953) is very like that of *Herpestes*, with the added complication that as the egg is thrown backwards, it is speeded on its way by a quick kick from one hind foot.

For the sea otter, *Enhydra lutris*, breaking open mollusc shells is a more difficult problem, since there is no hard object available against which they can be thrown and smashed. It is solved in a surprising manner. Fisher (1939) saw the otters dive and come up carrying a flat stone, together with some hard-shelled object, probably a small bivalve. The otter then turns on its back, places the stone on its chest and uses this as an anvil on which the mollusc is pounded until it is smashed open. The species being so treated could not be identified, but it was something small enough to be held in and largely concealed by one paw. The red abalone, *Haliotis rufescens*, which is one of the otters' favourite foods, is not, as reported by Bourlière (1955), treated in this way; since the shell is widely open below there is no necessity to smash it. Fisher, however, reports that the abalones are always brought up with a large piece broken out of one side of the shell. How this is done underwater is not known, but it is presumably incidental to whatever method the otter uses for prizing the mollusc loose from the substratum.

Perhaps the most highly specialised of all prey catching techniques is that of the insectivorous bats. Echolocation is used not only to guide the animal in its nocturnal flight, so that it does not fly into obstacles, but also to locate its prey. Most of the work on echolocation has dealt with the characteristics of the ultrasonic pulses emitted by different species and the problem of exactly what parts of the theoretically available information provided by the echos are actually used by the bats. Griffin and his co-workers have, however, found out a number of interesting details about the actual prey capture. When the bat is merely cruising, its pulses are emitted at a frequency sufficient to ensure obstacle avoidance, but when an insect is located the interval between successive pulses is shortened as the bat closes in on its prey, thus increasing its accuracy in keeping track of the moving prey (Griffin, Webster and Michael, 1960). Highspeed filming has shown that the bat's orientation does not have to be so accurate as to bring its jaws directly in contact with the prey—the latter may be "fielded" with a wing tip or with the inter-femoral membrane and then transferred to the mouth (Webster and Griffin, 1962). Griffin, Friend and Webster (1965) have shown that bats, *Myotis lucifugus*, which had been trained to catch mealworms thrown in the air for them can distinguish these from an inanimate object which gives back an echo of the same strength as a mealworm. Suthers (1965) has shown that the fishing bat, *Noctilio leporinus*, which takes small fish from the water surface with the enlarged claws of its hind feet, can locate a small piece of fish barely projecting from the water surface. A ripple made by a just submerged fish on an otherwise calm surface would also give back an echo. The experimental bats would eagerly strike at small disturbances deliberately created by the investigators, so it seems that in natural circumstances the disturbance caused by a fish without its actually breaking the surface may be sufficient to disclose its whereabouts to the hunting bat. Indeed, the bats Suthers studied were capable of such fine discrimination that they learnt to take a piece of fish held below the surface with its position marked by a fine wire projecting 5 mm from the surface.

Schnitzler (1967) has recently shown that rhinolophid bats utilise a technique which has not so far been recorded in vespertilionids. As they approach an object, the pitch of the sound emitted is altered in such a way as to compensate for the Döppler effect caused by the fact that they are moving towards the target. The result is that the return echo remains at the normal sending frequency for which the animal's ear is attuned. The bat, in fact, is performing a sort of auditory focussing on the thing that matters most, allowing the less vital background to become "blurred". Schnitzler's bats were able to keep focussed on a swinging pendulum and there is no reason to doubt their ability to do the same with a flying insect; this, however, has not yet been demonstrated. How many species other than the two studied by Schnitzler also make use of the Döppler effect in this way remains to be discovered, as do the details of how the bat's central nervous system computes distance from the sensory information received. Clearly bat hunting techniques are by no means "worked out" and may yet hold some surprises for us.

2 FOOD STORING

Food is, of course, not equally abundant at all seasons and many mammals have special behaviour patterns concerned with food storage; a few even process the food before they store it. Food storing, normally referred to as "hoarding", reaches its highest development among the rodents, but it is by no means confined to this Order, for Carnivora, Insectivora and one group of Lagomorpha also do it.

One may distinguish two main types of hoarding: storage inside the animal's home and storage elsewhere. The former is commonly practised by species that have a protected lair, a burrow or hole of some sort and is referred to as "larder

Plate I. Aardwolf (*Proteles cristatus*): piloerection in defensive threat. (*Photo* T. Taylor.)

Plate II. Ground squirrel (*Xerus erythropus*): typical eating posture. The white eye- and body-stripes are very striking but nothing is known about their function.

Plate III. Meerkat (*Suricata suricatta*): male cocks leg to mark the entrance to the sleeping box with the secretion of the anal glands.

Plate IV. Red kangaroo (*Megaleia rufa*): female cleaning pouch shortly before parturition. (*Photo* E. C. Slater.)

Plate V. RED KANGAROO (*Megaleia rufa*)

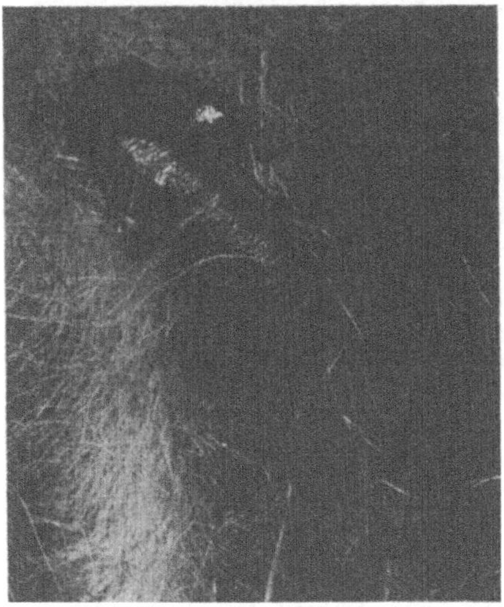

Newborn young making its way to the pouch.

Journey's end; note the disproportionately large fore limbs and nostrils (*Photos* E. C. Slater.)

Plate VI. EXAMPLES OF PLAY

Investigatory play: young meerkats (*Suricata suricatta*).

(a)

Substitute fighting play with inanimate objects: young Tasmanian devil (*Sarcophilus harrisi*) reaching up to seize a toy (*left*) and playing tug-o-war with a duster (*below*).

(c)

Plate VII. MATING BETWEEN BLACKFOOTED CAT (*Felis nigripes*) AND DOMESTIC CAT

Normal orientation: intromission impossible.

The male has worked his grip back to behind the female's shoulders and can now mate successfully.
(*Photos* P. Leyhausen.)

Plate VIII. DEFENSIVE THREAT BY KITTEN OF AFRICAN WILDCAT (*Felis lybica*)

(*Above*) Moderately high intensity.
(*Below*) Maximal intensity. (*Photos* Dr. C. K. Brain.)

hoarding". In other species, each individual load of food may be separately concealed and this D. Morris (1962) has called "scatter hoarding". An intermediate category does exist, where a number of relatively large caches, not inside the home, are made and repeated trips to the same store are required but this has not received any distinguishing name.

Larder hoarding, as mentioned above, is common amongst burrowing rodents. The hamster owes its name to this habit and it is also known in many other species, for example American ground-squirrels and chipmunks (*Citellus*, *Tamias*, *Eutamias*), pocket gophers (Geomyinae), kangaroo rats (*Dipodomys*), pocket mice (*Perognathus*), voles (*Microtus*, *Clethrionomys* and *Alticola*), the pouched tree-mouse (*Beamys*) and the giant rat (*Cricetomys*). Amongst the Insectivora, moles store earthworms and the short-tailed shrew, *Blarina*, stores snails and beetles in its runways (Ingram, 1942).

Although all these animals store food in their burrows, they do not all do so in an identical manner. In some cases the food is placed in special chambers separate from the nest; in some there is no special larder; while *Beamys* and *Cricetomys* (B. Morris, 1962, 1963) store the food in the nest, buried under the bedding. There is not, however, sufficient information to tell whether related species show the same methods of housing their stores. The brown rat, *Rattus norvegicus*, shows a slightly anomalous type of hoarding. According to Steiniger (1950) wild rats make three sorts of burrows. A colony has an extensive residential burrow system and small refuge burrows are also made which serve as temporary retreats if danger threatens. The third type is the food storage burrow, which is excavated somewhere near the main feeding area, or between it and the home burrow. Food is sometimes taken from the storage burrow to the home and eaten there.

Many of these hoarding species have cheek pouches, in which the food is transported, but details of the movements used in filling and emptying them are known only in a few cases. *Cricetomys* makes virtually no use of the paws; pieces of food are picked up directly in the mouth and the pouches are normally emptied without assistance from the paws, although occasionally, if there is difficulty in emptying out the last few grains of a load, the paws may be used to assist.

The hamster uses its paws in a highly characteristic manner to empty the pouches: the ipsilateral paw is used to push the back of the pouch forwards while the other paw scrapes out its contents. The pocket mouse (Bailey, 1939) uses its two paws simultaneously both in loading and unloading. Kangaroo rats load in the same way (Shaw, 1934) but I have not found a description of their unloading technique.

A number of species whose stores must last them through a long winter season of scarcity do more than simply accumulate food in the burrow. They process it first, drying it out before it is placed in the definitive cache. Shaw's (1934) work on the giant Kangaroo rat, *Dipodomys ingens*, provides the most complete account available of behaviour of this type. In the area he studied, rain fell only during December, fresh vegetation was growing in January and had produced dry seeds by May. During this short season when food was available, enough had to be stored to last the animals through until the following year. Investigations in March showed that the animals were collecting half-dry seed pods and storing them not actually in their burrows but in very numerous small shallow pits, each covered over with tamped down soil, all round the burrow entrance. One animal had 875 such pits round its burrow. Investigations in June showed that most, but not all, of the pits had been emptied. Excavation of a burrow disclosed two large caches containing between them 3¾ quarts (approximately 3½ litres)* of seeds, together with the remains of two old disused caches. A burrow dug up in December had 305 small pits round it, every one empty, and twelve underground caches. All this suggested the possibility that the unripe seeds might first be put in the surface stores until they had dried out and then be moved to the more permanent underground caches. The following year, towards the end of March, surface pits were therefore opened and seed pods dyed with mercurochrome and also some rice grains were put in. By May many of the surface pits were empty and on excavating the burrow, nine caches were found, four of which contained rice and dyed seed pods. The total

* Assuming the quarts are U.S. quarts, not imperial ones.

volume stored by the single animal in this burrow was almost 35 quarts (= 33 litres). Clearly Shaw's original idea was right: the food was being given preliminary outside storage that allowed it to dry before it was taken to the main stores.

Shaw followed up these observations by watching the collecting behaviour of the animals in May, when the seed pods on the food plants were already dry. He found that the seeds were now harvested and taken directly to the underground caches, so clearly the condition of the food determined how it was treated.

A somewhat different procedure is used by species whose stores consist not of seeds but of grasses and similar green vegetation. The food is first cut and left to dry before being carried in and stored in the burrow or stacked in a protected place nearby. This "hay-making" technique is best known in the pikas (Lagomorpha) but is also practised by rodents; it is recorded in the mountain beaver, *Aplodontia rufa*,* and several species of voles belonging to the genus *Alticola*. A surprisingly detailed parallel is shown in the behaviour of two of these animals, one a vole (*A. strelzovi*), the other a pika (*Ochotona pricei*). According to Formazov (1966) both make their final "haystacks" in niches under cliffs close to their refuges and both protect their hay against being scattered by the wind by building a wall across the entrance. Special collecting trips are made to gather the pebbles which are used to make the wall.

Not all types of food, however, have their keeping properties improved by preliminary drying out. Pine cones, if allowed to dry, open out and shed their seeds, but if kept moist will remain in good condition for a long time. Cones are a major constituent of the diet of the pine squirrel, *Tamiasciurus hudsonicus*, and the animal's behaviour is adapted to the unusual characteristics of its favourite food. Other types of food are stored in dry places but caches of pine cones are

* It must be pointed out that in the case of *Aplodontia* the significance of the hay-making is not fully clear. Wandeler and Pilleri (1965) quote Camp as believing that the grass is not used as food but solely as bedding. *Aplodontia* does, however, store other types of food and shows discrimination in the treatment of dry and moist materials.

preferentially made in moist ground. Indeed the pine squirrel's adaptability goes even further. Mushrooms are also stored but these are first laid out on a tree branch and allowed to dry before being cached (Shaw, 1936). It is not, however, clear whether one and the same squirrel can show both types of behaviour and clearly this species deserves a much more detailed study.

Storage of food in the wet, of course, reaches its highest development in beavers, *Castor fiber*, where the stockpiles of cut branches are actually in the water, close to the lodge.

Before leaving the subject of food processing, it may also be mentioned that the mole, *Talpa europaea*, treats its stored earthworms in a manner which improves their keeping properties. Many of the worms removed from a mole's cache are found to have the anterior end (or sometimes the posterior end) bitten off. This renders them almost immobile and so prevents their escape. A consideration of the mole's normal method of feeding shows that it is not necessary to postulate any great intelligence or insight to account for this behaviour. In finding food the tactile sense is the most important and the mole reacts very swiftly to anything that moves, biting at it instantly (Godfrey and Crowcroft, 1960). Since the anterior end of an earthworm is the part that moves most (and the posterior end is the next most active region) it is only necessary to suppose that as the mole collects the worm it gives its normal response and bites at the moving part.

Scatter hoarding shows many contrasts with the types of hoarding so far considered. The food, instead of being amassed, is dispersed as much as possible, only a single load being placed in each hoard. D. Morris (1962) has shown that the green acouchi avoids burying one hoard close to another. The most familiar example of scatter hoarding is the European squirrel's habit of burying individual nuts. It is also practised by the agouti, *Dasyprocta aguti* (Eibl-Eibesfeld, 1963), and by the African ground squirrel and brush-tailed porcupine (Ewer, 1965). This type of food storage is also found in Carnivora, and is best known amongst the Canidae, where the domestic dog's habit of burying a bone is a familiar example. Tinbergen (1965a) has recently made a study of the food storing behaviour of the European red fox, *Vulpes vulpes*, in the area round a black-headed gull nesting colony. One of the fascinating things about this study is that it was virtually all done by tracking. The sandy area round the gull colony provided an ideal surface on which the foxes' feet wrote the story of their night's activities for Tinbergen and his colleagues to read the next morning. Foxes were regular predators both of eggs and young gulls. When eggs were plentiful, the foxes scatter hoarded them. One egg is taken at a time and carried off in the mouth for a short distance (5–10 metres). The fox then digs a small hole with the fore paws, lays the egg in it and covers it by pushing sand over it with its nose. The fox then returns for another egg, takes it off in a different direction and buries it in the same way. Larger food objects may be hidden simply by thrusting them into deep vegetation. The eggs were not dug up until later in the year, after the gulls had gone and food was becoming scarce. The tracks showed that the fox comes from his home fast and direct, using regular highways along the beach and particular "passes" up through the dunes, to the general area where the eggs had been buried. He then starts to search slowly, quartering the ground carefully. He detects a buried egg from a distance of about 3 metres downwind and goes from there directly to it. It thus appears that he remembers the general area, but not the particular places where he buried an egg. This was further confirmed by the fact that the fox also found eggs buried in the same area by the investigators, of whose location he could have had no previous knowledge. Sooter (1946) describes very similar scatter hoarding of the eggs of various species of wild duck by the coyote. Wolves may kill in excess of their immediate requirements and when this occurs the remains are buried (Crisler, 1959). This does not occur in *Lycaon*, possibly because for them food is abundant all the year round (Kühme, 1965a).

It is virtually certain that scatter hoarding has been independently evolved in carnivores and in rodents and indeed has probably arisen independently more than once in the latter group. This is suggested by the fact that not all scatter hoarders use identical techniques. *Sciurus* and *Xerus* have burying patterns which are so nearly identical that it is difficult to believe they could be other than homologous (Eibl-Eibesfeldt

1963; Ewer, 1965). A small hole is scratched with alternating movements of the fore paws; the food, carried in the mouth, is placed in it and rammed down with the incisor teeth; it is then covered over and the soil tamped down with simultaneous movements of the two paws. Only in the details of this covering do *Sciurus* and *Xerus* show slight differences. Finally, the cache may be camouflaged by scraping a dead leaf or small stone on top of it. The pattern used by *Myoprocta* (D. Morris, 1962) differs in many details. The hole is dug as usual, but the food is then dropped in and rammed down with the fore paws; it is covered up using alternating paw movements and the camouflaging material is put in position with the mouth, not the paws. A very similar and probably homologous pattern is described in the agouti by Eibl-Eibesfeldt (1963). This all suggests that we are here dealing with two independently evolved patterns, one in sciurids and one in South American hystricomorphs.

If indeed scatter hoarding has been evolved several times, the habit must have considerable survival value, despite the fact that it may superficially appear to be rather inefficient. D. Morris (1962) concludes that its main value probably lies in protection against the stores being rifled by "thieves", whether conspecific or belonging to other species. A few caches may be found and stolen, but the dispersal policy makes it improbable that all will be found. Tinbergen (1965a) agrees with this and did actually find evidence of a hedgehog stealing eggs buried by a fox. In this connection it must be borne in mind that the question "is scatter hoarding advantageous?" is not simply a mathematical problem equivalent to the question "are more eggs likely to be found if they are all in one basket than if they are dispersed?". How the losses are distributed also matters. The regular loss of a certain proportion of the stored food may be of little importance, whereas the total loss of the reserve stocks, even if it only occurs as a rare event, could be extremely serious. Scatter hoarding could thus be advantageous even if it does not cause any reduction in the total losses as a result of theft. Storage away from the animal's home also has the advantage that as long as other food is readily available the store will not be drawn upon. If it were in the immediate vicinity of home the animal might consume its stores before a time of real need had come. Furthermore, in cases where seeds and nuts are cached, stores which are never rediscovered are equivalent to the animal sowing a crop of its favourite food plant.

While the function of food hoarding is obvious, the origin of this type of behaviour remains to be considered. Larder hoarding and scatter hoarding are so different that they must be dealt with separately. Bindra (1948) has suggested that larder hoarding originates from the animal's need to find a safe place in which to eat. This view has much to recommend it: I know of no rodent that will not, if at all alarmed, carry food to its burrow or to a familiar refuge of some sort to eat it and Bindra has found that in laboratory rats the tendency to carry food back to the nest to eat is increased by making the feeding place more exposed and decreased by making it more protected. There is, however, another factor which Bindra does not mention and which has an even more obvious effect on the behaviour of the black rat, *Rattus rattus*: this is the size of the pieces of food available. This I have seen very clearly in populations of wild animals that have learnt to come for food provided at regular feeding places. If there is no extraneous source of alarm, the boldest rats will devour small items of food (such as maize grains) just where they pick them up, the more timid will retire to the lowest branches of the nearest tree. If something larger (for instance a lump of cassava) is found, even the least timid of the animals will carry it right away to the shelter of the roof in which the nests are situated. As far as the evolution of hoarding is concerned, this seems to be an important factor, for the larger the piece of food, the greater is the chance that after the animal has had enough to eat, there will be something left over which can be used later—in fact an incipient hoard.

It must, however, be emphasised that if a complex hoarding pattern is evolved one would expect, from all that is known about emancipation, that it would during the course of its elaboration, acquire independent motivation of its own and would cease to be essentially part of feeding behaviour. In this connection I consider that experiments on rats in captivity are likely to be misleading. If we refer to the behaviour of the wild rat, as described by Steiniger (1950), then it is at once apparent that a rat in a cage, taking food from its feeding dish to its nest, is really doing something that corresponds with fetching food from the storage burrow to eat in the home burrow. It is not really hoarding at all, for the hoarding part of the wild rats' behaviour is the stocking of the storage burrow. Actually, in the cage situation, where true hoarding cannot occur, one would expect the behaviour of the animal to show a combination of the two components—hoarding in the storage burrow and removal of food to the home burrow to eat. The situation is thus highly confusing and the rat is clearly an unsuitable animal on which to study the factors influencing hoarding behaviour. Bindra (1948) from his studies on rats concluded that food collecting and feeding have the same motivation, but other workers (see Barnett, 1963 for details) have found some degree of independence. Such conflicting views are not unexpected in view of what has just been said.

On *a priori* grounds one would expect a hungry animal to be highly motivated to eat and that this would compete with and inhibit, the tendency to hoard. On the other hand, eating and hoarding have much in common, since in both types of behaviour the animal is directing its activities towards food. One might therefore argue that if a hungry animal finds abundant food its first response will be to eat, but that once hunger is satisfied eating might lead on to hoarding. This appears to be the case in the three hoarding species with which I am familiar. My tame *Xerus* would always eat first and bury food only after she had made a good meal and the same was true of a tame *Atherurus*. Eibl-Eibesfeldt (1963) says that the red squirrel also behaves in this way. My *Cricetomys* were kept in a cage but allowed to forage all round the laboratory during the evening activity period. They always fed first and emerged to collect food only when they were no longer hungry and would do so when there was still food in the nest. Clearly in these species, hoarding, whatever its origin, has in the course of evolution acquired its own motivation; it is not directly dependent on hunger, although the previous satisfaction of hunger may have a priming or facilitatory effect on hoarding.

In *Cricetomys* the most important external factor influencing hoarding is the quality of the food available. Making the food (maize grains) less attractive by mixing it with gravel caused a reduction both in the amount of food brought home and in the time the animal would go on collecting. Making access to the food more difficult had little effect on the amount collected, although the time required was, of course, increased (Ewer, 1967).

As previously noted, scatter hoarding is very different from larder hoarding and its original motivation cannot have been the same. This is very clear in the case of *Xerus*, which possesses a burrow, but does not normally hoard in it. This species shows the usual tendency to take the food to the safety of the burrow entrance to eat, but the one place it will not bury its hoards is close to the burrow. Similarly *Sciurus* buries food on the ground, but if alarmed takes it up a tree into safety to eat. From watching the behaviour of *Xerus* I came to the conclusion that its hoarding behaviour must have originated in food envy—its original function being to conceal food from a conspecific when the animal is itself not hungry enough to eat any more. The behaviour of tame fennecs, studied by Gauthier-Pilters (1962) suggests the same thing. Food remaining after a meal is commonly buried, but if a rival is present it is often dug up again and carried around for some time and then buried in a different place. Similarly with a pair of brush-tailed porcupines, if one was hiding a ground-nut and the other approached, the burying was always interrupted and the animal ran off and buried the nut elsewhere. The leopard's well known habit of taking the remains of its kill up a tree also seems to relate to protecting it from jackals and hyaenas. Since food envy is almost universal amongst mammals, it is not difficult to believe that it could have led to the habit of hiding the food being developed independently in different groups. However, as in the case of larder hoarding, the highly evolved patterns which many species possess have acquired independent motivation. Scatter hoarders kept singly will still perform their patterns, although there is no potential thief about. Moreover Eibl-Eibesfeldt (1963) has recorded both squirrels and agoutis performing their food burying patterns on a bare floor, where no concealment could be achieved, and I have seen the brush-tailed porcupine do the same thing.

If the basic motivations which have led to the evolution of the two types of hoarding are as suggested—food envy and the need to have a safe dining place—then they are both very simple and both very widespread. There is therefore no difficulty in believing that hoarding of either type might be evolved independently a number of times. Moreover, the motivations suggested are different. There is thus nothing mutally exclusive about them and there should therefore not be any particular difficulty in an evolutionary replacement of one type of hoarding by the other, should selective pressures alter in such a way as to make this advantageous.

Observations on *Xerus* and *Cricetomys* give some support to this suggestion. The female *Xerus*, predominantly a scatter hoarder, at one period, possibly connected with her sexual cycle, showed some tendency to bury food inside the burrow. This I am inclined to interpret as a possible first step in abandoning scatter hoarding in favour of larder hoarding, which seems to be a more efficient method for species possessing a burrow. The female *Cricetomys*, an extremely enthusiastic larder hoarder, would occasionally scatter hoard. On just a few occasions, when she had been free in the laboratory in my absence, I subsequently found individual loads of grain buried separately in a sand tray and an earth box at opposite ends of the laboratory. Although she readily carried food home to the nest in my presence, she showed the usual aversion to being watched while scatter hoarding and I only once actually saw her bury a load of grain. The circumstances that induced her to behave in this way remain obscure. The fact that the food carried home to the nest in normal larder hoarding is buried under the bedding, using a typical digging movement, does suggest the possibility that in this species we have an example of a more advanced stage of the same change from scatter hoarding to larder hoarding: a stage at which the latter is fully developed and the former persists only in a vestigial form.

The evolution of the two major types of hoarding, in their simple forms, thus present no great problems. The more specialised procedures, including those which involve processing the food, at first sight seem difficult to explain without attributing to the animals concerned a highly improbable degree of intelligence and foresight. In the cases where detailed information is available, however, there is in fact no need to invoke such mental powers and the evolution of the complex behaviour shown is actually explicable in very simple terms.

The beavers' stockpiling of branches is a complex process. According to Richard (1964) the store consists of an elongated raft of wood in which some sorting out of the materials is evident. Nearest the bank are placed the largest and straightest branches with their side shoots lopped off and beyond these, bundles of twenty to thirty smaller pieces with leaves and side branches intact are arranged in a radiating manner. Richard points out that there are resemblances in the constructional methods used for making the lodge, the food store and the dam and it may therefore be that the two latter patterns are derived from the former; in other words, they are modified nest building behaviour.

Haymaking must have been evolved independently in voles and pikas and its origin is therefore likely to be found in some type of behaviour that is not uncommon or unusual. The behaviour of the steppe lemming, *Lagurus lagurus*, closely related to *Alticola*, suggests a very simple origin for haymaking. In summer when food is plentiful, these animals eat at feeding places not very far from their burrows. Food storage is not highly developed: Formazov (1966) records that out of thirty burrows excavated, only three contained substantial stores. If food runs short during the winter, the animals may emerge from their burrows, collect from their feeding places the unfinished remains that have been left lying about and take them back to the burrows to eat. To cut down more grass than is eaten, when abundant food is available is a common enough habit. *Thryonomys*, for instance, does this and selects the most palatable pieces—ripe seeds and the tenderest parts of the stem—the rest is simply left lying. From such a simple beginning, the evolution of haymaking requires only two things: intensification of the tendency to cut more food than is immediately required in times of abundance, coupled with postponement of hoarding until later in the season, when the "hay" provides the most easily available source of food.

The seed drying habit of *Dipodomys ingens* probably also has a simple origin. The essential point here is that the seeds are treated differentially in relation to their moisture content. Dry seeds are taken directly into the burrow and stored, unripe ones are first buried in surface pits near the burrow entrance. In this connection it is of some interest to find that in *Xerus* also there is a relationship between the quality of the food and the treatment it receives. The female I studied scatter hoarded maize grains, but she discriminated between hard dry grains and fresh soft ones: the latter were eaten at once and only the dry seeds were stored (Ewer, 1965). It was possible to make this animal alternate between eating and burying by giving her alternately soft and hard grains. Whether this reflects an inhibition against the storing of soft grains or merely the fact that they were so tasty that the squirrel simply could not resist eating them is immaterial, for the latter alternative provides the simplest way in which a true inhibition against storing soft seeds could have begun. An inhibition of this sort is a very simple response and yet one which could have great survival value, since the stores would be much less likely to go bad. In the case of the kangaroo rat, such an inhibition could provide the basis for the evolution of the whole processing pattern. It is relevant here to note that in the related *D. nitratoides* food is stored in an enlarged chamber in the burrow. Little pits are excavated in the wall of the chamber to house the stored food, apparently very similar to the surface drying pits of *D. ingens* (Culbertson, 1946). If in the latter species, an inhibition of the type suggested does exist, then when abundant but unripe seeds are available, the animal is in a complex situation. The precondition for hoarding is that food should be available in excess of immediate need and this condition is amply fulfilled; but the food is moist and this inhibits storing. In this situation the animal may perform what is really a disrupted version of the full pattern. It collects the food and returns home with it, but is unable to complete the pattern and so digs its store pit not in the wall of the underground chamber, but outside its burrow. Digging at the burrow entrance could have started as a displacement action in the conflict situation and then have simply led directly into completing the normal sequence of unloading the pouches into the hole that has been dug. Once such behaviour had started, even if in an irregular manner, its advantages could reasonably be expected to lead to rapid evolutionary perfecting.

The pine squirrel's diverse methods of dealing with cones and with mushrooms still awaits an explanation, as does the wall building of *Alticola strelzovi* and *Ochotona pricei*. In none of these cases, however, is there sufficient detailed information to provide a guide to speculation. We know neither the full details of the behaviour itself nor sufficient about the habits of related species. When such information is forthcoming, it is to be expected that these forms of behaviour too will prove to have very simple evolutionary origins.

CHAPTER 4

Social organisation and territory

Although some mammals undertake large scale seasonal movements between feeding and breeding grounds, or between different feeding areas, the majority of them spend most of their time within a definite home range with whose topography they are familiar and within which they can find their way about with speed and assurance. It is convenient to designate this area as the "home range". The term "territory" has a different meaning, for in this case, a principle of exclusion is in operation. A territory, in its simplest form, is an area from which conspecifics are excluded by being attacked or threatened if they venture within its confines and it may or may not be coextensive with the home range.*

It must not be assumed, in an anthropocentric manner, that a home range or a territory necessarily means a continuous area enclosed within defined boundaries. When we buy a plot of land, or survey a tract of country, we think of it in these terms and for large species with good eyesight, living in open country, a home range may be of this type. In other cases, however, if eyesight is relatively unimportant or if cover reduces visibility, the home range may consist essentially of a number of places which are of importance—feeding places, drinking places, resting places, sunning or wallowing spots and so on—linked by a series of pathways. It is thus possible for neighbouring ranges to interpenetrate without significant overlap. Meyer-Holzapfel (1957) says that in bears territory is of this sort, and in the elephant shrew, *Nasilio brachyurus*, the territory consists of the burrow, one or more feeding areas and the trails linking them with the burrow (Rankin, 1965). In the domestic cat too, territories may interpenetrate. Leyhausen and Wolff (1959) found that neighbours might even

* Jewell (1966) discusses the concept of home range as applied to mammals.

use the same path—but not at the same time. By keeping to more or less definite timetables encounters were avoided and one cat did not interfere with another's freedom of movement.

Sometimes there is a part of the home range that is of particular importance because it contains the most favoured resting places and refuges, the best feeding places etc. This may be designated a "core area", without prejudice as to whether it is defended or not. The core area is the most frequented part of the range and if a core area exists, it is not uncommon to find that while the home ranges overlap, the core areas do not. This was the case, for instance, in the coatis studied by Kaufmann (1962) on Barro Colorado Island.

The advantages of remaining within a familiar area are obvious in terms of knowing where water is to be found and where are the best hunting or feeding grounds. Equally important is the ability quickly to find shelter in a known refuge if danger threatens. The difference which familiarity with the environment makes in this respect was graphically illustrated for me by the following incident. Cane rats are shy creatures, easily put to flight. When I first kept these animals and put them into an enclosure they fled blindly when alarmed, crashing violently into the wire netting and finally finding refuges more or less by chance. They soon became accustomed to their surroundings, but presently I decided that they required more space and built a larger enclosure. Not wishing to disturb them by catching them, I built a passageway, about 1 ft in diameter, from one enclosure to the other and allowed them to find their new quarters for themselves. The next day I accidentally frightened one near the corner of the new enclosure diagonally opposite to the passageway. Without hesitation it ran directly to the opening, along the corridor and "home". The passageway showed evidence that there had been heavy traffic along it and clearly the animals had spent the night passing to and fro and familiarising themselves with their new domain: the contrast between the resulting oriented flight to a known refuge and the original blind panic was very striking.

Use of the term "familiarity" implies that at some period

the animal has explored its area and has learnt its way about, but it implies something further. Unless its daily activities take it to every part of its range reasonably often, the less frequented parts will presently cease to be familiar, either because they have changed or because details have been forgotten.

E. E. Shillito (1963) in a study on voles, points out that in considering "exploratory behaviour" there has been a tendency to concentrate on the rather artificial situation where an experimental animal is suddenly placed in a new environment and its behaviour is recorded. She defines exploratory behaviour as "that behaviour which serves to acquaint the animal with the topography of the surroundings included in the range" and points out that in natural circumstances exploratory behaviour comprises two different components. Firstly there is investigation, which is behaviour oriented to anything unfamiliar and is the type of exploration studied in laboratory experiments. The pattern of investigation is characterised by extreme caution. The strange object is approached slowly and carefully, with much sniffing; at the slightest sound or movement there is a hasty retreat, followed by a new advance—indeed there may be withdrawals without any visible external cause—finally contact is made and the object is touched and smelt and its harmless nature established. In addition to investigation, however, there is also reconnaisance, in which the animal moves round its range in a fully alerted manner so that all its sense organs are used as much as possible, resulting in maximal exposure to stimuli from the environment. It thus "refreshes its memory" and keeps a check on everything in its area. If any change has occurred, it will be investigated and so cease to be strange. In her voles, Shillito found that this reconnaisance behaviour occurred as a regular activity in an already familiar environment and did not require the stimulus of a strange object. My *Cricetomys* also showed this type of behaviour.

Steiniger (1950) reports that wild rats, although normally living within a relatively small home range, may cover long distances to special food sources. He records a distance of over a kilometre being travelled to forage on the pieces of fish left in nets hung up to dry. To find such a feeding place in the first place suggests that periodic exploratory journeys may be made beyond the normal range.

If an animal is extending its range or has been forced by circumstances into a new area, then, of course, it shows investigatory exploration, with its attendant caution and readiness for instant flight. Once the animal has learnt its area, its demeanour changes. It moves about more rapidly, with greater assurance and is less readily put to flight. This correlation of readiness to flee with unfamiliarity and of assurance with familiarity is of great importance, because it is basic to true territorial behaviour. Once territories have been established, the animal within whose domain an encounter occurs will have the advantage, since he has the assurance based on familiarity, while the interloper is on unfamiliar ground. Without this sliding scale, the establishment of stable territories would be extremely difficult and costly in terms of the severity of fighting that would be necessary for an owner to maintain his area.

The primary function of territory is to ensure that the holder has at his disposal an area sufficient to provide for his needs in terms of food and shelter. The strongest animals in a population will be able to hold the best parts of the habitat, weaker ones will have to make do with slightly less desirable areas. In this way there will be a spacing out which ensures that the available habitat is fully exploited. In terms of natural selection, however, the survival of the individual is of importance only in so far as it ensures survival of his genes. Moreover, the rearing of a family increases the requirements of both food and shelter. It is therefore natural that territory should become associated particularly with reproduction and that its function should be extended to that of providing for the needs not merely of the individual, but of the family that is to be raised.

The simplest form of territorial organisation is one in which the animals are solitary, each holding its own territory, that of the female being sufficient to meet the needs of her family until they are old enough to fend for themselves. At least a brief truce to hostilities between the sexes must, of course, occur to permit mating. This type of territory is characteristic of squirrels (Eibl-Eibesfeldt, 1951b; Kilham, 1954) and here the male is permitted to enter the female's territory to mate and is driven out again shortly afterwards. The same happens in the hamster, *Cricetus cricetus* (Eibl-Eibesfeldt 1953a). In lemmings also (Arvola et al., 1962) and in the common shrew, *Sorex araneus*, (J.F. Shillito, 1963) the female holds a territory but permits the male to enter for mating. Whether the male at other seasons also holds an individual territory is not known in the latter two cases. Amongst the marsupials, the numbat, *Myrmecobius fasciatus*, is considered by Calaby (1960) to be solitary and territorial. The same individuals could be recaptured after as much as six months within a distance of 200 yards from where they were first found and on only one occasion did he find two adults together—presumably a reflection of the breeding truce.

In bears, the organisation may be rather similar. Meyer-Holzapfel (1957) regards them as being territorial solitaries, with a truce between the sexes during the breeding season. Krott, however, disagrees (Krott, 1961; Krott and Krott, 1963) and says that bears have no territories and are "socially neutral". It seems very improbable that any mammal is ever truly socially neutral; individuals of other species may be disregarded, but to have no response, either positive or negative, to conspecifics seems inherently improbable. It may be that Krott's opinions on lack of territoriality have arisen because in bears territory consists of a series of paths linking important places, not of a continuous area. Territories may therefore interpenetrate and individuals may ignore each other when, to an observer thinking in terms of a defined area, a violation of frontiers appears to have occurred for the very simple reason that, from the bear's point of view, there has not in fact been any infringement of territorial rights.

A further complication is pointed out by Leyhausen (1964). In moderately large species, with well developed distance receptors, there is probably no such thing as a truly solitary animal. Each individual periodically meets his neighbours at boundaries or, in the case of interpenetrating territories, at crossroads. Gradually the neighbours become known and familiar and a sort of social order in fact develops. Known neighbours are recognised and treated with a degree of tolerance that would not be permitted to complete strangers.

In cats, for example, Leyhausen describes how one male may wait at the approach to a crossroads for another to pass by "like a train held up at the signals". Such behaviour is in fact based on the recognition by each of the territorial rights of the other, but could easily give the impression that the animals are not territorial.

Even in the simplest type of solitary territorial social system, there is some sexual selection of males in operation: the male must be sufficiently strong and persistent to overcome the female's initial resistance to his entering her territory. A further complication is possible if, as in domestic cats, a female on heat attracts males from several nearby areas. When this occurs, then there may be competition between the males for access to the female and this imposes a further selection in terms of fighting ability and the degree of pugnacity that accompanies sexual motivation. In cats the test of endurance may be severe; the female usually becomes attractive a day or two before she will permit copulation and during this period the males may court her and threaten and fight each other almost continually, often without returning to their homes for food. Although the female may in the end copulate with more than one male, the dominant animal who achieves the first mating may have the greatest chance of fathering the resulting litter. In this way, territory becomes linked with competitive inter-male selection and so takes on a new function—that of ensuring that while all the females breed, the strongest males father the largest number of offspring; in fact the situation produces the natural counterpart of the stock-breeder's "upgrading".

In solitary species the predominant function of territory is the primary one—spacing out of individuals and ensuring an adequate area for the rearing of each family. The secondary, male competitive function, where it exists, is integrated with the primary one in a very simple manner; by means of some type of contest which decides who shall be allowed to enter the female's territory and mate with her. With social species, however, the picture becomes more complex. Here the major interactions between individuals are not restricted to those concerned with reproduction and care of the young; consequently the relations between social structure and territory and between the primary and secondary functions of the latter are complex, varying from species to species. To say that a species is social does not, of course imply that there is no intra-specific aggression and that all conspecifics are accepted and tolerated. On the contrary, it is usual for the social group to form a community within which aggression is reduced or controlled, while conspecifics not belonging to the group are treated as enemies. Without some such system of differentiating group members from strangers, the social unit could have no coherence and territory holding would be impossible. The ways in which animals succeed in having their cake and eating it—retaining the advantages accruing from social life without forfeiting those resulting from being territorial—are diverse. It will therefore be convenient to review briefly the type of social structure and territorial organisation found in some representative cases within a few of the major orders. It will then be possible to see some of the general principles that operate and note how the differences observed relate to general mode of life.

1 RODENTS

Within this order, various grades of social organisation exist. In wild rats, *Rattus norvegicus*, according to Steiniger (1950), colonies are formed as the result of expansion of a family group with very little aggression and great mutual tolerance. The group inhabits a large communal burrow system which acts as the living quarters and its members forage for food over a considerable area round about. Strangers are not tolerated, but viciously attacked. We have, however, little information about the interactions of independent groups when their members meet away from the home burrows. If strange adults are brought together, however, a colony will ultimately be formed, but the organisation is not the same as it is in a naturally established colony growing from a family group. On the basis of much fighting a type of rank order is established, and the weaker animals give way to the more dominant ones. It therefore seems likely that if individuals of two natural colonies frequently meet at a feeding place, the same thing will happen: a dominance order will be established, determining who has priority, and the subordinate animals will simply withdraw and wait until the dominants have finished feeding. Presumably if the colony outgrows its food supply it will split up, but we have no information about how this happens.

According to Trevis (1950) a beaver colony is also a mutually tolerant family group within which, surprisingly enough, there is little social integration except between the mother and her young during their dependent period.

The Mexican ground squirrel, *Citellus mexicanus*, is described as "colonial but not social" (Edwards, 1946). Burrows may be concentrated in a relatively small area, but each individual defends its own burrow and its immediate environs against any other. Each also has his own home range 50–100 yards in diameter, but strangers are permitted within this area, provided they do not approach the burrow too closely.

In the Uinta ground squirrel, *Citellus armatus*, the situation is similar, although here the territorialism is modified; each individual tends to drive off any other that approaches it too closely and is preserving a certain individual distance, rather than defending a particular area (Balph and Stokes, 1963). In both species the female is tolerant of the male for about a week during the mating season and then the two separate. The organisation is therefore essentially the same as in solitary squirrels; in fact the colony consists of a number of solitaries living at close quarters.

Wild guinea pigs, *Cavia porcellus* (Kunkel and Kunkel, 1964), are gregarious and occupy a home range within which there are no specific territories. There is, however, a strict rank order in which the subordinate individuals give way to the superiors whenever there is competition for food or for favourite resting places. In the males, the same rank order applies in relation to access to females, so that the highest ranking males are the ones that produce most progeny. The green acouchi in captivity establishes a strict linear hierarchy and any strangers introduced are attacked (D. Morris, 1962); it seems likely that the natural social system is much like that of the guinea pig.

Amongst the rodents, the most complex social system at

present known is that of the prairie dog, *Cynomys ludovicianus*, described by King (1955). Within a suitable habitat, burrows are not equally spaced but grouped to form a "town". This concentration alters the vegetation in a manner favourable to the animals, as their activities encourage the growth of herbs at the expense of grass and the former constitute the animals' favourite foods. A town may become very large, with over a thousand inhabitants. The larger towns are usually partially subdivided by topographical or vegetational features into a number of "wards", but there is visual and vocal communication between neighbouring wards. The latter are thus an incidental result of local topography, rather than an essential part of the social structure. Within each ward there is further subdivision into what King has named "coteries", which constitute the basic social units. Each coterie possesses its own territory and this is communally held against members of the neighbouring coteries, the male being more active than the females in defence. The coterie has a variable number of members, ranging from two to thirty-five, but a typical one will contain one dominant male and possibly a second who is subordinate to him, two or three females and about half a dozen immatures.

King's detailed study is one of the few investigations covering the entire annual cycle, the breeding relationships and the manner in which new social groups are formed. Mating occurs in early spring and at this period the adult males become more aggressive towards each other. King observed one case in which a male captured most of a neighbour's territory, together with his females and left the previous owner restricted to a very small area with only occasional access to the females. At this period, any coterie with more than one adult male may split up. An organisation is thus established in which the number of females and the area at the disposal of a male are decided on the basis of his strength and pugnacity.

During pregnancy and lactation the female establishes, within the burrow system of her coterie, a sort of individual sub-territory, defending her nest against any encroachment. This is, however, temporary and does not disrupt the coterie system. At about this time too, yearling males tend to become restless and more antagonistic towards their neighbours than previously. Many of them move away and establish themselves somewhere on the periphery of the town, where they may succeed in establishing a new coterie.

When the young first emerge from the burrow they are not attacked if they cross the boundary of the territory of their own coterie; soon, however, they begin to meet with increasing hostility and they then learn the limits of their area without ever being seriously attacked. Within the protected area of their own coterie's territory they complete their development with a very low juvenile mortality. During the latter part of the summer, adults, especially females, may gradually move out and start making themselves a new burrow, leaving the old area and established home to the growing young. This they do mainly in relation to two factors—finding a good feeding area elsewhere which attracts them to the new site, and being too much pestered by the growing young soliciting food, which tends to drive them away from the old home. As a result, the original territory does not become overpopulated. Furthermore, since it is the young who remain behind, they are assured of a good start in life, with all their needs for food and shelter provided. The whole organisation is thus one which gives all the advantages of group life: communal warning cries and an extensive burrow system increase the efficiency of defence against predators, while the growth of a favourable type of vegetation is encouraged. At the same time, the genetic benefits of inter-male selection are preserved and spacing out occurs in a manner which not only avoids overcrowding but ensures that the young grow up in protected and favourable circumstances.

The colony of marmots, *Marmota flaviventris*, studied by Armitage (1962) comprised a number of adult females and juveniles, together with a single adult male. Each animal had its own permanent burrow, in which it spent the night and in which the females reared their young, together with one or more auxiliary burrows, serving as temporary refuges. Each female occupied a home range which included the main burrow and pathways leading from it to feeding grounds and sunning places. Many of the ranges overlapped and three of the animals actually shared what was virtually a single area. Where overlapping occurred, the animals tended to avoid encounters with each other. Peaceful relations are thus achieved negatively, rather than by any positive attachment or cooperative behaviour between the individuals, but the members of the colony do benefit from each other's presence in terms of the increased protection afforded by the availability of more refuges and by their ability to react to each others' alarm calls.

My observations on captive *Thryonomys* suggest that here too the social unit comprises a single adult male, together with a group of females and juveniles. Rival males are driven away, but it is not clear whether they then form bachelor groups, exist as true solitaries, or remain in a loose peripheral relationship with their original unit. Within the group the females show mutual tolerance and, unlike the marmots, do not tend to avoid either each other or the male.

In all three cases, the basis of the social structure is a unit of the harem type, consisting of a single male, together with his females and their young. In the prairie dog, however, there is more social cohesion within the group than in either the marmots or the cane rats. In addition, the social units are not widely spread but closely grouped into the typical town. It is not possible to decide whether increased social cooperation preceded the formation of the complex colonies, followed it or was evolved *pari passu* with it. Shortage of suitable burrow sites might have tended to force groups together and so set a premium on cooperative behaviour or increasingly social habits might have facilitated aggregation to form compound colonies, with all the consequent advantages pointed out by King.

2 LAGOMORPHA

Surprisingly little is known about the solitary hares. Southern and Thompson (1964) merely report that the brown hare, *Lepus europaeus*, is said to travel long distances to feeding places. The home range, within which it has a number of frequently traversed pathways, may be a couple of miles in diameter, while that of the blue hare, *L. timidus*, is probably much smaller.

Investigations of social organisation have, however, been made on captive populations, kept in large enclosures, in both *Oryctolagus* and *Sylvilagus*. *O. cuniculus*, the common rabbit, has been intensively studied in Australia by Myers and Poole (1959, 1961) and also by Mykytowycz (1958, 1959, 1960). Myers and Poole studied three populations, each confined within an area of 2 acres. At the beginning of the breeding season, the populations divided up into a number of separate breeding groups, each comprising of two to six females and one to three males. A few males failed to become attached to any group. The way in which the groups formed is of interest. As the breeding season begins, the older females select certain areas and dig burrows. These digging sites become foci of social activity and both the males and the younger females are attracted to them. Adult males are mutually intolerant; as a result, only a few can coexist in a single group and amongst them there is a rigid dominance order. Females, however, are capable of developing a considerable degree of tolerance of others with whom association has made them familiar. Individual females seemed to differ in their aggressiveness, some being quarrelsome, others very peaceable. Indeed, in one group tolerance was so complete that no aggressive interactions between the females were ever observed. The females thus sort themselves out into groups of reasonably compatible individuals and to each group a few males attach themselves. Myers and Poole (1961) suggest that the limit to the number of females in the group is set by the inability of each either to recognise or become intimately associated with more than three or four others and the only group containing as many as seven females was unstable and presently divided into two.

Each group acquires its own territory; the boundaries are clearly defined with little overlap. Although each individual spends almost all its time within its own territory, periodic exploratory and foraging excursions are made over a wider range, so that an area larger than the territory is actually known and if better conditions are available elsewhere, a move can be made. This happened on two occasions when a hole was deliberately opened in the fence separating two of the enclosures. The populations were not equally dense in all three, and when communication was made between enclosures A (containing 53 adults) and B (containing 39) this was promptly noticed and there was a shift of 6 animals from the former to the latter, resulting in an almost equal distribution of 47 and 45 animals. When subsequently B enclosure was linked with C, the former contained 52 and the latter 115 animals. Again there was one way transfer from the more to the less heavily populated area, with near equality again reached at 81 and 86 animals.

Mykytowycz (1958, 1959, 1960) did not find the same division into breeding groups and in his colony the females as well as the males appeared to establish a linear dominance hierarchy. His experimental situation, however, was complicated by the fact that the rabbits were provided with two artificial warrens. These they adopted and did not dig burrows elsewhere. In view of the important role played in the organisation of the breeding groups by the first outburst of digging by the older females, this difference in the environmental situation may account for the disparity between his findings and those of Myers and Poole.

Marsden and Holler (1964) have made comparable studies on two species of *Sylvilagus*, *S. floridanus*, the cottontail and *S. aquaticus*, the swamp rabbit. The two species occur within the same geographical region, but the cottontail occupies the more open uplands whereas the swamp rabbit is restricted to low-lying wooded or marshy areas. Unlike *Oryctolagus*, they are not burrowers, but the young are protected in a nest constructed by the female. Although the populations studied were much smaller, the swamp rabbit showed essentially the same social organisation as Myers and Poole found in *Oryctolagus*. The males were mutually intolerant and formed a linear dominance hierarchy while the five females were mutually tolerant but divided up into two breeding groups of two and three. Two males attached themselves to the larger group and one to the smaller, while two failed to become members of either. Although the females of the two groups tended to restrict their movements to separate ranges, the males did not defend these areas as such, but attacked rivals only in the vicinity of the females.

In the cottontail too, the males formed a linear hierarchy

but the females were less mutually tolerant than in the swamp rabbit and also tended to establish a hierarchy. They showed some degree of division into two groups occupying different areas, but the subdivision was less marked than in the swamp rabbit and the males did not show any territorial behaviour.

These observations suggest that in the wild it is the females in all three species who set the pattern for the organisation of the social group. By selecting areas suitable for breeding and feeding they effectively determine the location and extent of the range or territory. According to their degree of mutual tolerance they determine the number within the group and this, in turn, decides the number of males that can become attached to the group. In the burrowing *Oryctolagus*, where the existence of a more or less permanent refuge makes the tie to a definite area essential, territorialism is well developed; in the other two species, less so. The habitat of the swamp rabbit is more interrupted than that of the cottontail and Marsden and Holler point out that the absence of territorial behaviour in the latter is another aspect of its generally more free ranging behaviour.

3 CARNIVORA AND PINNIPEDIA

Apart from true solitaries, the simplest type of social organisation found in the Carnivora is one in which the mated pair stay together and defend a joint territory until the young are old enough to care for themselves. This occurs in the fox (Tembrock, 1959), in the raccoon dog (Bannikov, 1964) and probably in the black-backed jackal (van der Merwe, 1953).

In wolves and in *Lycaon* the social group is larger and hunting is cooperative. The *Lycaon* pack studied by Kühme (1965a, b) consisted of six males, two females and their fifteen young, and it may have originated as a family group. In the wolves, the male and female have slightly different status: the males normally take the lead in hunting, while the individual remaining behind to guard the young is a female (usually the mother). In *Lycaon*, however, there is little evidence of social inequality of any sort within the group: the females are as active in hunting as the males and the animal that stays behind to guard the young may be of either sex.

An area of approximately 1 square kilometre immediately round the burrows in which the young were cared for constituted a defended territory, from which the *Lycaon* pack attempted to exclude any other carnivores: even lion were barked at within this area. Round this lay the hunting range of 100–200 square kilometres, which was not clearly marked off from the ranges of the other packs in the area. On the only occasion when two packs were seen to meet in the hunting area, there was barking but no actual fighting and the smaller group soon withdrew.

Amongst the Felidae, the lion is the only social species. Prides appear to originate as family groups. According to Stevenson-Hamilton (1947), the size attained depends upon the type of terrain. In open country there may be up to thirty—indeed he recorded one pride of thirty-five in the Kruger National Park—but in thick bush anything more than a family party of seven or eight is unusual. The female does not give birth more often than once in two years, and shortly before parturition, she abandons the young of the previous litter. They may then either form an independent hunting group or stay with the pride and the female herself may rejoin the pride with her new cubs. Although large groups may be formed in this way, there is generally only one adult male with the pride, so presumably the juvenile males must leave or be driven off when they become sexually mature. Small prides of a few young adult males are sometimes formed when this occurs and sometimes a group of young animals of both sexes may form a hunting unit of their own and leave the pride.

The extent of the hunting range will, of course, depend both on the size of the pride and on the availability of prey. According to Guggisberg (1960) a group of two females and their eight young which he studied in the Kruger National Park hunted over an area of 33 square kilometres. It was his impression that the ranges of different prides overlapped without this leading to hostilities but he also reports a fight, which may have been territorial, between a group of six females and another pride composed of females and young.

Lions are not prone to indulge in unnecessary activity and it is probable that the degree of vigilance shown in defence of the hunting ground is very much influenced by the relative densities of predators and prey. Moreover, if a pride splits up, it is likely that the resulting groups may be more tolerant of each other than they would be towards complete strangers. Carr (1962) is of the opinion that this is the case.

Schenkel (1966a) has studied three prides in the Nairobi National Park for periods spread over three years. The pride to which most attention was devoted comprised two adult males, one very old female and a pair of females of reproductive age. In agreement with other observers, he found that the females always went apart to give birth but used to return periodically and make contact with members of their pride only a few days after the birth of the young. The cubs, however, at first remained without any such contact and were introduced to the pride at the age of approximately ten weeks. All the adults, including the males, were tolerant, or even "maternal" towards the cubs and the young of the two females developed mutual attachment. Later on, when the females bred again, they became increasingly intolerant of the juveniles, now in their second year of life.

Within the home range, the members of the pride showed hostility to strangers and attacked any trespasser vigorously. Strange females were attacked by the resident males, as well as by the females of the pride. True territoriality thus appears to exist and Schenkel notes that the demeanour of a lion is very different according to whether it is within its own domain or that of a foreign pride. Both scent marking and roaring appear to function as means of indicating occupancy of a territory.

It is much to be hoped that these observations will be continued and extended, for a full unravelling of the complexities of lion social relationships requires a study lasting over a long period and embracing several different areas. The fact that their organisation appears to be rather looser and more facultative than is the case in the social Canidae probably reflects the fact that a single lion is capable of making a kill and a female can care for her young single handed, so that although cooperation may be advantageous, it is much less essential than it is for the social Canidae. Tigers are usually considered to be solitary, but Schaller (1965a) has found that although they hunt individually, a group of females within an area are mutually tolerant, meet frequently and will even share their kills. The females share their area with a single male, to whom they do not show hostility, but the adult males are intolerant of each other and there is only one in association with each group of females.

The spotted hyaena, *Crocuta crocuta*, is the only hyaenid whose social structure has been studied. Kruuk (1966) has found that the organisation varies to some extent according to the habits and availability of the prey. In the enclosed area of the Ngorongoro crater, the crocutas form "clans" which may have anything from 10 to 100 members. Eight clans occupied the crater, each with its own territory, within which the dens were situated and inside which hunting was carried on. When a pursuit happened to cross a boundary and the kill was ultimately made inside the range of a neighbouring clan, Kruuk found that the owners generally chased the intruders off their kill and appropriated it. In Serengeti, although the same organisation was also found, many of the clans did not have a fixed abode but migrated, following the movements of the wildebeest, on which they principally preyed. There was also a third category, the "commuters", who had permanent dens to which they regularly returned but from which long excursions were made to wherever the main concentration of game might be. These hunting trips might take them a distance of as much as 50 miles and they were often away from home for several days at a time.

In the coati, *Nasua narica*, the social unit is a band composed of one or more adult females with their offspring of up to two years of age (Kaufmann, 1962). Each has its own home range but there is some overlapping and little hostility is shown if two bands meet. The females are intolerant of adult males and the latter live as solitaries for most of the year. During the breeding season this alters and, for a period of about a month, each band permits a single adult male to join them, but when mating is over he again leaves. The solitary males each have a home range but are less bound to it and more inclined to wander than are the bands. The females leave the band temporarily to give birth to their young, but rejoin their fellows, together with the new litter, as soon as the latter are sufficiently mobile. While the advantages of foraging together as a group are obvious both in terms of communal care of youngsters, communal vigilance against enemies and of efficient exploitation of food resources, the advantage of male solitariness is not clear. It would be interesting to have information about the social structure of the other species of the genus, as it is quite possible that the organisation just described represents a stage in evolution from a solitary habit towards a completely integrated social unit including adults of both sexes.

Many of the smaller viverrids are highly social, but nothing is known of their communities. In *Suricata* in captivity the male plays a very definite role in defending the young, which suggests that the family may form the basic social unit. Communities, however, may be extremely large and there must be some form of organisation integrating a number of families into a larger group.

From these data it thus appears that in the terrestrial Carnivora territory and social organisation are mainly concerned with the problem of ensuring an adequate food supply. The advantages of cooperation lie in increased efficiency in hunting and in defending the young. In the smaller species communal protection against larger predators is also probably important. In the Pinnpedia, however, feeding is aquatic, but breeding occurs on land. Territory is therefore related primarily to breeding and its secondary functions have become predominant. Typically, the mature males come ashore before the females and each attempts to secure a territory. When the females appear each male attempts to herd as many as possible of them into his territory. Those who have managed to stake a claim close to the favourite landing places are at a considerable advantage and competition for these sites is therefore very keen. Having acquired his females, the male then has to keep them. This involves both defence against neighbouring males and herding the females to prevent them leaving his territory. Parturition occurs within the territory and copulation with the owning male follows shortly afterwards during the post-partum oestrus. The number of young fathered by any one male is therefore a function of his strength and pugnacity, tested by his ability to preserve his territory and his harem. The number of females in the individual harems is highly variable from species to species. In the sea-bear, *Arctocephalus gazella*, Paulian (1964) records a maximum of thirteen to fourteen with the mean around seven, but in the northern fur seal, *Callorhinus ursinus*, the numbers are much larger—up to 100. In the Galapagos sea-lion, *Zalophus wollebaeki*, the usual number is about twenty (Eibl-Eibesfeldt, 1955). In the latter species and also in Steller's sea-lion, *Eumetopias jubata*, juveniles are included in the social unit but in most species this is not the case. Orr (1967) notes that in the two species mentioned a female is sometimes seen suckling a large juvenile as well as her newborn pup. Presumably the former is her own yearling young, but in the absence of marking experiments this is no more than an assumption.

This type of harem is typical of the eared seals. It also occurs in the Phocidae, where the Grey seal, *Halichoerus grypus*, has a small harem and the Elephant seal, *Mirounga*, a large one. More typical of this family, however, is simple pairing of an individual male and female. Since the Phocidae are not in any way a primitive group and since the usual harem arrangement is shown by certain species within it, it is possible that the simple pairing situation is secondary, rather than primitive. Its significance is by no means clear but it may be related to the relatively poor terrestrial locomotory powers, characteristic of the Phocidae.

4 THE LARGE HERBIVORES

(a) *Artiodactyla*

Within this group the simplest organisation is found in the Suidae. Here, typically, the animals live within a home range which includes favourite resting, feeding, drinking and wallowing places. There is little evidence of territorial defence and the animals may move to a new area, if seasonal vegetational changes make this necessary.

In the Warthog, *Phacochoerus africanus*, the social unit is a small group, often composed of two or more mature females with their half grown young, sometimes accompanied by an adult male. The female's habit of going off on her own when about to give birth to a litter appears to be the main factor responsible for preventing the formation of large groups. In the animals studied by Frädrich (1965) in the Nairobi National Park, each group had its own home range, including the regularly used sleeping holes, but there was no territorial defence and the ranges of the groups overlapped considerably. Frädrich found that in this area females were more numerous than males and he notes that in births recorded in European zoos there were thirty-one females to seventeen males. This figure is significant at the 5% level. In the Hluhluwe Game Reserve, recording only full grown animals, I did not find females preponderating, in fact there was a slight excess of males, but the counts made were not large enough for this to be statistically significant. Whether this difference between the two localities (supposing it to be real) has any effect on social organisation is not known.

In the hippopotamus, *Hippopotamus amphibius* (Verheyen, 1954) the amphibious mode of life has produced some complications. In the water, the females and young form a group occupying some favourite resting and basking place, such as a sandbank. Round this area the males each have a resting place and a rank order determines which may occupy the places closest to the females. Presumably this also implies a rank order in access to the females for mating. According to Laws and Clough (1966) the group (school) may contain a few adult males along with the females and juveniles. In the Queen Elizabeth National Park in Uganda, where the latter authors worked, the average number of animals in a school was nineteen, while the largest recorded number was 107. Close to the resting place of a male is usually to be found the opening of a path leading inland to a feeding area. The feeding areas and the paths linking them to the river appear to be owned by individual males and defended against rivals, but it is not clear from Verheyen's account how access to them of the females and juveniles is determined.

In the grazing ungulates, there is protective advantage in the formation of large herds and in most species this occurs. The area occupied by a herd may not always be the same, as seasonal migrations in relation to changes in vegetation may occur. This, although it may modify the picture, does not alter the fact that large numbers of individuals remain gathered together in relatively close contact with each other. We have already seen how, in the Pinnipedia, close aggregation during the breeding season has led to the development of a harem system, with individual males attempting to monopolise as many females as possible. It is not surprising to find that in the ungulates, although the primary factor responsible for the formation of large groups may be different, the consequences of close grouping are similar in relation to the breeding system. As in the Pinnipedia, local concentration favours the development of male competition for females and the resulting organisation is generally of the harem type.

The details of social structure and its relationship to territory vary from group to group. The simplest type of organisation occurs in the Camelidae. Koford's (1957) study of the vicuña, *Vicugna vicugna*, is the most complete that deals with undomesticated animals in their natural habitat. Here each male defends a territory within which he has a number of females (usually about eight) and their young. The area is large enough to provide food for the group and Koford found that its size was inversely related to the richness of the pasture. The younger males formed bachelor troops which were excluded from the richest areas and occupied the peripheral part of the locality inhabited by the animals. The organisation thus served for spacing out and ensuring an adequate food supply for the females and their young, as well as for ensuring selective breeding of the males.

Inter-male rivalry seems to be the rule in the Camelidae and with the domesticated species it is the practice during the breeding season to keep only a single male with a group of females (Pilters, 1956).

In many of the Bovidae, the organisation is not very different and the same two functions of spacing out and of ensuring that while all the females breed, there is selection of males, are fulfilled in almost the same way. In Grant's gazelle, *Gazella granti*, for example, (Walther, 1965) the breeding group consists of an adult male, with a group of females and juveniles. The group lives within a territory defended by the male; the territory is moderately stable, although there may be some degree of seasonal shifting of ground. The groups Walther studied averaged about twenty animals and the territories were approximately 300 metres in diameter. As in the Camelidae, the younger unmated males form bachelor herds on the periphery. Although the latter groups tend to stay within a definite home range, they do not defend it and their ranges show some overlapping. In contrast to what occurs in the vicuña, however, the parturient female leaves her group and remains apart with her new born young for a few days. After this she will join other females with young and form a temporary association with them, usually close to the territory of her mate. Later on, she and her young will rejoin the male in his territory. Peripheral to the main breeding herd in charge of the territorial male, there are thus two other groupings, which are less stable and have more floating populations. These are firstly, the bachelor herds of young males who have not yet acquired a territory and a group of females; secondly, the group composed of the recently calved females and their young, who have not yet rejoined the male in his territory. The organisation in Thomson's gazelle, *Gazella thomsoni*, is essentially similar, although the females and young tend to stay apart in their own group for a longer period before they return to the territorial male (Walther, 1964a).

In the impala, *Aepyceros melampus*, studied by Schenkel (1966b), there is also an organisation into breeding herds consisting of a single adult male, with a number of females and juveniles. Here, however, there is no distinct territory: the male herds his females, attempts to prevent their becoming scattered and attacks or threatens any rivals who approach his group, but he does not defend a specific area. The male yearlings are driven out and join with bachelor herds, the older members of which attempt to secure for themselves a group of females. This a young male may do either by challenging another already in possession of a group or by herding off for himself a few females from a breeding herd. The size of the herds is extremely variable; a breeding group may contain anything from two to over a hundred females, together with their young and a bachelor herd from two to about sixty males.

In these cases, although the organisation is very like that of the Camelidae, there is some degree of separation of the functions of ensuring adequate feeding and selective breeding, in that the females and young have a temporary feeding area outside the breeding territory. In the Uganda kob, *Adenota kob*, this process has gone much further; here male territorialism is concerned purely with breeding and is unrelated to feeding (Buechner and Schloeth, 1965). Within the home range occupied by the females and juveniles a central "territorial breeding ground" contains the mating territories of the mature males. These territories are extremely small, only 15–30 metres in diameter, and there may be thirty or forty of them in the territorial ground. Each male defends his territory and here the females come to him for mating. Territories are not all equivalent in value: the central ones are the most hotly contested and the ones in which most breeding occurs. In the Toro Game reserve, where the observations were made, breeding occurs all the year round and any particular male may hold a territory for a very variable period—anything from a couple of days to what is described as "semi-permanent". In the latter case, the male periodically leaves the territory to feed or drink and then returns to it, which emphasises the fact that here the territory is concerned solely with mating. Leuthold (1966) has recently extended Buechner and Schloeth's work and has found that in addition to the males occupying the central territorial ground, there are always a number of others scattered about the periphery, each holding a larger territory some 100–200 metres in diameter. There is a rough gradient, with the territories in general smaller towards the middle of the range until finally they form the central clump which constitutes the territorial breeding ground. Although the peripheral males do defend their territories, there is little competition for them and the attention of the females is concentrated on the central area, where almost all the mating occurs. Bourlière (quoted by Buechner, 1961), studying the same species in the Congo, failed to find this type of social organisation. Leuthold's observations now make it clear that this was because the population density was too low to produce the characteristic clumping into a central territorial breeding ground.

In the waterbuck, *Kobus ellipsiprimnus*, the need for access to water affects the social organisation. In the study made by

Kiley-Worthington (1965) in Nairobi National Park the males each held a territory one of whose boundaries was formed by a length of river bank. The region next to the river, which included the best cover for sleeping and was close to the favourite grazing grounds, was the most strenuously defended and here the territorial boundaries were sharply demarcated. Passing further away from the river, territorial behaviour became less intense and the distal parts of the boundaries were often not very clearly defined. Within the same area, the females have their home ranges. By day they congregate in groups of up to thirty individuals, but they separate into small family groups at night. The home range of a female group overlaps those of several males and the latter attempt to herd as many females as possible within their territories. The system is thus rather similar to that obtaining in Thomson's gazelle, modified in relation to the linear arrangement of the male territories which, in its turn, is dictated by the need to include a strip of river frontage within the territory.

In many artiodactyls, seasonal movements of the herds in relation to changing food supplies make the holding of a stable breeding territory impossible. This, however, need not abolish "territorial" behaviour on the part of the males; instead of a definite area, the group of females itself is defended, regardless of the fact that they are constantly on the move. We have already had an example of this in the impala, where the area over which the animals feed is too large, in relation to the degree of cover, for its defence as a spatial unit to be possible. The same principle operates even more strikingly in the case of the blue wildebeest, *Connochaetes taurinus*. In the animals studied by Talbot and Talbot (1963) in Masailand, mating occurs during the period when the animals are on the move from the open plains towards the bush. The large herd splits up into temporary breeding units consisting of a group of females with the young of the previous year together with the male, who herds them together and drives off other adult males. The size of the breeding groups is highly variable and may be anything from two or three to over 150 individuals. If the group of females and yearlings is large, then two or even three males may share it and cooperate in defending it against others, although showing no animosity to each other.

This form of organisation is adapted to areas where seasonal changes in vegetation make a more or less nomadic existence necessary. The wildebeest is, however, capable of leading a settled existence if the habitat provides permanent water and grazing. The social organisation is then modified and very much resembles what has been described in the waterbuck. The large aggregations, including males and females of all ages which characterise migratory life, are replaced by subdivision into the usual three categories; relatively small groups of females and juveniles, territory-holding solitary adult males and bachelor herds of non-territorial males. The female herds occupy home ranges that overlap the territories of several males. Each territory holder attempts to retain as many females as possible within his area while the bachelor herds are excluded and have therefore to occupy the peripheral and less favourable parts of the range; an arrangement which automatically reserves the best grazing for the females and young. Estes (1966) has made a study of this species in the Ngorongoro crater, where, owing to the peculiar environmental situation, both types of organisation coexist. A minority of the total population shows the sedentary social organisation just described, the rest form large herds with the usual migratory structure. The behaviour of the latter animals, however, is not entirely typical since the boundaries of the crater restrict movement and the high productivity of the area permits the population to reach an unusually high density. These factors result firstly in there being little of the long range movement to which the social organisation into large herds is primarily adapted; the migration of only a very small proportion of the animals takes them beyond the crater and back again. Secondly, the aggressive activities of the permanently territorial males, who would not be present in the typical migratory habitat, tends to disrupt the large herds. This frequently results in young being separated from their mothers and falling prey to predators much more easily than would be the case in more normal circumstances.

Browsing species living in woodland or forest, where poor visibility makes communication more difficult, are in general much less social than the grazers of the open plains and tend to be solitary or to live in small groups. In the bush buck,

Tragelaphus scriptus, for instance, the male and the female with her young each have their own territories. These are arranged in a radiating manner around a common grazing ground, to which the animals repair in the evening (Verheyen, 1955). On one occasion Verheyen saw ten animals together in such an area. Here contests between males may occur and some form of bond between individual males and females may be established. Verheyen does not describe mating, but he reports that one male of particularly splendid appearance had only to show himself on the grazing ground to be at once joined by one or other of a pair of females. He would then return to his territory, accompanied by the female.

In domestic cattle the normal practice of keeping the sexes separate and castrating excess males does not permit observation of the natural social relationships. Schloeth (1961), however, has studied the behaviour of the semi-wild Camargue cattle. Here, during the winter, when food is scarcest, the animals tend to scatter, but in spring they unite to form a more stable breeding herd. This includes both sexes and within the herd a rank order is established, in which adult males are superior to females and they in turn are superior to juveniles of both sexes. There does not appear to be any tendency for the males to gather a group of females together and defend them, but mating is related to the rank order and the superior male will drive away an inferior from a female on heat. It is not clear whether the absence of territorialism should be regarded as primitive, or as secondarily derived from an originally more complex system, possibly of the type found in the wildebeest.

Amongst wild sheep and goats, the most complete study is that of Geist (1964) on the mountain goat, *Oreamos americanus*. The study was made in British Columbia over a period of two seasons. The main rutting time is in November and Geist found that the associated behaviour was greatly affected by the weather. In 1962 there was little snow and the animals were free to move over a large area. Each male then formed an association with one or more females and defended them against any rival but no attempt was made to herd them within a fixed area. In 1961, conditions were very different: heavy snow kept the animals within a relatively small area.

In this situation males and females formed a single group in which the males did not appear to restrict each others' movement and they did not defend individual females.

Amongst the Cervidae there are many parallels with the Bovidae but, in general, their habitats are such as to involve more seasonal changes of feeding ground and nothing approaching the complexity of organisation shown by the Uganda kob is known. In this family, Fraser Darling's (1937) classical study on the red deer, *Cervus elaphus*, still remains one of the most thorough and complete. In this species, outside the mating season, the sexes tend to remain separate. The female herds, which include juveniles of both sexes, each occupy a home range, within which there is frequent separating into small family sub-groups. There is a seasonal shift down from the higher ground to more protected areas during the severe winter weather and back again in spring. The males form smaller groups and their ranges tend to be grouped peripherally round those of the females, although there is some overlapping. Towards the end of September, the rutting season begins. The adult males move in on the female home ranges and there each attempts to establish a territory and herd within it as many females as possible. The territories are not very stable and boundaries are subject to continual adjustment, as the number of competing males first increases and then decreases with the onset and termination of the rut. The picture is not very clear, but it appears that here, as in the wildebeest, the group of females is of more importance to defend than is a particular area.

The same principle is more strikingly shown in the reindeer, *Rangifer tarandus* (Espmark, 1964). Here, during the rut, the male herds a group of females together and chases away any other male that attempts to approach them. There is no spatially determined territory and the group may shift around over a considerable area.

(b) *Perissodactyla*

In the Equidae, Antonius (1937) distinguishes two main types of social organisation; one characteristic of the true horses, or *caballus* group, the other of the asses. The former are highly gregarious and a large herd of females and juveniles is formed. This is dominated by a single male, who both defends and directs the movement of the group. In the Przewalski horse, *Equus przewalskii*, Dobroruka (1961) reports the same behaviour in captive herds and notes that the dominance of the male is also shown by his having priority in feeding and drinking. The herd male drives off the juvenile males as they start to mature and they must then live a peripheral existence until they can succeed in winning a group of females to form a herd of their own.

In the asses, however, the sexes remain separate for the greater part of the year (Antonius, 1937). The females and juveniles form groups, each under the leadership of an experienced female, while the males are solitary, or form smaller groups. In the breeding season the males approach the female herds and each attempts to mate with as many oestrous females as possible. Since courtship in the asses is extremely violent and involves fighting between male and female, the competition between males is indirect, rather than direct. The number of mates a male will get depends more on how many females he has the strength and determination to subjugate than on how many rivals of his own sex he can defeat.

Amongst the zebras the position is much less clear. According to Antonius (1937), in Burchell's zebra, *Equus burchelli*, there are large loosely knit herds, tending to divide into sub-groups and then rejoin. Fighting between males is not very severe and he saw no signs of any type of harem formation. This apparently somewhat casual arrangement Antonius regards as primitive, and he suggests that the situations found in the true horses and in the asses are the result of divergent evolution from such a condition. In Grevy's zebra, *E. grevyi*, males appear to be more aggressive and there is some tendency for the sexes to separate.

A recent detailed study of Burchell's zebra by Klingel (1967) considerably modifies the picture given by Antonius and makes clear that there is, in fact, a very definite social structure. Klingel studied the animals in a variety of parks and game reserves in eastern and southern Africa, covering most of the present range of the species. He found that the basic social unit is a small family group, consisting of a single adult male, together with a few females (from 1 to 6) and their foals. Although the stallion is dominant, the group on the move is usually led by one of the mares, the others following her in order of their dominance. The stallion may occasionally lead but more commonly he brings up the rear, or moves parallel to the group. These family groups are remarkably stable, the same females remaining in their group for the whole of their adult lives while the stallion is replaced only if he becomes weakened by sickness or old age. The young males leave the group of their own accord during their second year and join together to form bachelor herds within which no dominance appears to be shown. Young stallions may later establish a group of their own by abducting young females from an existing family unit.

When favourable grazing conditions bring many groups together to form a large concentration of animals, there is no alteration of the social structure. Both family groups and bachelor herds maintain their coherence and the large herd is merely an aggregation not a social unit. Presumably it was the lack of social organisation within the large herd as such that impressed Antonius and he failed to detect within it the structure of the smaller constituent groups. The organisation described by Klingel closely resembles what has been found in many artiodactyls but with this difference: the coherence of the group appears to depend more upon mutual ties between the females than upon herding activity by the male. The result is a greater long-term stability than is usual in artiodactyl harem groups.

In the Rhinoceroses, the black rhino, *Diceros bicornis*, essentially a browser, is largely a solitary whereas the grazing white rhino, *Ceratotherium simum*, is gregarious. In the latter, small family parties are common but larger groups of up to twenty-four are recorded by Heppes (1958) in Uganda. Each group has its own favourite grazing and watering places and its own pathways linking the two. The mating system, however, has not been described.

(c) *Proboscidea*

In the African elephants, *Loxodonta africana*, which were studied in Uganda by Quick (1965) breeding occurs throughout the year and the social structure is adapted to facilitate this. Most of the time the sexes are separate. Females and juveniles form herds of their own, separate from those of the males and each group remains within its own home range. Periodically, however, they aggregate together to form an enormous breeding herd, which may number a thousand or more. This gathering remains together for a period of three or four days and then breaks up again, the male and female groups returning to their various home ranges. During one of these breeding aggregations, Quick observed that mating was occurring between some of the females and the older males. Fights between males were also in progress on the periphery of the group, but these seemed to be between the younger adults. This suggests that some type of rank order is established and that once this has been settled further fighting is not required and the males recognised as dominant are permitted to enter the herd and seek a mate. It seems likely that any female on heat will mate during the aggregation period. The formation of breeding herds was observed in every month of the year, but it is not clear whether every group in an area participates in a breeding aggregation each month. The "dance of the elephants", so vividly described by Kipling in his story *Toomai of the elephants*, is presumably based on the occurrence of a similar type of breeding aggregation in the Indian elephant.

The principles operating in the social systems of all these large herbivores seem to be the same. For large grazing animals living in open terrain and depending for safety on swift flight, there are advantages in being gregarious. If herds are formed, inter-male competition for females is likely to occur and this will automatically produce selection for male intolerance since the fiercest male is the one who leaves most progeny. Any advantage so gained will, of course, be lost if the male kills his own sons before they can breed. Since his own offspring are amongst the individuals with whom contact is likely to be greatest, there will therefore be at the same time strong selection for avoiding lethal contests. The various types of social organisation that have been evolved all represent solutions to the problem of how to preserve the advantages conferred by male aggression and at the same time avoid its potentially deleterious effects. The simplest and basic one is based on territory. If the individual male holds and defends a territory then the juveniles can be excluded from it but not pursued beyond its boundaries when defeated. They will thus not be lost but remain as a breeding reserve, ready to replace any breeding male that is killed, injured or is past his prime. As a secondary modification, spatial territory may be abandoned and replaced by defence of the females, wherever they may happen to be. Possession of a group of females then functions in exactly the same way as possession of a territory, both in conferring psychological advantage on the owner and in limiting pursuit of a defeated rival.

As will be discussed later, intense inter-male competition favours the development of a type of contest in which threat may be as important as actual fighting, and even when fighting does occur, it is of such a type that there is no great danger of a fatal outcome. Once such a stylised method of deciding who is the victor in a contest has been evolved, the way is open for further changes in social organisation. It then becomes possible to form a rank order which permits the males to remain in each others' company without perpetual warfare, as we have seen occurs in the Camargue cattle.

5 PRIMATES

The habitats occupied by Primates range from dense forest to open savannah. It is therefore to be expected that there will be some variation in social organisation, since the forest environment is less conducive to the formation of large groups than is more open country, where visual contact is more easily maintained. Moreover, in many tropical arboreal species, abundant food appears to be readily available all the year round—a factor which reduces the importance of territory in its primary role, and all observers are agreed that in the higher primates there is very little territorial behaviour. In general, in natural conditions home ranges frequently overlap and when groups meet there is little or no evidence of aggression. This is true both of highly arboreal species such as the howler monkeys, of more terrestrial forest dwelling species such as the langurs (Jay, 1965) and the gorillas (Schaller, 1963) and chimpanzees (Goodall, 1965) and also of the more open-country baboons (Hall, 1962a, b, Hall and De Vore, 1965). In the howlers, *Aloutta palliata*, when two groups meet, they may howl at each other, but this does not lead to fighting and there is no evidence of any clearly defined territorial boundaries (Carpenter, 1934, 1965). In general, it appears rather that habitual routines tend to keep groups largely separate without the need for any strict territorial delimitation or defence of boundaries. In most species of monkeys, large groups may be formed with little or no fighting or threat. A dominance order usually appears to exist, but is often maintained with very little obvious aggression. In the langur, *Presbytis entellus*, for instance, dominance is shown in terms of priority of access to favourite foods and drinking places and to oestrous females, but seems to be maintained virtually without enforcement (Jay, 1963, 1965). Macaques are at the other end of the scale and in them aggressive behaviour is much commoner.

It must, however, be noted that although in natural conditions serious hostilities, particularly between groups, are rarely observed, the potential for extremely aggressive behaviour is not necessarily absent. In restrictive captivity conditions, where avoidance is impossible, fighting is much more common and the introduction of a stranger is almost certain to lead to bloodshed. Hall (1965) quotes a case where an alien adult male and female were introduced to a group of seventeen chacma baboons in the Bloemfontein Zoo. Violent fighting broke out and many of the animals were killed or subsequently died of the injuries they received.

One of the interests in studying the social organisation of the higher primates lies, of course, in the light they may cast on the origins of human society. From this point of view, two groups are of particular relevance: on the one hand, our closest living relatives, the great apes; on the other, those species that are not arboreal or forest dwelling but live predominantly in open savannah, as we believe our early ancestors to have done. None of the great apes lives in this manner and for information on primate social structure in this type of habitat, we must turn to the baboons and the patas monkeys, *Erythrocebus patas*.

The two groups have adapted in different ways to the dangers of life away from the refuge of the trees. In both, sexual dimorphism is extreme, the adult males weighing approximately twice as much as the females, and in both, the security of the group depends on the adult males. The roles they play are, however, very different. In the baboon, their role is active defence, whereas in the patas, the male protects his group by his vigilance, not his fighting strength. Long limbed and lightly built, the patas have been described as the greyhounds of the monkey world (Hall, 1966). Like the grazing ungulates whose habitat they share, their safety depends on their speed and it is the responsibility of the male to be constantly on the watch for danger and to ensure that no predator takes his group by surprise. Another characteristic of the patas is that they are much less noisy than most other monkeys. Hall regards this as related to the need to avoid attracting the attention of predators and refers to it as "adaptive silence".

The social structure of the two species reflects this difference in mode of life. In the patas, the group consists of a single adult male, several females and their young. Males who have not succeeded in becoming leaders of a group are met with as solitaries and Hall (1966) records one small all male party. The usual number in a group is about fifteen; the largest recorded by Hall was thirty-one. Groups are widely spaced and mutually intolerant. When one group was observed to sight another, the male of the larger band threatened and ran towards the others, who beat a hasty retreat. A solitary male was similarly chased off.

In the chacma baboon, *Papio ursinus*, (Hall and De Vore, 1965) the troop generally numbers some twenty-five to thirty individuals and comprises several adults of each sex, together with juveniles and infants. Very large troops of up to 200 have been recorded. Within the group there is a dominance hierarchy amongst the adult males, and young males, although belonging to the group, may remain somewhat peripheral to it. Mating is not according to an exclusive harem system but the dominant males have priority of access to the females at the time when they are at the height of oestrus. Upon the dominant males too falls the main responsibility for defence of the group, although the peripheral males also play an important part in giving early warning of danger. When on the move, the positions taken up reflect the social structure.

In the centre are the females and the youngsters, accompanied by the dominant adult males, while the less dominant males are on the periphery. From whatever quarter danger may threaten, the females and young are thus assured of protection. According to Hall and De Vore (1965) hostile encounters between groups are uncommon. Groups living within the same area generally tend to follow habitual routines that keep them apart and if they sight each other, they usually simply move away. Within the group aggressive incidents maintaining the dominance structure are not uncommon. Although these may be rough and noisy, serious damage is rare.

The populations of *Papio anubis* studied in Uganda by Rowell (1966) were much more arboreal than is *P. ursinus*. In correlation with this habitat, where visual contact is less easily maintained and where long movements across open spaces are less frequent, Rowell found that the social structure was more relaxed and friendly and a male hierarchy was not obvious. In general, organised defence by the males was less in evidence and the "order of march" typical of the chacma was not shown.

In the hamadryas baboon, *P. hamadryas*, the picture is different again. Here, although large aggregations are formed each night at the sleeping cliffs, these split up for the day's activities into groups consisting of a single adult male, a number of females and their young (Kummer and Kurt, 1963). The male prevents his females from straying into any other group, so that here something approaching a closed harem system does appear to operate.

Amongst the apes, the highly arboreal, brachiating gibbons, *Hylobates*, live in small family groups, consisting of an adult male, an adult female and their young—rarely more than half a dozen individuals in all. There is little sexual dimorphism and male and female appear to be equally involved in leading and guarding the group. Neighbouring groups may mingle for a time without fighting, but this appears to be relatively rare (Schaller, 1965b). In the other three apes, the situation is more complex. In the gorilla, *Gorilla gorilla*, the group is larger and generally includes more than one adult of each sex, together with young of various ages. Hostility between groups is rare and different groups may even join together for some hours; more usually, they simply ignore each other (Schaller, 1963, 1965b). In chimpanzees, the tendency for groups to mingle is even more marked, so much so indeed that it is difficult to characterise any stable unit, apart from the tendency of the young to stay with their mother. About sixty to eighty animals living within an area constitute a sort of major social unit which continually breaks up into sub-groups: these may be either males only, or females and young, or may include adults of both sexes. The sub-groups are not stable but mix and change composition frequently (Goodall, 1965; Reynolds and Reynolds, 1965). Little is known about social structure in the Orang-utan, *Pongo pygmaeus*, but there is frequent meeting and parting of sub-groups in the same sort of manner (Schaller, 1965b).

The very free and easy relationships that characterise the social life of the great apes are a reflection of equally easy conditions of existence: food is readily available and dangerous enemies are few. The less secure open savannah demands some more stable system. Groups of females and youngsters cannot roam at will unprotected; some organisation combining vigilance with readiness for flight or for active defence is required. The same must have applied to early man, but the three different solutions found by the patas monkey, the hamadryas and the two other baboons do not go very far in suggesting how our ancestors met the same problem, for in early human organisation a new factor was in operation—the emergence of man as a predator and furthermore a predator not dependent on tooth and claw alone but one wielding weapons. This must have had repercussions on social structure in at least two ways. Firstly, as predation developed from the incidental taking of small and relatively defenceless prey to the deliberate hunting of larger species, cooperation must have become increasingly important. This would favour the retention of sub-adult males as full group members, not relegated to a peripheral position. Secondly, it would become more and more difficult for the females to remain with the hunting group. Early hominids are unlikely to have had a shorter childhood than the living apes and in modern man the period is extended still further. Had the females taken an active part in hunting, there would have been selective pressure against prolonging the dependent childhood stage: it therefore appears likely that some system involving leaving the females behind with their young, and possibly some adult male guards, must have arisen at an early stage of the development of true hunting. This would at once rule out any harem system under the control of a single adult male. These considerations suggest that early hominid social structure must have had a greater stability than is shown by chimpanzees and gorillas but it must have shared with them a greater degree of mutual tolerance and less marked dominance relations than are found in baboon societies.

This survey, although by no means exhaustive, is sufficient to illustrate how social organisation is related to habitat and mode of life and to give some idea of the role of territory in maintaining social structure. Most of the principles that operate have already been mentioned. The simplest and most basic is the degree to which the habitat facilitates or hinders contact keeping between individuals; large societies are only able to exist in situations where the members can easily keep in contact. The second relates to the type of food; large societies can exist only if adequate food can be found for all, within an area easily covered in the normal feeding period. This is generally simpler for herbivores than for carnivores. Prey must always outnumber predators and, in addition, the activities of one hunter alert the prey and so make things more difficult for another—a factor which obviously does not apply to vegetable food. The advantages conferred by group vigilance and group defence will also favour gregariousness in species that are preyed upon but are of lesser relevance to predators. For all these reasons one will therefore expect large social units to be characteristic of herbivores rather than carnivores. There is, however, another factor that works in the opposite direction. Obtaining vegetable food rarely requires cooperation but the killing of large prey may be greatly facilitated by it. Moreover, providing sufficient food for the young without leaving them unguarded for a dangerously long time presents a problem for the predator, whose hunting may be rather a slow business. Predators, especially those killing large prey, may therefore be expected sometimes to form groups although these are not likely to be very large. Herbivore herds may number hundreds but predator packs are likely to be an order of magnitude smaller.

Terrain and type of food are thus the two external factors which have most influence on the evolution of social relations. The latter, however, are directly dependent on the behavioural characteristics of the animals themselves and it is therefore necessary to consider the factors which tend on the one hand to keep individuals together and, on the other, to drive them apart, and then see how these may be expected to interact with the relevant external factors.

The most important cohesive force is the tendency of the young to stay with their mother. This is, of course, essential during the suckling period and when this has ended there may be a further interval before the young are capable of making their own living. Even when they reach this stage, if there is nothing to disrupt the family, there is a strong tendency for the young to remain where they are, in company with the individuals to whom they are accustomed and in the area that has become familiar to them. A second type of bond is, of course, formed between male and female during the mating period. This, however, is frequently only a temporary association during the time when the female is on heat. It is more enduring where there is strong selective advantage in male cooperation in the rearing of the family. For the reasons already mentioned, this is most likely to occur in carnivorous species. The need for cooperative hunting will also tend to cause selection against prolonged male antagonism. In the herbivorous species, however, these considerations do not apply and in them male competition has free play. It is therefore in the large herbivores that selection will favour on the one hand, the formation of large associations and, on the other, the elaboration of male competition. It is thus not surprising that some type of harem arrangement has been evolved independently in many unrelated herbivore groups. In carnivores this is found only in cases where feeding and breeding do not occur together, as in the Pinnipedia, and, to a lesser extent, in the lion, where cooperation between males in hunting is not particularly vital. With the harem system one may contrast the extreme "democracy" shown by the *Lycaon* pack, resulting from strong selective pressure in favour of fully cooperative hunting and guarding of the young.

Where a harem system is in operation, territorial behaviour may be modified. Territorial defence, in these circumstances, may become primarily an agent of male selection: weaker rivals are excluded and the dominant male alone has access to the females inhabiting the territory he defends. In these circumstances, the territory as such may lose its importance and the male, instead of defending a specific area, simply defends his group of females. Rivals are attacked not when they transgress a spatially fixed boundary, but when they approach the females. Examples of this type of behaviour have been cited amongst the artiodactyls and the lagomorphs.

The attainment of a stable social organisation with a high level of male aggression is, however, possible without subdivision into individual harems. This is attained if the aggression is not abolished, but merely controlled by the establishment of a rank order, in which each individual accepts the dominance of his superiors but is ready in turn to penalise any inferior who infringes his rights. This solution has been independently evolved in artiodactyls, rodents and primates. It should, however, be noted that in primate societies dominance hierarchies are complicated by the fact that individual personal relationships are often of great importance. Certain animals may support each other in aggressive interactions, forming a sort of "alliance". The result of an encounter between two animals may therefore depend not only on their individual positions in the hierarchy, but on whether an ally is or is not close at hand. Hall and De Vore (1965) report a case of this sort in chacma baboons, where a pair of males supported each other so effectively that the true top ranking male rarely had the opportunity to demonstrate his superiority. Similarly, working with a caged group of six rhesus monkeys, *Macaca mulatta*, Varley and Symmes (1966) found that the animal who occupied second place in the hierarchy did so only by virtue of her relationship with the top ranker; when the latter was temporarily excluded, she sank one place and at the same time the previously bottom ranking animal rose one step in social scale.

One final point must be made. It is not easy to avoid anthropomorphism when thinking about animal societies. Our own organisations have a certain degree of stability and include various homeostatic mechanisms designed to keep them running and fulfilling the purposes for which they were designed. Animal societies, however, are not based on a planned conscious recognition of the functions they serve but on a series of behavioural patterns and responses to various situations. The social organisation is therefore not something the species "has got" in the same way as it has got characteristic anatomical and behavioural attributes. If the external situation changes, we may therefore expect that the end result of the animals' behaviour may be a change in the social order. This is most clearly shown in the way restrictive captivity conditions may produce a more aggressive society than is met with in the wild. It also becomes clear when we compare populations living in a tropical area and breeding all the year round with others living at higher latitudes and showing defined breeding seasons or dense populations in one area with sparser ones elsewhere. Once we recognise that the social order is not a thing-in-itself, but the secondary result of a complex interaction of the animals with each other and with their environment, variations of this sort cease to appear as distressing discrepancies and become themselves a subject worthy of study.

Such studies are still in their infancy but the large ungulates provide a number of examples illustrating how the social structure within a single species may be modified in relation to the environment. In the Uganda kob, for instance, we have seen that population density—which itself must be related to the carrying capacity of the habitat—determines the presence or absence of concentrated central territorial breeding grounds. The case of the blue wildebeest is even more striking. Here, according to the nature of the area occupied, behaviour may be either migratory or sedentary and each has its own characteristic social organisation. The males are able to show sufficient mutual tolerance to remain together in a single herd only provided the habitat precludes their remaining in one place and acquiring territories. Reproductive behaviour will be dealt with in more detail in Chapter 10, but it is relevant to note here that it is of a type most suited to the migratory life and that there is little alteration when this is replaced by the sedentary territorial habit. Their general distribution suggests that the wildebeest are primarily adapted to the migratory type of life and are facultatively sedentary, rather than *vice versa*, a view which their reproductive behaviour strongly supports. Estes (1966) mentions that in the Queen Elizabeth National Park in Uganda, the topi, *Damaliscus korrigum*, shows a similar ability to adopt either a migratory or a sedentary habit, with changes in social organisation paralleling those of the wildebeest. He does not, however, mention their reproductive behaviour and it is therefore not easy to be certain which should here be regarded as the primary and which as the modified mode of life.

As might be expected, the effects which the unusual environmental features of Ngorongoro exert on wildebeest social organisation are not restricted to that species but have further consequences. Crocutas prey very extensively on young wildebeest calves and the ease with which these can be killed in the crater may be the main factor responsible for the large crocuta population which the area supports. This in turn has led to the reversal of the usual relations between the latter and the lions, to which reference has already been made, the lions driving the crocutas off and appropriating their kills rather than the latter scavenging on the remains of the kills made by the lions. This set of interactions provides a striking example of the complex way in which environment affects ecological relationships and of the important role which behaviour plays in determining these relationships. It is likely that continuing studies in the crater will bring to light still more far-reaching consequences of the peculiar characteristics which make Ngorongoro one of the few terrestrial habitats that in actuality exemplifies what is implied in the theoretical concept of an ecosystem.

CHAPTER 5

Scent marking

Of all the forms of behaviour concerned with maintaining a territory, the most direct are those used in actual physical combat with an intruder. Before dealing with fighting however, it is convenient to consider a different type of behaviour, more directly related to the animal's home ground and less directly to trespassers upon it. The essential thing about a home is that it is familiar. To us, highly visual creatures, the important thing is the sight of familiar objects in their accustomed places and it is almost entirely by visual cues that we find our way about. It is therefore not altogether easy for us to appreciate that for other species olfactory cues may be at least as important as visual ones and that familiarity may be as much a matter of the known smells as of the known sights.

By virtue of the fact that he moves about within a certain area and that his urine and faeces are distributed within it, an animal's home range will inevitably become loaded with his own scent. It thus seems reasonable to suppose that the presence of "own smell" should become important in the recognition of home. Very many mammals, however, are not dependent on such casual methods of smell distribution, but have distinctive forms of behaviour concerned with marking their homes by setting their own smells upon suitable objects. The evolution of this type of behaviour, and of the special smell-producing structures often employed, may have begun purely in relation to the marker himself—by setting down his own smell he makes an area safe; he has been here before; he belongs here. This alone will tend to restrict straying, since the animal in an area lacking its own smell marks will be more timid, more ready to flee, than one reassured by their presence. Extension of the responsiveness to marks may be such that the presence of a foreign one increases the readiness for flight—in fact, constitutes a sort of threat. This will make the system even more restrictive, particularly if population density is such that straying usually takes an animal not only beyond its own normal range but into that of a neighbour. The two functions, reassurance and threat, are to some extent complementary, since anything increasing the confidence and pugnacity of the animal on home ground automatically reduces the interloper's chances of victory. The evolutionary transition from the individual function of self-assurance to the social one of threat is therefore easy and has probably been made independently many times. It is also easy to see that in territorial species, the importance of marking behaviour will be greatly increased and it is therefore in them that we should expect it to be most highly evolved. In view of the way in which territory becomes linked with reproduction, it is not surprising to find that where this is the case, marking may be most in evidence during the breeding season and may be restricted to one sex, or at least be very much more important in one sex. Since it is usually the male who is most active in direct defence of the territory, it is usual for him to be also the partner more active in marking.

Before attempting to consider in any detail the functions of scent marking, it is convenient to deal first with the structures and modes of behaviour that are involved. Their number and diversity attests the importance of marking and at the same time shows that this type of behaviour must have been evolved independently many times over. It is only to be expected that urine and faeces should frequently serve as marking substances—they are odoriferous products ready to hand, requiring only the evolution of special behaviour in relation to their deposition to increase their value as marks. The male hippopotamus marks his pathways between aquatic resting place and terrestrial feeding ground by the deposition of dung at special places along the trail, usually close to some conspicuous object (Verheyen, 1954). The white rhinoceros also marks his paths in this way (Hediger, 1951) and polecats too deposit urine and faeces at particular places along their trails (Lockie, 1966). The hippo has a special method of making his droppings more obvious: as he defaecates, he vibrates his tail rapidly from side to side, so that the faeces are flung out and some get lodged in vegetation above ground level and thus, placed at nose height, are made more noticeable. The male *Cricetomys* sometimes places his droppings above ground level by standing on his hands, with his back legs supported against some solid object as he defaecates, which suggests that the droppings may have some role in marking (Figure 5). The little polecat, *Poecilogale albinucha*, backs up against a solid object so that

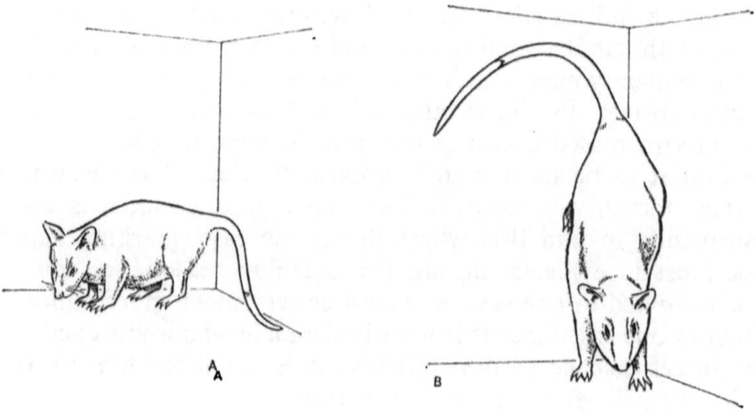

Figure 5. Young giant rat (*Cricetomys gambianus*). (a) Normal defaecation. (b) Defaecation in the hand-stand posture.

the droppings are plastered up against it as the animal defaecates (Alexander and Ewer, 1959). *Aonyx capensis*, the Cape clawless otter, voids urine and faeces together in a somewhat similar manner (Eyre, 1961) while some lemurs mark by smearing their faeces on branches (Petter, 1965).

The use of urine in marking is familiar to us all in our domestic animals. The dog, cocking a leg to urinate, is placing his mark where it is likely to persist and to be noticed. According to Lorenz (1954) he is also leaving a record of his own size since the larger the dog, the higher the mark. Amongst the Canidae, this type of urine marking is widespread. There is, however, variation from species to species in such things as whether the marking is done only (or mainly) by the male or by both sexes and whether marking is restricted to or intensified during the breeding season (see Table 1 page 128 for details). The bush dog, *Speothos venaticus*, is anomalous, in that while the male cocks a leg in the normal canid manner, the female urinates in a hand-stand posture (Kleiman, 1966a). From Kleiman's description, it seems likely that in this case the animal may be marking with tail-root or anal glands as well as urinating.

The tomcat also marks his territory with urine, but uses a different method to place it at nose level. He selects some suitable object, backs up to it, extends the back legs and raises the tail. As the urine is passed, the tip of the penis is curved backwards, so that the urine emerges as a fine jet between the hind legs. A characteristic wriggling movement of the tip of the tail often accompanies the marking. In lions, marking of the same type is used to sprinkle urine on bushes. The male marks in this way much more commonly than the female, and he may also sometimes adopt a squatting position and then wipe his feet in the urine (Schenkel, 1966a). Apart from these special marking procedures, felids normally bury their urine and faeces. Lindemann (1955), however, notes that in both the European lynx and the wild cat, *Felis silvestris*, burying is done only well within the animal's territory; near the boundaries, not only is burying dispensed with, but particularly obvious and salient places are chosen for the deposition of the urine and faeces.

The Poglayen-Neuwalls (1966) describe the use of urine as a marking substance by the olingo, *Bassaricyon sp*. The marks are most commonly made on tree branches, a projecting stump being particularly favoured. The animal crouches slightly over the object to be marked and rubs its genital region over it, at the same time depositing some urine. *Nasua narica* marks in a very similar manner (Kaufmann, 1962). In bears, the habit of marking a tree by scratching and chewing the bark is well known: in addition, they frequently urinate on the marking tree (Meyer-Holzapfel, 1957). Amongst rodents, urine marking occurs in rats and mice, in the red squirrel, *Sciurus vulgaris* (Eibl-Eibesfeldt, 1958) and in the mountain beaver (Wandeler and Pilleri, 1965). In *Cricetomys* a urine trail is deposited only when the animal is anxious or uncertain and its significance is not clear. Amongst marsupials, marking with urine is recorded in the sugar glider, *Petaurus breviceps* (Schultze-Westrum, 1965) and the brush-tailed possum, *Trichosurus vulpecula* (Kean, 1967).

Amongst the lower primates, urine is extensively used in marking and there is special behaviour to ensure its effective deposition on branches. In lorises and in some New World monkeys, the urine is rubbed on the hands or feet and thus transferred to the branches on which the animal climbs. In African lorises the movements used are extremely rapid and stereotyped: one paw is held cupped in front of the urinary opening and filled with urine and this paw and the ipsilateral foot are then rapidly wiped across each other two or three times, the animal supporting itself the while with the limbs of the opposite side. The technique used by the capuchin monkey, *Cebus apella* (Nolte, 1958) is similar but the slow loris, *Loris tardigradus*, described by Ilse (1955), showed a slightly different technique. Here, although the urine is caught in the cupped hand, the hind foot is first brought into contact with the urinary opening and apparently used to squeeze out a little urine. The bush baby, *Galago crassicaudatus*, is peculiar, in that the urine is put directly on the foot (Eibl-Eibesfeldt, 1953b). Amongst the lemuroids, *Microcebus murinus* uses urine for marking and applies it in the same way; some species smear it on branches with their hands (Petter, 1965) while others simply wipe the urinary opening on a branch as they micturate.

Many species, however, are not dependent on urine or faeces, but have evolved special glands producing the marking substances. These have the advantage that their use can be restricted to occasions when marking is necessary and, furthermore, development of the glands may be linked with the reproductive cycle in a way that is not possible with urine or faeces. The commonest smell producing glands are anal glands but in some species scent glands are associated with the genitalia. The behaviour involved in applying the secretions is, however, very similar and smell glands situated in the anal and genital regions will therefore be considered together. The usual procedure in applying such secretions is to evert the gland and dab or wipe the hindquarters on the ground or whatever object is to be marked. Sometimes this anal drag is prolonged and the animal may pull itself forwards for a short distance in a sitting posture.

Marking of this type occurs in monotremes, marsupials and a variety of placental orders. The spiny ant-eater, *Tachyglossus aculeatus*, everts the cloaca and rubs it on the ground, leaving a smelly mark (Hediger and Kummer, 1956). The brush-tailed possum and the sugar glider have two types of cloacal gland and the dasyurids, *Sminthopsis crassicaudata*, *Dasycercus cristicauda*, and *Sarcophilus harrisi* also mark with cloacal glands, using the anal drag to apply the secretion to the substratum.

In *Antechinus flavipes*, the yellow-footed marsupial mouse, the cloacal secretion may be rubbed on twigs or branches but in captive animals it is also liberated in another way. My animals were kept in a glass fronted cage and I found that the glass gradually became spattered with little drops of a milky fluid which was quite distinct from the urine. How this happened remained a mystery for some time, but in the end I was lucky enough to see the male, as he ran head first down a tree trunk, emit the fluid as a little squirt from the cloaca. Whether this is a normal form of behaviour or a captivity artefact is not known. If it is normal, it is peculiar in that the secretion is not being placed on a definite object, but distributed in a rather general manner over the areas frequented by the animal.

Amongst rodents anal marking occurs in the dormouse, *Glis glis*, (Koenig, 1960), the Muskrat, *Ondatra zibetica*, (Darchen, 1964), the guinea-pig (Kunkel and Kunkel, 1964), the agouti (Roth-Kolar, 1957), the acouchi (D. Morris, 1962) and I have seen it in *Atherurus* and *Thryonomys*. The anal glands of the marmot, *Marmota marmota*, produce a strong-smelling secretion which Müller-Using (1956) believes is used in territory marking. Koenig (1957), on the other hand, considers that the glands are not concerned with territory but have a purely defensive function.

In the Carnivora, anal gland marking is characteristic of viverrids and mustelids. Amongst the latter, the secretion is applied by a simple anal drag in the badger, *Meles meles* (Eibl-Eibesfeldt, 1950), the polecat, *Mustela putorius*, and pine marten, *Martes martes* (Goethe, 1938, 1940), the honey badger, *Mellivora capensis* (Sikes, 1964), and the tayra, *Eira barbara* (Kaufmann and Kaufmann, 1965). The same method is used by the viverrid, *Mungos mungo* (Dücker, 1965). Many herpestines, however, although they may use this method if a flat surface is to be marked, usually set their marks on objects above ground level. The male *Suricata* raises one hind leg, much as though about to micturate and wipes the everted anal glands downwards along the surface to be marked (Plate III). The Indian mongoose, *Herpestes edwardsi*, may use the same method, but is also capable of going one better and setting the mark as high above the ground as the length of the body permits by rearing up on both fore paws and then wiping the anus downwards. Anal marking in this hand-stand position is also found in *Atilax paludinosus*, *Crossarchus obscurus* (Dücker, 1965) and the dwarf mongoose, *Helogale undulata* (Zannier, 1965). Amongst the viverrines, the same method is used by genets, *Genetta*, but the civets, *Viverra* and *Viverricula*, mark with the secretion of glands situated near the genital opening which they press or rub against a suitable surface; first backing up to it, very much like a tomcat about to urine spray. The fossa, *Cryptoprocta ferox*, is exceptional amongst viverrids in that marking is done by glands which are not situated in the anal or genital region but on the chest. Both sexes mark, but the glands are larger in the male than in the female and are maximally active during the breeding season (Vosseler, 1929). A number of lemurs also mark by rubbing the anal or genital region against branches.

Saliva, like urine, is a secretion available to all mammals and it is not surprising to find that it too has been pressed into service as a marking agent in some cases. A number of marsupials deposit saliva on twigs or other suitable objects by mouthing or chewing them. Such behaviour is recorded in the sugar glider (Schultze-Westrum, 1965) and I have seen it in *Sminthopsis crassicaudata*, *Dasycercus cristicauda* and *Antechinus flavipes*. A much more complicated method of using saliva is described by Eibl-Eibesfeldt (1965) in a tenrec, *Echinops telfairi*. The animal first salivates copiously on the object to be marked and then scratches with one paw alternately in the saliva and on its own body, working from head back to hindquarters. Presumably this serves to impregnate the saliva with body odour—at any rate, Eibesfeldt found that after one animal had spent a quarter of an hour marking his hand in this way, the smell of tenrec transferred to him was very strong. The curious self-anointing behaviour of hedgehogs, in which the animal covers itself with its own saliva, may bear some relationship to this type of marking, but does not itself appear to constitute marking since the saliva is transferred directly from the mouth to the animal's own body. Burton (1959) and Herter (1957) discuss this habit more fully.

Marking with saliva alone does not seem to be very common in placentals, but glands at the angle of the mouth occur in many rodents (Quay, 1962, 1965) and in *Cricetomys* licking of the object to be marked is combined with rubbing the sides of the face upon it. Despite the commonness of glands near the mouth or on the face, their use in marking has been described in relatively few cases. Some ground squirrels, including both American (Balph and Stokes, 1963) and African species (Ewer, 1966), mark by rubbing with the sides of the head and so does the American porcupine, *Erethizon dorsatum* (Shadle, Smelzer and Metz, 1946); I have also seen *Thryonomys* do so. Glands on other parts of the head are present and used similarly in a number of species. In the marmot, for instance, they occur in the area between eye and ear (Koenig, 1957). In the Australian hopping mice of the genus *Notomys* there are glands under the chin in some species and further back, in the sternal region, in others. Amongst the lagomorphs also the glands lie under the chin in the common rabbit and the swamp rabbit, *Sylvilagus aquaticus*, while in the pika, they are situated just below and behind the eye. Amongst the Carnivora, *Herpestes edwardsi* (Dücker, 1965) and *Helogale undulata* (Zannier, 1965) mark by rubbing with glands situated on the face as does *Atilax*. In the domestic cat there are hypertrophied sebacious glands on the chin, along the lower margin of the mouth. A cat will rub this region against his owner's face in greeting and in its natural context this behaviour is related to courtship, rather than to territorial marking.

It is, however, amongst the artiodactyls that the development of marking glands on head or face reaches its greatest development. Many species possess preorbital glands which range from a sort of gutter-like prolongation of the inner corner of the eye in the axis deer, *Axis axis*, to a deep pit completely separate from the eye in many antelopes. Glands on the forehead or near the base of the horns are also present in a number of species. Two ways of applying the secretions of these glands have been described. Where the object to be marked is something firm, such as a tree trunk or a large branch, the head is rubbed against it. This pattern is used not only by species possessing antorbital glands such as the Wildebeest (Talbot and Talbot, 1963), but also by species of *Tragelaphus*, where such glands are absent. Here the action is possibly effective in applying the secretion of small apocrine glands on the cheeks (Walther, 1964b). When grass or small yielding shrubs are being marked a pattern which Walther calls "weaving" is used. The animal lowers its head and swings it gently from side to side, so that the face is repeatedly wiped across the vegetation. This movement is used by Grant's and Thomson's gazelles (Walther, 1964a, 1965) and the mountain goat also uses it to apply the secretion of its supra-occipital glands (Geist, 1964). In the camel, glands situated on the neck are rubbed against the object to be marked (Pilters, 1956) and the bear also rubs its neck on its marking tree.

Apart from glands situated in the region of the anal or genital openings and on the head, a number of species have developed scent glands on other regions of the body. Their location is usually such that the animal's normal activities will serve to transfer the secretion to its surroundings. Pedal scent glands occur in species as diverse as marsupials, mice and artiodactyls. Glands on the sides or flanks occur in a number of small rodents that either inhabit burrows or make runways through the vegetation. Shrews, also runway users, have similarly situated glands. In some of these cases, in addition to incidental transfer of the scent to the surroundings, there is special behaviour for its deposition at specific times and places. The hamster (Eibl-Eibesfeldt, 1953a) wipes the secretion of its lateral glands onto its feet and the vole, *Arvicola terrestris*, first wipes its feet across the glandular area and then stamps on the ground (Frank, 1956). In the peccary a large gland is situated mid-dorsally on the hindquarters and the secretion must be transferred to any vegetation overhanging the pathways frequented by the animal; the same may be true of the supra-caudal gland of the male guinea pig (Martan, 1962). As long ago as 1923, Lang pointed out that amongst the elephant shrews the location of glands in the pectoral region or on the lower surface of the tail is related to ease of transfer of the secretions to the substratum in the various types of terrain inhabited by the different species. The tail gland touches the ground as the animal lands from a jump and is characteristic of species living in open rocky areas while the secretion of the pectoral gland is likely to be rubbed off when traversing vegetation and is found in species that frequent grassy areas. Possibly also concerned with incidental transfer are the glands that occur on the dorsal surface of the tail in many of the Canidae (but not the domestic dog). Their position suggests that the secretion is rubbed off on the roof of the entrance to the den or burrow, but no detailed study of their function has been made. Mid-ventral glands are present in many cricetine rodents (Quay and Tomich, 1963) and their location may possibly be related to the deposition of scent on frequently used trails. The hamster has such a gland in addition to the lateral glands and Quay and Tomich (1963) also mention its occurrence in the shrew, *Blarina*. Glands in the sternal region are present in a number of arboreal species; those of *Cryptoprocta* have already been mentioned. In arboreal marsupials sternal glands are common and occur in the koala, the phalangers, *Trichosurus* and *Petaurus* and the dasyurid, *Antechinus*, but they are also present in the terrestrial red kangaroo.

Amongst the primates, sternal glands occur sporadically in a number of species that are not particularly closely related. Their presence has been recorded in the tarsier, *Tarsius* (Hill, 1955) and in two New World monkeys; the spider monkey, *Ateles* (Hill, 1957) and the titi monkey, *Callicebus moloch* (Moynihan, 1966). Amongst the Old World monkeys, the drill, *Mandrillus leucophaeus*, possesses a functional sternal gland (Hill, 1944) and amongst the apes, the gland occurs in the orang-utan but is very small (Schultz, 1921) and whether it is of any functional importance is not known. In the sifaka, *Propithecus verreauxi*, the male has a gland situated more anteriorly, in the throat region, while the douroucouli, *Aotus trivirgatus*, has a gland at the base of the tail. In the gentle and ringtailed lemurs, *Hapalemur griseus* and *Lemur catta*, there are two specialised glandular areas, one on the forearm and one near the axilla. In *L. catta* the use of these glands in an aggressive encounter involves a complicated series of actions. The forearm gland is first wiped briefly across the opening of the shoulder gland and the tail is then pulled down between the forearms several times, which must impregnate it with the scents of both glands. Facing his opponent, the animal then stands with his tail arched forwards over his head and quivering violently in an up-and-down movement, so that dissemination of the scent is combined with a visual display (Jolly, 1966).

The types of marking known to occur in marsupials and in three of the major placental orders are summarised in Table 1 (page 128). From this a number of points are apparent. Firstly, there is no uniformity of marking methods within the larger orders and furthermore, it is obvious that similar methods of marking have been developed independently in unrelated species. It seems clear too that in mammals specialisation of skin glands into scent glands capable of being used in marking may occur virtually anywhere on the body surface. One cannot imagine this occurring otherwise than in relation to the animal's existing habits, selection placing a premium on intensification of smell production in those areas where the secretion is most likely to be transferred to the surroundings. The location of sternal glands in tree climbers and of lateral glands in the burrowers or runway users reflects this relationship and the widespread occurrence of glands near the anus or genital opening may stem from the common habit of wiping this region of the body on the substratum in response to any irritation. Glands on the face may similarly be related to the same type of rubbing in response to irritation. Once a smell-producing organ has been evolved in a particular location, however, there will be selective advantage in changes in behaviour which increase the efficiency of scent deposition. The pattern of wiping the feet on the lateral glands in the hamster and in *Arvicola* is an example of this and in the latter, the pattern has been further elaborated by the addition of foot stamping. In the Herpestinae too we can see evidence of a series of modifications of the simple anal drag method of scent application from the leg cocking of *Suricata* to the full hand-stand posture. The latter method is used by species that live in forested regions to apply the scent to tree trunks or to branches that overhang their pathways. There is thus a reciprocal interplay between behaviour and structure: behaviour may decide where the scent glands are developed in the first place, but the position of the glands then decides what new forms of behaviour will be adapted to improving the efficiency of smell deposition. The animal's general structure and habitat are of course also important in deciding what form these patterns take. This is most clearly seen in the case of the use of urine as a marking substance. The primates, with highly mobile limbs, have developed various patterns employing hands and feet in urine application but the arboreal carnivore, *Bassaricyon*, merely squats over the object to be marked. Leg cocking in the Canidae may have started as a movement whose original function was to ensure that the urine did not soil the animal's own coat and the same may be true of the retromingent habit of the Felidae: the behavioural difference may be related to the differences in penis structure in the two families. Amongst the viverrids, the male *Suricata* also cocks a leg to micturate, but there is no evidence that the urine has any marking significance in this species.

It is, of course, one thing to observe and describe marking behaviour but quite another to determine the significance of the marks in the life of the species. A visitor from Mars might find it a rather slow business to work out the meanings of the visual signs we set up and respond to, but with patience and application most could gradually be deduced by correlating the details of the signals and their locations with the responses shown to them. In dealing with olfactory signals, however, we are in a more difficult situation. We have virtually no means of comparing the detailed characteristics of one signal with those of another. Our noses may suggest that cats respond to the smell of valerian because it has something in common with their own odour but mine gives me no help in explaining why fresh tortoise droppings should trigger off a frenzy of anal marking in a male kusimanse but have no such effect on a female marsh mongoose. Frequently we cannot detect a mark unless we see it made and the fact that the marks are not permanent adds another difficulty. A further complication is that comprehension of marking behaviour involves two considerations —the motivation of the marker and the subsequent effect of his action upon the animal that responds to the mark. In general, since both are the products of selection, we must expect a considerable degree of correspondence between the two. If a mark has the social function of acting as a threat, this will be most effectively fulfilled if it is made in appropriate circumstances, which is equivalent to saying that the motivation for marking must here be largely aggressive. The making of a mark whose main function is self-reassurance, on the other hand, is likely to be more defensively motivated, since the appropriate time for it to be made is when the animal is more ready to retreat than to attack. Methods of study therefore include both analysis of the circumstances in which marking occurs and of the responses shown to the marks. If the species studied is one with a rich vocabulary of expressive movements, these can act as a sort of behavioural Rosetta stone, for in them the animal, whether making his own mark or smelling a foreign one, will express his mood in a language which we can learn to read.

Studies of this type have made it clear that marking has not one, but several functions. These overlap and intergrade considerably, which means that even where a species has more than one method of marking, it may not be possible to ascribe to each a perfectly clear and distinct role. The complex interplay of reassurance and threat has already been mentioned: the two are not mutually exclusive but complementary, and the same mark can function in both ways, the relative importance of the two varying from case to case. This is shown by comparing the agouti studied by Roth-Kolar (1957) with the Dieterlen's (1959) golden hamsters, *Mesocricetus auratus.* In the agouti Roth-Kolar judged reassurance to be the more important and the animal's behaviour showed that any new object was a source of unease until it has been made safe by marking. On one occasion this cost an animal its life. A sleeping box was removed for cleaning and returned after being thoroughly washed and so transformed from "home" into something unfamiliar and dangerous. Rather than risk entering it, the animal stayed out all night and was found dead of cold in the morning. In the hamsters (both *Cricetus* and *Mesocricetus*) when the animal marks, its ears and tail are in positions indicating readiness to fight; moreover, when presented with the smell of a stranger, the male will frequently not only mark in reply but also threaten as though the rival were actually present. Dieterlen's animals were raised from a very early age (four to ten days after birth) in an environment devoid of any smells of biological significance, but they still showed this type of behaviour when adult. Here, clearly, the threat element in marking is of considerable importance. The same appears to be true of cloacal gland marking in two species of possums, *Trichosurus vulpecula* and *T. caninus*. Thomson and Pears (1962) found that males responded with violent scolding and threat to a piece of tissue impregnated with the cloacal gland secretion of another male, whether of his own or the other species. On the other hand Jordan et al. (1967) record a lone wolf cowering and showing signs of anxiety on smelling the urine marks left by the main pack. In the rabbit, the motivation for territorial marking with the chin gland probably includes both agressive and defensive tendencies (Mykytowycz, 1962, 1964, 1965).

Any form of behaviour which discourages a rival may, at the same time, increase the self-confidence of the animal that performs it and a smell which acts as a threat to a stranger may also function in this way. Two cases in which the smelling of its own glandular secretions appears to act thus as a form of reassurance behaviour are known. Kühme (1963) records that the African elephant, when facing a rival, will sometimes turn his trunk round and smell his own temporal gland. This appears to increase the animal's confidence and make him more likely to attack. Sharman and Calaby (1964) record that in the red kangaroo, *Megaleia rufa*, rival males, when threatening each other, will lick or bite at their pre-sternal glands. No comment is made on the significance of this behaviour but the similarity of the situation in the two cases suggests that the glandular secretion is being used to increase aggressive motivation in the same way as Kühme believes the elephant uses its temporal gland. In neither case is anything known about any other function of the glands concerned. In the coati, when rival males face each other and threaten, they may also indulge in bouts of urine marking (Kaufmann, 1962). Their other behaviour does not suggest that the marks of a rival have any great threat significance, so here too, the main function of the marking accompanying a hostile encounter may be that of increasing the aggressive motivation of the marker.

The complex process of tail anointing which has been described in *Lemur catta* (page 114) very probably also acts as a form of reassurance. The complex series of actions is difficult to account for on any other hypothesis, for if the object were only to direct the scent towards the opponent, merely waving the arms at him would surely be more effective. Possibly the peculiar performance involving the use of the sternal gland which Hill (1944) has described in a captive drill is also a form of reassurance behaviour.

Jolly (1966) describes the behaviour of *Lemur catta* as a form of "self-advertisement", by which I judge her to mean very much the same as I do by the term "reassurance behaviour". I prefer the latter name, since it does not imply quite such complex mental processes. "Advertisement" implies an appraisal on the part of the advertiser of the probable effects of his behaviour on its recipient; "reassurance" only that he "feels better" as a result of the performance.

A foreign mark can, however, convey information without necessarily constituting a threat. Where territories interpenetrate, the same marking places may be used by a number of individuals who thus leave a record of their comings and goings to be read by subsequent visitors. Such information may be of importance in the type of social life described by Leyhausen (1964) as characterising many solitary species. The lynx's habit of depositing urine and faeces on conspicuous places near territorial boundaries has already been mentioned. These signposts are regularly visited and inspected by neighbours (Lindemann, 1955) and thus each animal can keep a check on his fellows without actually seeing them; a newcomer in the area will at once be detected and may be challenged and required to fight for his right of admission to the community. Possibly even more important, the vanishing of a familiar mark will advertise a disappearance from the locality and constitute the equivalent of a "to let" sign. The significance of "bear trees" has been much disputed, some authors believing that they are of importance in territorial marking, others denying that they have any such role. If, in fact, they do not constitute threats so much as "general information disseminators", most of the conflicting observations can be reconciled. In domestic dogs urine marks can act in this way and the behaviour of hippos and of beavers suggests that the same is true of their marks. In the hippo a number of males have a system of paths which are partly distinct but cross or join here and there and may have sections which are shared. The dung piles at crossings or on common sections are used by all the animals frequenting the track and each will smell the communal pile before adding his contribution to it (Verheyen, 1954). Beavers also use set pathways and treat each other's anal gland markings in the same sort of way (Grassé, 1955). Leyhausen and Wolff (1959) suggest that in such cases the marks are used as a means of avoiding encounters. A very fresh mark implies that to proceed is likely to lead to a hostile encounter and is equivalent to a railroad "section closed" signal; a slightly older mark merely means that some caution is desirable, while the absence of any recent mark is a go ahead signal. Before going on its way, the animal deposits its own mark, which will then "close the section" for the next comer.

It must, however, be borne in mind that a stable organisation of mutually tolerant neighbours using their sign-posts in this way is something that has gradually been built up and any alteration may lead to a temporary outbreak of hostilities. The basic motivation of the marking is therefore presumably the one usual for marking with a territorial significance—a combination of aggressive and defensive tendencies in proportions which differ in different cases. Once an animal has established his place, however, the situation has changed. He has learnt to know his neighbours, as they know him; the system has stabilised with each accepting the presence of the others within limits which, although they may have been determined originally by fighting, now no longer require hostile action for their maintenance. This learning process affects the marking behaviour too: the animal habituates to the marks that he knows, so that their threat value wanes; he also learns the locations of the signposts. Marking then becomes an action associated with a particular place and not strictly tied to a particular mood. This was perfectly clear in the male *Suricata* which I kept as a domestic pet. In this species, one of the social duties of the male is defence of the family and, when a litter of young was present, the sight of strangers coming to view them was enough to set the male marking. He would also usually mark after an encounter with a dog. In both cases the marking occurred in association with threat behaviour. This animal also had regular signposts: a particular corner of a wall some distance from our home, for instance, was marked whenever our walks took us near it. Here, however, there was no visible change in mood, no sign of threat: his whole demeanour showed that he recognised the place and was marking simply because he had come to the place to mark.

In the cases just discussed, the function of the sign posts is the preservation of the territorial *status quo*. In rats, on the other hand, odour trails seem to act as communal guide marks for all the members of a colony and even strangers tend to use a marked path rather than to avoid it (Barnett, 1963). In shrews, their function appears to be the opposite—to enable an animal to avoid a path recently traversed by another. The glands are most highly developed in males, less in anoestrous females and least of all in oestrous females. Crowcroft (1957) points out that this may indirectly assist the female in finding a mate since, when she is in oestrus, her "keep away" signal is turned off.

The information transmitted by marks is obviously not available only to conspecifics. The scents of predators are of biological significance to potential prey and *vice versa*, but the degree of responsiveness to odours of foreign species is highly variable. Roth-Kolar's agoutis, for instance, although disturbed by the absence of their own smells, would unconcernedly carry in dog hairs to line their nest boxes, whereas Dieterlen's golden hamsters, reared in olfactory isolation, gave immediate defensive responses to the smells of dog and polecat. My *Suricata* also growled at the smell of dog droppings. This emphasises a point of some importance, to wit that whether a mark does or does not act as a threat is not a function of its own intrinsic chemical characteristics but of the responses of the animal that smells it. This means that in the course of evolution alterations in the significance of marks may be brought about by selection of the behaviour of the receiving animal without any change in the chemistry of the message sent. In the case of the agouti, the hamster and the meerkat there may be a perfectly sound adaptive reason in terms of their respective natural predators why the latter two should be alerted by the smell of a dog but the former have no such response. It would be extremely interesting to test the agouti's responses to the smells of indigenous South American predators, to which a response might have been evolved.

It is not uncommon during courtship for the male to mark the female. In *Cricetomys*, for instance, after initially repulsing him, the female permits the male to approach and he then grooms her round the face and ears and usually also marks her with his cheek glands. In this species the reassurance role of marking seems to predominate and it is very easy to see how courtship marking comes about. The male, in his first direct contact with the female, must, after his initial repulses, be in a mood of anxiety conducive to the performance of any behaviour which increases his confidence. In marking, cheek rubbing is accompanied by licking, so that the transition from the licking involved in grooming the female to cheek gland rubbing is a very simple one. The male rabbit also may mark his female with his chin gland during mating (Mykytowycz, 1965) but here its significance is more difficult to assess: it may be no more than a reflection of the linkage of sexual motivation and aggression that characterises the male of many territorial species.

In the case of courtship marking, the smell transferred to the female does not appear to have any further significance. Sharing of smells may, however, be very important in social species where outsiders are violently attacked and driven off but behaviour is friendly towards members of the group. If smell is one of the main characteristics used in recognising a foreigner, then clearly everything will be much simpler and mistakes rarer if smells are shared within the group. It is then not necessary for each animal to remember individually the odours of every one of his fellows but only to recognise the presence or absence of a group odour. Any form of close bodily contact, such as sleeping together or mutual grooming will not only familiarise the individuals with each other's odours but will result in some smell sharing. A number of species have additional forms of behaviour which serve to increase the efficiency of smell sharing. In meerkats, the male, after having marked an object with his anal glands, rubs his body along the mark and impregnates his own fur with the smell. After he has done this, the odour on his fur is quite strong, even to the human nose. Since the family group sleeps together closely piled on top of one another, his smell is transferred to all his sleeping companions. One of my tame male's favourite marking places was the entrance to the communal sleeping box—the equivalent of the entrance to the burrow. I originally interpreted this behaviour as having the function only of informing any visiting stranger that the burrow is occupied (Ewer, 1963), but I have since come to wonder whether it may not also assist in smell sharing, since all the users of the burrow brush against the marks every time they go in or out.

The crawling under and over each other described by Steiniger (1950) in rats will also result in smell sharing. Steiniger regarded this behaviour as of mixed origin—partly a form of social amicable behaviour, comparable with mutual grooming, but also as having some sexual significance; regardless of its immediate motivation, however, one of its results will be smell sharing. In this connection it may be noted that some species of *Rattus* possess a mid ventral gland (Rudd, 1966). It would be interesting to know whether in them crawling over and under are more frequent than in *R. norvegicus* and whether the occurrence of this behaviour parallels the changes which occur in the gland in relation to the breeding season.

The male rabbit marks the youngsters of his own group with his chin gland secretion (Mykytowycz, 1965); young treeshrews, *Tupaia*, are marked with the sternal gland (Martin, 1966) and young *Helogale* with the cheek glands (Zannier, 1965). Peccaries (*Tayassu*) will interrupt a session of mutual grooming to rub their throats and shoulders on each others' dorsal glands and amongst the bovids, marking of a fellow member of the group by rubbing with the facial region is quite common (Walther, 1966). Smell-sharing behaviour however, reaches its highest development in the sugar glider (Schultze-Westrum, 1965). In this species the male has scent glands on the forehead and on the chest. Within a group, the males mark the females with the secretion of the frontal glands. The procedure is quite complicated: clasping her neck on either side with his paws, he twists his head so as to face upwards and rubs his forehead on her chest with a circular motion. An adult female will sometimes rub her head in this way on the chest of a male, which must result in her head becoming impregnated with the secretion of his sternal gland. This head rubbing behaviour is not related to courtship and mating and its importance in relation to group recognition is attested by the fact that if a female is removed from a group and kept apart for a day or two, she will be promptly marked on her return. If, in the meantime she has been marked by a foreign male, she is at once attacked. When strange males fight, the importance of the foreign odour in arousing aggression is shown by the fact that bites are directed mainly at the region of the opponent's gland.

In these cases, although its function in preventing attack between members of a group is clear, the motivation causing the marking is not easily deduced. In the case of the rabbit, the primary function of chinning appears to be territorial. Marking reaches its peak during the breeding season, when direct aggressive territorial behaviour is also most in evidence amongst the males. Moreover, the non-territorial cottontail does not possess marking glands. One would therefore judge the motivation for marking to be largely aggressive; but, if so, why should a male mark a youngster? If he is aggressive towards it, why does he not just attack it? Possibly there is something about the juvenile—a special smell or a visual sign—which inhibits attack and the chinning results from aggression which is not permitted to take the form of overt attack: its causation may therefore include a displacement element.

The complex behaviour of the sugar glider is much more difficult to understand. Here we have a secretion which clearly functions as a releaser of aggression being applied to a partner by means of what appears to be a piece of amicable behaviour. Lorenz's (1963) analysis of the greeting ceremonies of geese suggests that the solution to this apparent paradox may not be as difficult as it at first sight appears. By a careful study of related species he has shown that the complex "triumph cry" ceremony of the grey lag goose, whose function is to create the bond between the sexual partners, has actually been derived from aggressive actions originally aroused by any conspecific, including the potential mate. The first stage of evolution consists of turning the hostile action away from the partner and directing it towards any other goose within sight. The message now conveyed signifies "I am ready to attack anything and anybody—except you". Further evolution dispenses with the need to direct the originally aggressive display against a conspecific and it finally carries the simple message "You are the one, the only one for me". The grey lag's display can thus be described as ritualised redirected aggression (Lorenz, 1964). The sugar glider's frontal marking may have had a similar history, but a study of related species will be necessary to lay the basis for any analysis of its origins and evolutionary history.

Now, however, we must return to consider cases where the emphasis in marking is mainly on the threat element. In many species this is underlined by the fact that marking is often associated with or follows a fight and is then most often performed by the victor. Hamsters actually mark a defeated rival after a fight and the marked individual subsequently retreats if he comes on the smell of his erstwhile conqueror (Dieterlein, 1959). Mice and rats similarly mark a defeated opponent with urine. Often marking associated with fighting is accompanied by vigorous scratching or digging movements. These may have originated as displacement behaviour but may become so linked with the marking that the whole constitutes a sort of intimidatory display. Uinta ground squirrels when about to fight merely extrude the papilla of the anal scent gland as an accompaniment to threatening (Balph and Stokes, 1963). Two species of *Sylvilagus*, the cottontail and the swamp rabbit, scratch with the paws when facing a rival (Marsden and Holler, 1964) but in the latter, where the chin gland is used in marking, this activity does not appear to be associated with the scratching. In the common rabbit however, the two frequently occur together. A dominant male studied by Mykytowycz (1965) chinned sixteen times in an hour and in four cases the chinning was accompanied by digging. In guinea pigs a fight is usually preceded by marking and, if the general excitation is high, the animal first scratches the ground and then marks with its anal gland on the scraped area (Kunkel and Kunkel, 1964); at the end of a contest, the victor generally marks. In *Thryonomys* too, the victor in a fight frequently "scratch-marks" in the same way, and if there is something suitable nearby—a stick or stone or the stems of growing vegetation—he may also rub his cheeks on it. Such a display following a victory may have the function of further intimidation of the opponent, making sure that he does flee, while at the same time it ensures that he is free to do so without further attack.

In the cases just mentioned, the behaviour of the marker intensifies the threat element of the scent mark by the addition of a visual component. The latter, however, is transitory and vanishes with the completion of the action. In a few instances, particularly amongst the artiodactyls, more complex behaviour has been evolved, resulting in the production of a lasting visual signal which remains effective even when the animal that makes it has gone away; it may thus constitute a semi-permanent "trespassers prosecuted" sign. We owe to Graf (1956) the realisation that two sorts of behaviour shown by deer have this significance—scraping with the antlers against trees and slashing with them at bushes or shrubs. The former has traditionally been regarded as no more than an attempt to remove irritating strands of velvet. Graf, however, points out that there is no correlation between antler rubbing and velvet-shedding: an animal may go about with masses of half-stripped velvet hanging from the antlers, making no attempt to rub them off, and rubbing is in evidence long after all the velvet has disappeared. In the elk he studied, *Cervus canadensis*, the adult males during the rut perform quite a complex ritual, in which bark is first scraped from a tree with the burr at the base of the antlers and the nose is then rubbed up and down the scraped area. This may then be followed by the animal rubbing the sides of the face and chin on his own flanks. Two males on sighting each other, usually indulge in a vigorous bout of this "sign posting" as a prelude to any hostilities and a single male will also do so when approaching a group of females and juveniles. The latter signpost their favourite "loafing grounds" by a somewhat similar process, but scrape the bark from the tree with their incisor teeth. De Vos (1967) notes that there is some tendency for males to select the same trees for signposting in successive years.

Thrashing of vegetation with the antlers has been interpreted as combat practice but Graf again points out that the facts are not in accord with this interpretation. An animal may continue to thrash a single shrub until he has reduced it to a mere cluster of stubs, leaving untouched another next door which would make a much better sparring-partner substitute. The actual process as he describes it in black-tailed deer, *Odocoileus hemionus*, does not suggest fighting practice, but is strongly reminiscent of the "weaving" used by antelope to apply scent marks. Graf describes how a male he watched selected a shrub 3–4 feet high and then, with his antlers turned forwards, swept them to and fro across it, vigorously but not violently. He would pause after 10–20 seconds and carefully nose over the twigs that he had been whipping and repeated the process five or six times in succession. Graf's interpretion, that the animal is thus producing a visible record of his presence in the area seems the most plausible and in view of the frequent sniffing of the twigs, it is likely that a smell is also being applied. The number of shrubs so treated and the degree to which they are mutilated will automatically give a newcomer a measure of the vigour of the marker. A good deal remains to be learnt about the significance of these various marking procedures since in the elk, at least, the frequency of male bark scraping differs in different parts of the animal's range in a manner which may be related to population density. In one area, where the concentration of competing males was highest, there was very little bark scraping and intruders were driven off directly by actual fighting. It may be that above a certain density such visual signals are not effective and the sign-posting behaviour may be inhibited if meetings with rivals rise above a certain frequency. In the related white-tailed deer, *Odocoileus virginianus*, "pawed circles" are produced by a combination of scraping the ground with the fore foot and thrashing at overhanging branches (Pruitt, 1954). The male mountain goat, digging "rutting pits" during the breeding season (Geist, 1964) is also producing an outward and visible sign of his presence and vigour. The animal squats down on his haunches, with his neck arched and head held low and paws vigorously first with one fore foot and then with the other, throwing the dirt on his own belly and flanks. Periods of scraping may be interspersed with marking of surrounding vegetation with the secretion of the supra-occipital glands, so that the rutting pit carries the olfactory signature of its author as well as the visible signs of his prowess.

The same principle applies to the bear's chewing and scratching of the bark of a marking tree: signals which can be seen as well as smelt are produced. According to de Leeuw (quoted by Lockie, 1966), wild cat "claw-sharpening" trees also act as markers and carry the scent of a foot secretion as well as the visible mark. The brush-tailed possum marks trees by chewing (which may apply a saliva odour) and scratching off the bark, usually within two feet of the ground (Kean, 1967) and a urine mark is often also set on a projecting root, just below. The trees chosen for such treatment are usually rather large and the marks may be renewed on the same place throughout the life of many generations of possums. Presumably the marks designate favourite pathways and may be involved in some type of priority system governing their use.

What has been found out about responses to scent marks, with only slight modification and elaboration, falls into line with the simple principle that an animal's own mark is a socially positive signal, whereas a strange one is socially negative. The information may be summarised in tabular form (see Table 2 page 133).

The types of effect listed in Table 2 are very simple, even if quantitative assessment of the relative importance of the different factors in particular cases is not easy. The problem of the factors which evoke marking, however, remains much more obscure, particularly in the cases where a number of different sorts of mark are used by a single species. *Trichosurus*, for instance, in addition to marking trees as described above also uses the secretions of two sorts of cloacal gland and a sternal gland, while the olfactory vocabulary of *Petaurus* includes all these together with a frontal gland and smaller glandular areas near the ears and at the angle of the eye (Schultze-Westrum, 1965). Even the much simpler question of the roles of the two sorts of marking practised by some viverrids remains an unsolved problem, so it seems likely that the complexities of possum olfactory communication will remain beyond our microsmic comprehension for some time to come.

Table 1—*MARKING METHODS IN FOUR MAJOR ORDERS*

MARSUPIALS

DASYURIDAE

Antechinus flavipes
1. Saliva—mouthing and biting branches
2. Cloacal glands—cloacal drag and squirting
3. Sternal glands—chest rubbed on branches (often combined with 1) gland larger in ♂ than in ♀

Dasycercus cristicauda
1. Saliva—mouthing objects
2. Cloacal glands—cloacal drag

Sminthopsis crassicaudata
1. Saliva—mouthing
2. Cloacal glands—cloacal drag. ♀ does not mark if a ♂ is present

Sarcophilus harrisi
1. Cloacal glands—cloacal drag

PHALANGERIDAE

Trichosurus vulpecula
1. Urine—zig-zag marks on tree roots or simple drip trail
2. Cloacal glands—cloacal drag: usually follows disturbance
3. Sternal glands—rubbed on branches (most common type of marking)
4. Saliva (?)—bite and scratch marks on special trees

Petaurus breviceps
1. Urine—drip trail
2. Saliva—gnawing of branches
3. Cloacal glands—cloacal drag
4. Sternal glands—on branches and ♀ rubs her head on ♂'s chest
5. Frontal glands—present only in ♂, rubs his head on ♀'s chest
6. Foot glands—incidental transfer
7. Ear glands

Phascolarctos cinereus
Sternal gland—much more developed in ♂ than in ♀

MACROPODIDAE

Megaleia rufa
1. Sternal gland in ♂—bites at his own gland in encounters with a rival

---

CARNIVORES

CANIDAE

In this family urine marking is the rule and it is usual for the male to mark more intensively than the female. The normal posture is to lift one leg and the male usually lifts a leg whether directing the urine against a vertical surface or on to the ground. The female normally either squats or raises one leg slightly, but does not usually adopt the full vertical leg-cocking posture. Species which have been investigated are listed below according to the intensity of marking behaviour shown by the female. Species in which the female shows the full leg-cock are marked with an asterisk. A gland is present on the dorsal surface of the tail in many Canidae and possibly serves for incidental scent transfer to the den entrance.

♀ *not seen to mark*

Canis lupus; Canis mesomelas; Vulpes vulpes (glands at root of tail rubbed on entrance of refuge); *Lycaon pictus*

♀ *Marks only during or around the time of oestrus*

*Canis familiaris; Canis latrans; Alopex lagopus**; *Nyctereutes procyonoides** (glands at root of tail may also be rubbed on roof of lair); *Otocyon megalotis.*

♀ *marks all year round*

*Canis adustus**; *Dusicyon gymnocercus; Speothos venaticus* (♀ marking is intensified during oestrus; ♀ marks in hand-stand posture, possibly rubbing glands on surface as well as urinating)

URSIDAE

Ursus arctos

1. Urine: voided against tree
2. Neck glands: rubbed on tree
3. Saliva?: rubbing neck on tree is often combined with scratching and chewing

PROCYONIDAE

Bassaricyon sp.; *Nasua naria*

1. Urine
2. Glands in ano-genital region: squats over object to be marked and rubs ano-genital region on it

MUSTELIDAE

Mustela putorois

1. Urine and faeces deposited on trails
2. Anal glands: anal drag

Martes martes; Eira barbara; Mellivora capsensis; Meles meles

Anal glands: anal drag

Poecilogale albinucha

Faeces: backs up to vertical surface

Aonyx capensis

Urine and faeces: backs up to vertical surface

VIVERRIDAE

Genetta

Ano-genital glands: hand-stand posture

Viverricula; Viverra: Ano-genital glands: backs up to vertical object

Suricata

1. Anal glands: ♂ cocks leg: ♀ rarely marks and then by anal drag
2. Urine: ♂ cocks leg to urinate

Herpestes edwardsi

1. Cheek glands: rubbed on objects
2. Anal glands: leg cocked or hand-stand posture

Helogale parvula

1. Cheek glands: rubbed on objects
2. Anal glands: hand-stand posture

Atilax paludinosus

1. Cheek glands: rubbed on objects
2. Anal glands: ♂ uses hand-stand posture; only anal drag seen in ♀

Mungos mungo

Anal glands: anal drag

Crossarchus obscurus

Anal glands: hand-stand or anal drag; only anal drag seen in ♀

Cryptoprocta ferox

Sternal gland: rubbed on branches

FELIDAE

Commonly retromingent: backs up to vertical surface
Lion also known to squat and wipe feet in urine

Felis sylvestris; Lynx lynx Urine and faeces deposited on prominent places near boundaries of territories

---

RODENTS

APLODONTOIDEA

Aplodontia rufa

Urine

SCIUROIDEA

Sciurus

Urine

Xerus

Cheek glands: rubbed on objects

Marmota

1. Cheek glands: rubbed on objects
2. Anal glands: anal drag

Citellus

Cheek glands: rubbed on objects

CASTOROIDEA

Castor

Anal gland: rubbed on ground

MUROIDEA

Cricetus; Mesocricetus; Lateral glands: transferred to ground by rubbing foot on gland; more developed in ♂ than in ♀

Arvicola

Lateral glands: rubbing of foot on gland followed by stamping

Ondatra

Anal glands: anal drag

Rattus

Urine: drip trail and may mark defeated rival
Mid-ventral glands in some species

Mus

1. Urine: drip trail and may mark defeated rival
2. Foot glands

Notomys

Ventral glands on chin or chest: usually larger in ♂ than ♀

Cricetomys

1. Urine: drip trail—only when disturbed
2. Faeces: hand-stand posture seen occasionally when disturbed and then only in ♂
3. Saliva
4. Cheek glands

(3 and 4:) Licking combined with rubbing cheeks on prominent objects. Commoner in ♂ than ♀: ♂ may mark ♀ during courtship

GLIROIDEA

Glis

Anal glands: anal drag

HYSTRICOIDEA

Atherurus

Anal glands: anal drag: seen only in ♂

ERETHIZONTOIDEA

Erethizon

Cheek glands: rubbed on objects

CAVOIDEA

Cavia

Anal glands: anal drag may be combined with scratching. Commoner in ♂ than in ♀

Dasyprocta

Anal glands: anal drag may be combined with scratching. Commoner in ♂ than in ♀

Myoprocta

Anal glands: anal drag may be combined with scratching. Commoner in ♂ than in ♀

OCTODONTOIDEA

Thryonomys

Cheek glands: rubbed on objects, may be done after a victory
Anal glands: anal drag, may combine anal marking and scratching the ground after a victory

(both:) Seen only in ♂

PRIMATES

TUPAIIDAE

Tupaia sp.

Sternal gland: used in marking young

LEMURIDAE

Microcebus murinus

Urine: voided on foot

Hapalemur griseus

Glandular areas on forearm and front of shoulder

Lemur catta

1. Perineal glands: rubbed on branches
2. Palm of hand: rubbed on branches by ♂
3. Glandular areas on forearm and upper arm, larger in ♂ than in ♀—secretions rubbed onto tail which is then waved at opponent: used together with 2 in intra-group disputes

Propithecus verreauxi

1. Urine and faeces deposited on branches
2. Genital region: rubbed on branches
3. Throat gland in ♂ only: rubbed on branches. All types of marking often associated with inter-group hostilities

LORISIDAE

Most lorises

Urine: voided on hand and wiped on foot

Loris tardigrada

Urine: voided on foot as well as hand

TARSIIDAE

Tarsius sp.

Sternal gland

CEBIDAE

Cebus apella

Urine: voided on hand and wiped on foot

Aotus trivirgatus

Gland at base of tail

Callicebus moloch

Sternal gland: rubbed on branches

Ateles sp.

Paired glandular patches on chest

CERCOPITHECIDAE

Mandrillus leucophaeus

Sternal gland

PONGIDAE

Pongo pongo

Sternal gland: possibly non-functional vestige

TABLE 2—*RESPONSES TO SCENT MARKS*

(a) OWN MARK

(In social species may be extended to include the mark of any member of the group).

Increased confidence or reduced anxiety.

In social species may also evoke increased friendliness or decreased aggression —especially important where there is smell sharing.

(b) FOREIGN MARK

Increased aggression, readiness to fight Decreased confidence, readiness to flee	} Relative importance may depend on whether responder is or is not on home ground.

Where there is recognition of individual marks, habituation may reduce the effect of familiar marks so that, in their accustomed places, they evoke almost no response.

Where the setter of the mark and the responder are of opposite sex, a foreign mark may have a positive effect, causing sexual arousal and producing appetitive behaviour for mating.

In social species the usual negative effect may be combined with, or replaced by a positive effect so that even the marks of members of a foreign group may not be avoided but act as pathway signals.

CHAPTER 6

Fighting

When it actually comes to fighting, it is rare in any species of mammal for two individuals simply to fall upon each other at sight and attack with all their might. What happens is usually much more complex. As examples, species belonging to three different orders may be taken—a shrew, a guinea pig and an antelope.

The following description of shrews (*Sorex araneus*) fighting is taken from Crowcroft (1957), who studied animals in captivity. If two male shrews share an enclosure each will presently establish himself as owner of part of the area. Since the territories are necessarily small, encounters are relatively frequent. When the two meet, the first to notice the other's presence raises his nose and squeaks loudly, whereupon the other may either retreat or squeak in return. In the latter event, the two now stand face to face, a few inches apart and "scream" at each other. Many, indeed most, contests go no further than this: one animal apparently becomes intimidated, retires and is not pursued by the victor. If neither will give way, one will suddenly rear on its hind legs and at this point the opponent may beat a retreat, possibly in response to the sudden apparent increase in size of his adversary. If they are very evenly matched, however, both may rear up and a genuine fight may start: striking and clutching with their paws, they bite at each other, continuing to squeak the while. The biting does not appear to be very effective and serious damage is rare. After a few moments, one may break away and retreat but if neither does so then one will suddenly throw himself on his back, still screaming and striking out with his legs. At this, the opponent turns away and leaves him. If it happens that both throw themselves on their backs at the same moment, they may lie thus for a few seconds, screaming, until either one makes off or they rear up and start again at the "boxing and biting" stage. Sometimes it happens that as the fight is resumed, each may bite the other by the base of the tail and for a moment, with jaws firmly gripping, they spin round and round, before managing to break loose and resume the more orthodox fighting position.

Bunn (1966) is of the opinion that territorial defence is distinctly more vicious than Crowcroft's observations suggest. His findings are, however, not really relevant. His method was to allow one shrew to establish ownership of an enclosure and then suddenly put in a stranger. The latter was, of course, defeated and although he attempted to flee, he had no way of removing himself from the other's territory—a totally unnatural situation. As far as the territory owner was concerned, the intruder was behaving quite improperly: although thoroughly defeated, he was obstinately refusing to obey the rules and withdraw. In such circumstances it is hardly surprising if quite unnaturally violent and persistent attacks are provoked.

The Kunkels (1964) have described what happens in encounters between guinea pigs. Territorial aggression is well marked in this species. An adult male, if he meets a cowering newcomer in his domain, will run at him and simply bite him anywhere that offers without hesitation; the bitten animal promptly flees, squeaking loudly and usually with a bleeding wound. The course of a contest between more evenly matched rivals is very different. As the two approach each other, they pause and start to gnash their teeth. They may continue this for some time, while their hair rises, first on the nape of the neck, then spreading down along back and flanks. At this point, one may flee but if he does not, they continue to face each other and soon a slight change can be seen. One is directly facing his opponent, with hair bristling and open mouth; the other, with back slightly hunched but legs fully extended, so that he appears short and high, has turned his body round through almost a right angle. He still faces his opponent, so his neck is sharply curved. As he takes up this position, his testes are suddenly protruded and rapidly retracted again and he may emit a high screaming sound. Since the wild type colouring includes a brightly coloured scrotum, the testes make a striking optical signal during their brief exposure and a smell may possibly be emitted from the anal glands. If we have already seen a number of fights, we know that this animal is getting the worst of it: in a moment he will turn and flee, often first treading with his hind feet and swaying his hind quarters from side to side. He is usually pursued and may get a bite on his rear from the victor. Only rarely does a true fight develop, with the two animals springing at each other, each attempting to bite the other on head or shoulder. Even then, one bite that finds its mark is usually sufficient to decide the issue. The bitten animal flees and hides, emitting a long-drawn cry, while the victor usually marks with his anal glands.

With these, we may compare Walther's (1965) description of fighting between a pair of male Grant's gazelle (Figure 6).

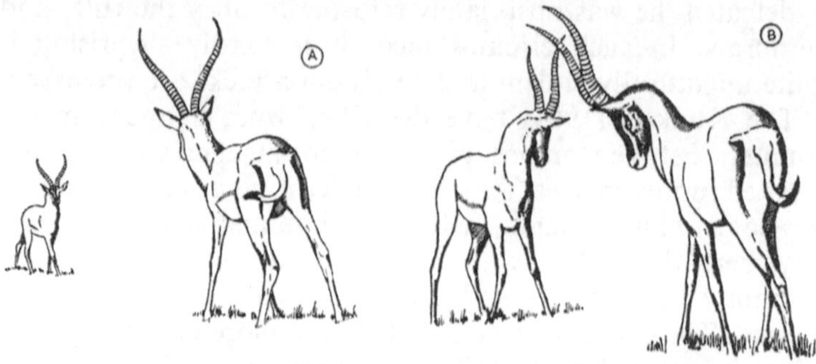

Figure 6. Grant's gazelle: preliminaries to a fight. (a) Challenge and first approach. (b) Heads turned aside as the two draw near. (c) Horn nodding as they come alongside. (d) Head-high throat display. (After Walther, 1965.)

The contestants "square up" to each other at a distance of 50–100 metres; they then advance towards each other with heads held high, ears turned forward and noses up, so that the horns slope backwards slightly. The march is not aimed directly at the opponent but a little to one side, so that the two will not meet head on, but pass close beside each other. As they draw near, each turns his head aside and as they draw level he turns it forward again and nods his horns towards his opponent. When the nose of each is level with the other's tail, they pause, stretch up their necks to their fullest extent, turn their heads towards each other and raise the nose so that the white patch on chin and throat is fully displayed. The white throat display may be repeated again and again, as many as thirty times. And now, the contestants may part, without a blow having been struck. Their positioning makes this very easy: they simply move onward and neither need withdraw. If they fail to separate at this stage, then the true fight begins. It may start with a sort of "measuring up": the two touch each other, either nose to nose or horn against horn, apparently getting an estimate of the opponent's height. From this they may go over directly into the fighting position (Figure 7), with foreheads together and horns interlinked.

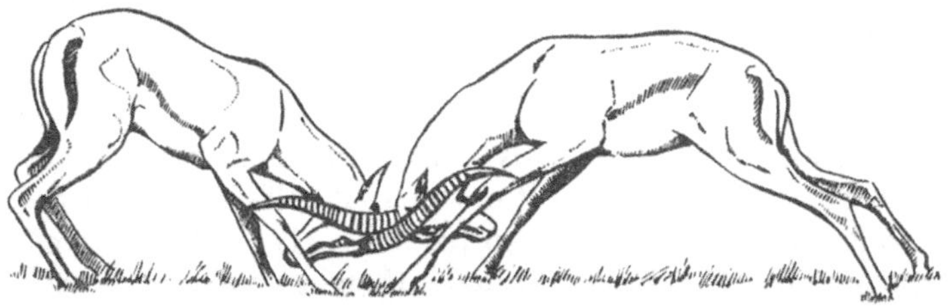

Figure 7. Grant's gazelle: fighting—pushing with foreheads, horns interlocked. (After Walther, 1965.)

More usually they draw back a trifle and then hurl themselves together, horn interlocking with horn. Now it is a trial of strength: each trying to thrust the other back, the contest may swing back and forth or round in a circle. Sometimes one, forced to give ground a little, may leap back, freeing his horns for an instant, only to leap forward and re-engage them with a clash. If one is clearly outmatched, after such a leap back he will wheel and flee and the victor will pursue him only briefly. If the disparity is less, he may withdraw at a walk and be followed also at a walk: in fact, he is being "escorted off the premises", rather than chased off. If neither can overpower the other, then in the end they will desist, turn their heads aside from each other, scratch or shake themselves and slowly move apart, often grazing as they go. Walther found that in only a trifle more than one in ten of all the contests he watched did a prolonged fight develop and in almost half the encounters, the opponents separated after the preliminary display without any physical contact at all.

These three examples serve to illustrate the main principles that operate in aggressive encounters. The first and most basic is that the objective is not to kill the opponent, but to drive him away. The function of the contest is to decide who is the stronger, not to kill off the weaker. In terms of natural selection it is easy to see why non-lethal contests should have evolved. It is to the advantage of the stronger animal if he can get rid of his rival without himself being hurt or wounded. It must be borne in mind that in the natural situation a wound that causes temporary disablement may be as good as a death sentence. To be slow off the mark may spell death at the hands of a predator for the prey or death by starvation for the killer. In addition, it has already been pointed out how in this type of circumstance kin selection may also be in operation. It is genetically disadvantageous for a male to kill his own offspring and, if the species is sufficiently long lived, his sons are likely to grow up to be his rivals. It is therefore doubly to his advantage to drive them away without either eliminating his own genes, which they carry, or running the risk of serious hurt himself. If group selection should also be in operation then there are further obvious advantages. The youngster making his first challenges may be weaker now than his older and more experienced rival—but he may grow to be a top ranker if not cut off in his youth. In any case, peripheral males form a breeding reserve, from which the death of any territory holder can at once be made good. As far as the group goes, it is thus an advantage not to eliminate the defeated males.

Bearing these things in mind, there are clearly two desiderata. Firstly, if one of the contestants is totally outmatched, it should be possible for this to be demonstrated and for him to withdraw without ever actually fighting. Secondly, even if the rivals are approximately equally matched, it should be possible for them to fight in a manner which measures strength and pugnacity but is so arranged that they are unlikely to kill each other. Natural selection has proved equal to the occasion. The retreat of the weaker without fighting is made possible by means of threat contests as a preliminary to genuine attack, while the development of stylised methods of fighting limits the damage done if the two actually come to grips with each other. The former are of general occurrence: I do not know of any mammal that has been shown to be quite devoid of any threat vocabulary. The three examples chosen show a variety of types of threat; mainly vocal in the shrew, visual in the gazelle and combining both auditory and optical signals and possibly olfactory ones too in the guinea pig. Stylised fighting, is, however, much less common. It certainly occurs in the horned and antlered artiodactyls, as is illustrated by the case of Grant's gazelle, and in general discussions on the subject, mammalian examples are normally drawn from this group. It is therefore of some interest to examine intra-specific fighting methods in as many groups as possible and find out in how many of them stylised fighting methods have been evolved.

In the majority of small mammals, once the opponents have actually come to grips, fighting is relatively unstructured. Unless the limbs are specialised in some manner making them unsuitable as grasping organs, the contestants usually clutch and strike at each other and fall to the ground in a wrestling, writhing confusion, each attempting to bite the other. If an upright posture is a reasonably stable one, this stage is often preceded by rearing up on the hindquarters and "boxing" the rival off with the paws, at the same time attempting to bite him (Figure 8). This initial boxing and biting stage is characteristic of the fighting of many species of rodents, but it also occurs in other orders. We have already seen it in the shrew and it is also found in the Dasyuridae. *Sminthopsis* fights very much as a shrew does, rearing up biting and striking out with its paws, often squeaking the while, sometimes ending with the two falling over and wrestling. In *Dasycercus* the rearing up is very brief and the contestants close with each other almost at once and fall to the ground wrestling and biting. In rodents like the guinea pig, which are so completely

Figure 8. (a) to (d): Successive stages in a fight between wild rats (*Rattus norvegicus*). (After Eibl-Eibesfeldt, 1963.)

quadrupedal that the paws are never used as hands, there is no rearing up. In species such as the kangaroo rats, where the hind limbs are very elongated and the fore limbs very reduced, the fighting method is modified. Although a fight may end with the usual wrestling and biting, the preliminary rearing up is followed not by boxing but by each attempting to leap over his opponent, kicking at him on the way (Allan, 1946; Culbertson, 1946).

In these cases, although there is close in-fighting, the animals are always somewhat inhibited in their efforts to bite each other by the likelihood of receiving a return bite and "bites at" are much commoner than true bites. The fighting is not stylised, but it might be described as "restricted". Since the animals do not possess very dangerous weapons, such restriction is sufficient to make fatalities very rare. It is amongst the species that do possess weapons whose unregulated use could easily result in a death that stylised fighting is most likely to occur, for it is in them that its selective value will be greatest. This principle is illustrated by the case of *Thryonomys*. This rodent has evolved a highly stylised method of fighting. As we have seen, the social organisation appears to be a harem system, based on male intolerance. Moreover, *Thryonomys* is also a species characterised by the possession of incisors which are extremely powerful and razor sharp. If the teeth were used in the normal rodent manner, the animals would cut each other to shreds. The efficacy of the teeth is attested by the fact that a rump bite administered to an inferior who does not move away fast enough usually leaves a bleeding gash. I have seen only one fight, but it was of such a surprising nature and presents such an unexpected parallel with artiodactyl frontal fighting that it seems worth describing. The contest was a ranking fight between two males and was already in progress when I came on the scene, so that it is not possible to say whether it was preceded by any form of threat duel. One night, my attention was drawn by a rustling sound from the enclosure in which the animals were kept. As I approached to investigate I could see something elongated writhing about in the grass. There was not much light and for a moment I thought one of the animals was being eaten by a large python; then I made out what was really happening. The "snake" was the bodies of two *Thryonomys* engaged in a snout to snout head pushing duel. Nose against nose, all four feet firmly on the ground, they heaved and strained and the battle wavered back and forth over a distance of three or four metres as first one, then the other, was forced to give ground a little. As they shoved, one would occasionally rotate his head to one side and back again, with a motion reminiscent of someone trying to force a gimlet into hard wood. Finally, one gave way, broke loose and ran but was not pursued very far. A tame youngster subsequently showed exactly the same pattern in play. He would snout push against my knuckles and twist his head in the "boring in" movement; he would also now and then interpolate a very rapid flip of his body so that his hindquarters struck my wrist and he was back again in shoving position almost before I was clear what was happening. In the real fight I did not see this movement, but it was clearly an attempt to throw the opponent off balance and presumably could be used only if pressure were relaxed for an instant. In litters of young reared naturally, play fighting of this type is extremely common and the "backsiding off" movement is frequently used.

The case of *Thryonomys* accords very well with the idea that stylised fighting is most likely to be found in species possessing potentially lethal weapons. Such weapons are also found in elephants, rhinoceros, hippopotamus, pigs and the horned and antlered ruminants. Their most obvious possessors however are the predatory carnivores, whose livelihood depends on their efficiency as killers. Of these, the felids take first place but unfortunately there is a dearth of information about their fighting methods, particularly the larger species. As far as lions go, this is clearly partly because serious fights are rare. Carr (1962), for instance, speaks of his pair of young animals at first retreating in face of a superior. As they became adult, however, they grew bolder and he finally left them to fend for themselves after having seen them "successfully driving off" another adult and so showing that they could carve out a territory—he never mentions having seen a real fight although the youngsters were under observation during the period when they changed from being the losers to the winners of the encounters. The only felid whose fighting methods are well known is the domestic cat. In a tomcat fight, the adversaries bite at each other and rake each other with the claws of the fore paws. If defence predominates over attack, the animal throws itself on its back and strikes at its opponent with hind as well as fore feet. The need for each to protect his own face tends to restrict biting; claw wounds predominate and the really lethal weapons, the canine teeth, come into play very little. The noise accompanying a fight is impressive but out of all proportion to the actual damage done and wounds are concentrated on the non-vital shoulder region. The location of the lion's mane suggests that here too the attack is oriented in the same way. The mane may have some protective value but fighting is not restricted to males and killings do sometimes occur (Schenkel, 1966a).

Although many of the smaller carnivores are noted for

their pugnacity, little is known about their intra-specific fighting. I have not seen fights between two male *Suricata*, but when a strange male and female were being introduced to each other, two patterns were shown which served to minimise injuries. The first of these is a curious form of behaviour, which I have referred to as "inhibited biting" (Ewer, 1963). The animals make as though to bite at each other but the jaws are held firmly closed and the head is rotated first to one side then the other. Darting their heads at each other in this way, they resemble a pair of fencers, sparring for an opening. If contact is made with the partner's body, there is still no bite; the jaws remain closed and the rotating, wriggling head movement is repeated so that the snout is rubbed against the other's fur. When I first saw this behaviour I was unaware that marking with cheek glands occurs in related viverrid species and suggested another possible derivation of the pattern. I now believe that this was quite erroneous and that inhibited biting is in fact derived from cheek gland marking. The second pattern is "back attack", but could equally well be called "back defence". As one animal approaches the other, instead of biting, he spins round and backs into his opponent, so that contact is not made with the head but with the hindquarters. The attack may be parried in the same way, so that the two animals merely bump their rumps together and biting is avoided. This back attack/defence was also often seen in squabbles between youngsters over food. Curiously enough, an almost exactly similar fending off with the hindquarters is found in the hairy-nosed wombat, *Lasiorhinus latifrons* (Wünschmann, 1966). In both cases the animals studied were of opposite sexes, so it is not known whether the same patterns are shown in fights between males. In the case of *Suricata* the female was much more inclined to bite the male than he was to make a genuine attack on her, and his behaviour may therefore reflect the very common carnivore inhibition against biting a female. His treatment of a strange member of his own sex may well be much more violent.

Kühme (1961, 1963) describes fighting in the African elephant. The two animals meet head on, but fighting is carried out mainly with the trunk and the tusks serve mainly as holding devices, very much as bovid horns do. Each animal tries to encircle the base of the other's trunk from above and push him back and down; this trunk wrestling may pass over into forehead pushing. When one is forced to give way, he flees and is pursued by the victor, who may strike him on the side or flank with his tusks. Kühme's animals were captive ones, kept in a large enclosure: it is quite possible that in the natural situation more violent fights may occur, in which the final tusk attack has serious consequences. According to Carrington (1958) this is the case with Indian elephants, *Elephas maximus*. The preliminary stages of trunk fighting and forehead pushing are similar to those of the African species, but Carrington says that as the weaker breaks away, the tusk attack on his now exposed flank may be lethal.

The encounters involving three male black rhinoceros described by Goddard (1966) occured in the presence of an oestrous female. In spite of this, the fighting was not very violent and threat formed an important part of the duels. One of the males was accompanying the female and met the two others in rapid succession. The first promptly retreated when the male in possession of the female charged him with flattened ears and emitting "a ghastly puffing shriek". The second rival was less easily deterred and responded to being charged by charging in return. The charges were not pressed home and the two stopped short about five yards from each other. The initiative, however, remained with the male in possession. He repeatedly attacked with short charges, always with the accompaniment of vocal threat; his opponent met the charges defensively and in silence and the two then struck at each other with their anterior horns "trying to club one another on the side of the head". Goddard did not see the animals attempt to gore each other with the point of the horn and from his description the contest appeared to be highly stylised.

Hippopotamus fights are described by Verheyen (1954). The area he studied was heavily populated and aggressive encounters between males were not uncommon. Frequently these were limited to threat duels, but in a genuine fight the two rush at each other and meet head on with a clash of tusks: a shoving contest with much grunting and roaring ensues.

When the weaker flees, he is usually pursued and speeded on his way by being rammed in the rear. On land, the weaker, being usually the lighter and more agile, generally makes good his escape, but if a fight occurs in the water he has more difficulty in doing so and Verheyen says that sometimes the loser is killed. Much more commonly, even after prolonged battles, there is no fatality and the two separate with wounds on heads, necks and flanks.

In the Suidae, two types of fighting occur (Frädrich, 1965): lateral and frontal. Lateral fighting is characteristic of species not possessing very long tusks. The opponents stand shoulder to shoulder, facing in opposite directions and either push sideways against each other or strike with the head. The European wild boar, *Sus scrofa*, fights in the latter manner and the thickened skin of the "shield" on the shoulder is used to parry the blows. Fighting in domestic pigs appears to be simply an attenuated version of the same method. At the beginning of the fight the two do not strike, but merely push shoulders against each other, but in the closing stages they strike with open mouths upwards and sideways at each other. In species with larger tusks, such as warthog and bushpig, frontal fighting replaces lateral. The opponents approach each other head on, hurling themselves at each other with considerable violence if the fight is a really serious one, and then push each other, head to head. In the warthog, the forehead is broad and flat and provides a large area of contact, so that there is little danger of slipping. The heads are held symmetrically, with the noses pointing downwards and the two are in contact from the forehead down along the length of the snout. For *Potamochoerus*, with its long narrow skull, this position is not suitable, as the heads would be very liable to slip off each other. Their technique is modified and instead of attempting to maintain a symmetrical forehead contact, the heads are turned slightly sideways so that the snouts cross each other and so are less likely to slip apart. In a *Phacochoerus* fight, the animals are like two warriors pushing against each other, shield to shield; in *Potamochoerus* they attempt to bear each other down, sword crossed against sword.

Fighting in the forest hog, *Hylochoerus meinertzhageni*, has not been described, but its extremely broad, slightly concave forehead suggests that, like *Phacochoerus*, it must use a head-to-head bull-dozing technique. In both warthog and bushpig, as the weaker turns to flee, he may be struck at by the victor. Frädrich records that in captive warthog, where escape may be difficult for a cornered animal, such a blow may be fatal. According to game rangers' reports in the Hluhluwe Reserve this may also occasionally occur in natural conditions.

The fighting techniques of the Camelidae are of special interest, since they may give some indication of the state of affairs in the ancestors of the horned and antlered Bovoidea and Cervoidea and hence give some basis for speculation about the evolution of the stylised fighting methods found in the latter groups. In both Old World and New World camelids, fighting includes biting, kicking and neck-wrestling (Pilters, 1956; Gauthier-Pilters, 1959; Koford, 1957), in which each animal tries to get his neck over that of his opponent and, bringing all his weight to bear, force him down and strangle him. Usually one succeeds in breaking away before serious damage is done but Gauthier-Pilters describes an exceptional case in which two dromedaries, *Camelus dromedarius*, succeeded in bringing each other down with their necks so entwined that both were suffocating. They were almost at their last gasp and both might have perished, had not their Arab owners come on the scene and extricated them from their predicament. Amongst the New World camelids an attack is often started by one animal leaping at the other and ramming him with his head as a preliminary to neck wrestling. Although the attack is usually amidships, Pilters (1956) notes that sometimes the two may attack simultaneously in this manner and meet head on.

Amongst the horned and antlered artiodactyls, although there are differences in details, fighting techniques, with only a few exceptions, are based on a single principle: the fight has been transformed into a trial of strength by head pushing. With horns or antlers in contact, the contestants exert all their strength against each other until the weaker is forced to give ground. When he yields, pursuit is brief and he usually escapes without damage. The main function of the horns or antlers in these fights is thus not to slash or gore the opponent but to provide a method of maintaining head contact, so that there is no danger of foreheads slipping aside as the animals strive against each other. Branched antlers interlock effectively but unbranched horns may tend to slip: the rings on the horn bases of many species are non-slip devices, serving to minimise this tendency. Depending on the intensity of the fight, the head to head position may be taken up quite gently or with extreme force, the animals hurling themselves at each other with great violence. In low intensity fights, or in the initial stages, the animals may rotate their heads slightly in opposite directions, so that the horns do not interlock, but cross (Walther, 1965).

Variations on the basic theme are shown by different species and can be correlated with the general body build and the form of the horns. In some species, the animals go down on the carpal joints of the fore limbs and head push in a "kneeling" position. This is characteristic of the heavy shouldered wildebeest (Talbot and Talbot, 1963) and Walther (1964b) also records it in one of the Tragelaphines—the sitatunga, *Tragelaphus spekei*. In the genus *Tragelaphus* as a whole, fighting with the horns crossed, rather than interlocked tends to be emphasised: the same pattern is found in the sasin, *Antilope cervicapra*, and in the genus *Oryx* (Figure 9), which, although not closely related, have rather similar horns (Walther, 1964b). In all these species the characteristics of the horns are more suited to this type of fighting than to simple head pushing, in exactly the same way as the long narrow head of the bushpig is to its entirely comparable "sword crossing" technique. In yet a third variant, characteristic of wild sheep and goats, there has been concentration on the initial impact, rather than on the pushing contest that follows it in the antelopes and deer. Here the fight consists essentially of a series of charges, in which the opponents leap at each other and "ram" head against head with maximal violence. The horns do not interlock but have wide bases and there is exaggerated development of all the other anatomical specialisations strengthening the forehead in adaptation to withstanding the shock of the impact.

Highly stylised fighting, in which both partners adhere to a formula which results in making serious damage to either very rare, is typical of the effectively head-weaponed species. There are, however, some exceptions. Walther (1964b) notes that *Tragelaphus scriptus* (the bushbuck) may attack in an unrestricted, unoriented manner. This is the smallest-horned species of the genus and it is of interest to see how behaviour and structure match each other, both being primitive, compared with those of congeners. Walther (1958) also quotes a description of an eland fight (*Taurotragus oryx*) in which there appeared to be little "keeping to the rules". Both parties

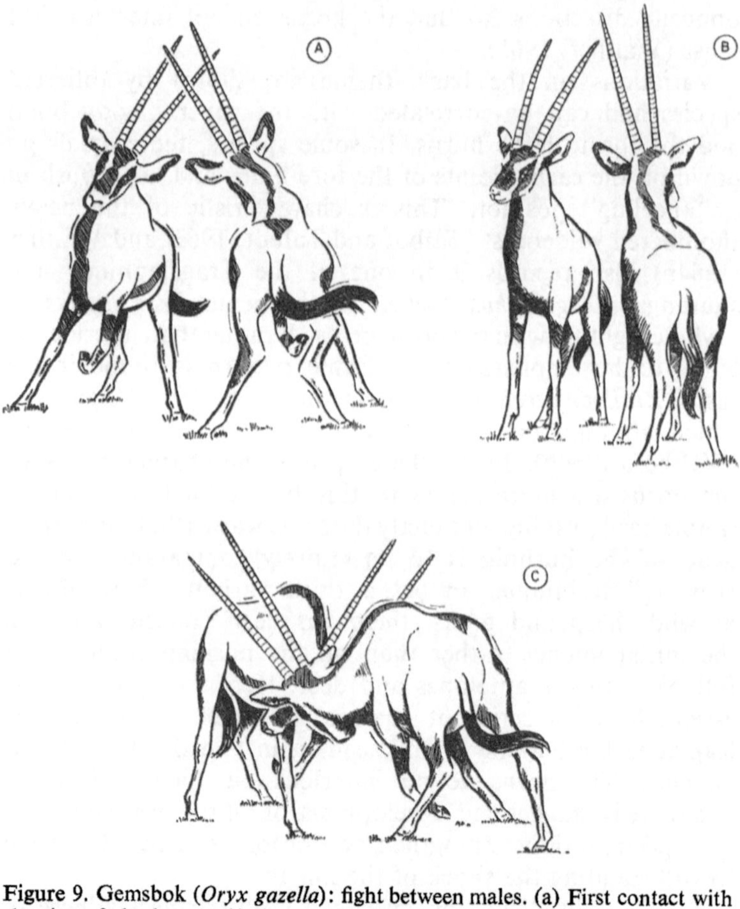

Figure 9. Gemsbok (*Oryx gazella*): fight between males. (a) First contact with the tips of the horns. (b) Pause with disengagement. (c) Maximal intensity; pushing with foreheads in contact. (After Walther, 1958.)

were severely injured and as one finally turned to flee, he was gored in the side, receiving a lethal wound. Geist (1966) gives other examples of damaging or fatal encounters.

In giraffes, *Giraffa camelopardalis* (Innis, 1958), as is only to be expected in view of their exceptional body build, the fighting technique is atypical. The heads are used to administer violent blows against the opponent's body and legs. Orientation is not critical and the animals may face each other or stand side by side. Delivered with a wide swing of the neck, the force of the blows is considerable: Innis notes that the impact of a heavy blow on the chest was clearly audible to her at a distance of 80 yards and quotes one case where the recipient of the blow was "knocked senseless and lay stretched out on the ground for 20 minutes before he struggled to his feet and wandered off". Bouts of this type of striking with the head may be interspersed with neck fighting in which the two "wrap their necks together" and push against each other with their bodies. Here, too the correlation of fighting technique and horn type is clear: the blows administered could not fail to be lethal if the horns, instead of being short and blunt were longer and sharp. The atypical fighting technique, the unusual body form and the peculiarities of the horns, when seen in relation to each other, thus form an adaptive unity. The same, however, is not true of the mountain goat, studied by Geist (1964). In this species fighting is lateral, not frontal, but there is no obvious correlation with structural characteristics. The opponents come together broadside on and thus oriented, circle about, attempting to strike upwards and sideways at each other with their horns. The blows are mainly aimed at chest and forequarters and, despite the thick hair, serious wounds may result. Since related species fight in the conventional frontal manner, the mountain goat's method might possibly be secondary. It is therefore of some interest to enquire into the origin of this unorthodox and apparently maladaptive technique. Geist's observations are sufficiently detailed to suggest an answer. According to him, damaging fights of this type are rare. A threat duel usually suffices and the older and more effectively armed the animals are, the less willing are they to strike a genuine blow. "Broadside threat" is extremely common amongst adult males; the threatener stands sideways on to his opponent, with legs stiffly extended, back humped and head turned slightly away and may emit a deep harsh roar. Two males may circle each other thus for some time, until usually one withdraws. It may thus be that concentration on this type of threat has led to abandoning of frontal fighting. If neither partner will yield but both persist in threat with ever increasing intensity, they have got themselves into a situation which precludes a change over to the head on position; they are virtually trapped by the efficacy of their own threat behaviour and in the threat position they may be forced ultimately to go over into attack.

In kangaroos, as is only to be expected, anatomical peculiarities are matched by a specialised form of fighting. When two males fight, they do so reared high on their hind feet. Grasping and clutching at each other with their paws and, whenever the chance offers, administering a kick with one foot, each attempts to down the other. The objective appears to be simply to bring the rival down, rather than to injure him. The kicks are delivered with a thrust not a downward rake of the claws, so that their effect is to throw the recipient off balance, not to disembowel him. The fighting is thus a distinctly stylised trial of strength. The kangaroos are often described as the ecological equivalents of the grazing artiodactyls. It is thus interesting to note that there is also a behavioural parallel to the stylised fighting methods of the latter.

There are many species in which no detailed study of fighting methods has yet been made but even from this brief survey, a number of points emerge. Firstly, in almost every case where some degree of stylisation has been evolved it has taken a basically similar form: the contestants come together in such a way as to permit them to measure their strength by pushing against each other. This is hardly surprising; in a normal quadrupedal mammal it is virtually the only possible method. The detailed mutual adaptation of fighting techniques and form of weapons is, however, very striking. In the course of evolution there must have been a continual interaction between the two, with each influencing and being influenced by the other. Usually it is not possible to say which took the lead at any particular stage—did the sasin evolve its peculiar horns in adaptation to its fighting method or *vice versa*? In the case of the giraffe, however, it is reasonable to assume that the elongated neck and fore limbs are a primary adaptation related to feeding habits. These must have determined the type of fighting that was possible and this, in turn, must have determined the selective forces in operation in the evolution of the horns.

Geist (1966) regards the lateral fighting of *Oreamos* as primitive; indeed, he considers lateral fighting to have been, in general, an early stage from which frontal techniques were subsequently evolved in other groups. This seems to me improbable; in view of the fact that the unarmoured camelids do use butting as a mode of attack and do sometimes indulge in simultaneous head ramming, there seems to be no difficulty in visualising frontal fighting as having originated directly a number of times. It could have done so without a preliminary lateral stage and would then provide the initial selective advantage leading to the evolution of head weapons. One might then interpret lateral fighting techniques as an early aberrant development, first appearing at a stage when the weapons were still relatively small and possibly arising, much as has been suggested for *Oreamos*, from a concentration on lateral threat. While frontal fighting led to elaboration and increase in size of the weapons, lateral fighting, in which the wounds they inflict may be serious, has had the reverse effect and prevented the evolution of exaggerated weapons. This shows itself as clearly in the relatively small tusks of *Sus scrofa* as compared with those of *Phacochoerus* and *Hylochoerus* as it does in the small horns of *Oreamos*.

It is, of course, also possible that any attempt to decide, in general, whether lateral fighting is derived from frontal or *vice versa* is misguided. It may be that the ancestral species at first used a variety of techniques in which striking with the head was involved but orientation was no more critical than it is in the extant giraffe. From this undifferentiated stage there may have been independent concentration in different groups on the frontal or the lateral orientation with all the attendant consequences for subsequent evolution of the weapons and of the techniques of using them which have been outlined above. Whether a fuller knowledge of the behaviour of extant species will make possible a decision between these possibilities remains for the future to show.

From what has been said above, it is apparent that even where a stylised form of pushing contest has been evolved, this is not necessarily sufficient to prevent bloodshed. There is always the moment of danger when the weaker yields and must turn to make good his escape. In elephant, hippopotamus and pigs, we have seen that the victor may seize this moment to deliver a lethal blow. The fighting methods of these species are therefore incompletely stylised. Amongst the Bovoidea it is not always clear whether the rarity of attacks of this sort reflects an inhibition of attack by the victor or is mainly due to the agility with which the loser disengages and springs away. The cases of the bushbuck and eland show that such inhibition, if it exists, is certainly not universal. Schloeth (1961) also notes that wild cattle attempt to gore each other in the side, while Innis quotes a case in which a defeated giraffe was delayed in his retreat by becoming entangled in fence wire: his opponent rained blows mercilessly on his unprotected rear. It is not really to be wondered at that even stylised fighting methods do not suffice completely to eliminate injuries. If the fight is to fulfil its biological function, it must test the pugnacity and determination without which strength alone is of little avail. These qualities may be decisive whether it is a matter of defending a territory or of rivalry for a mate, whether it is a question of a predator attacking potentially dangerous prey or of the latter defending itself or its young against such attack. It is therefore too much to expect that fighting will reach the point of being a good natured affair, like a friendly sporting event. Stylised though the attack may be, the adversaries in a serious fight are highly incensed and well-nigh blind with fury and one should not exaggerate the "gentlemanliness" of the encounter. There is, however, another factor which helps to limit the injuries inflicted in intra-specific fighting: this is the readiness with which the loser accepts the verdict of a single contest. This may be important in species that do not have a highly stylised method of fighting as well as in species that use stylised techniques. I have, for instance, seen a male *Sminthopsis* put up very little resistance, once he had been worsted in the first round of a contest with a rival, and in two cases where male *Thryonomys* fought, a single defeat was enough to send the loser into hiding.*

Those species in which fighting is highly stylised are all possessors of potentially lethal weapons, but the converse is not true: the weaponed carnivores appear to rely on restricted rather than highly stylised fighting. This is probably to be accounted for by the fact that their livelihood depends on their efficiency as killers. They cannot afford to develop any characteristics which would reduce this efficiency; the weapons cannot be modified in relation to a stylised fighting technique and some restriction on their use may therefore be the best compromise that is possible in the circumstances.

It is, however, characteristic of all the species studied that fighting is a last resort. It is preceded by threat, which functions in two ways. In the face of threat by a clearly superior animal, a weaker one may withdraw and so avoid launching into a fight in which he would be totally outmatched. Prolonged threat duels, however, also characterise encounters between equally matched contestants and these duels often end by the two separating without actually engaging in combat. It thus seems that here the threat provides a means of discharging aggressive tendencies which obviates the necessity for fighting; a situation which may be to the advantage of both parties. Threat therefore forms an essential part of aggressive behaviour and to it we must now turn.

* It should also be borne in mind that reproductive fighting reaches its full intensity only gradually, as the animals come into breeding condition. Preliminary encounters in which there is little ferocity may thus give the weaker animals a chance to "learn their place" without ever being involved in a contest with a superior at the height of his rut.

CHAPTER 7

Threat and appeasement

A threat may be defined as a signal denoting that, contingent upon some act or failure to act on the part of the recipient of the signal, hostile action will be taken. One may distinguish an offensive from a defensive threat: the former carries a message whose equivalent is "If you do not retreat, I will attack you"; the latter means "I am not about to launch an attack, but if you take the offensive, I will retaliate". The function of the threat is to deter the opponent; to drive him away in the first case, to prevent him making an attack in the second. Obviously a threat may be more effective if its recipient is given a good view of the threatener's size, strength or weapons. It is therefore not surprising to find that threat frequently includes actions which serve to draw attention to such features. In some cases, a demonstration of size or weapons may precede and be quite distinct from the signals which directly presage an attack; such preliminaries may then be referred to as an intimidation display.

The use of such terms as *intimidation* and *threat* must not be taken to imply a conscious effort to influence the behaviour of a conspecific in a particular way; still less that the evolutionary origin of the actions which function as threat is to be sought in any such deliberate calculation. Threat displays are, of course, not a mammalian prerogative: threat behaviour is shown by all classes of vertebrates and also by many invertebrates. In other groups it has been possible to show that threat behaviour can be derived from intention movements and displacement activities which are performed in a situation where there is conflict between tendencies to attack and to retreat. It is therefore desirable to examine mammalian threat behaviour from this point of view and see whether it is in any way exceptional or whether it shows all the hall marks of a similar evolutionary origin.

The first question that arises is whether the situations in which threat is shown are ones in which motivation is not purely aggressive, but is mixed with opposing tendencies. This, however, is less simple to demonstrate than might at first sight appear and there is some danger of a circular argument developing. One cannot say that if an animal does not attack at once but first threatens, this shows that there must be some conflicting tendency interfering with the attacking motivation to produce the threat and therefore threat is characteristic of conflict situations. The very nature of the case precludes any direct demonstration of the correlation between conflict and threat. It is, however, possible to demonstrate the reverse correlation, to wit one between absence of threat and absence of conflict. Although attacks not preceded by threat are relatively rare, it is a fact that when they do occur, the attack is never hesitant or half-hearted. It is carried out to the fullest extent and with great rapidity; there is no sign of motivation for anything other than attack. This was particularly clear in the case of the meerkats. These are extremely pugnacious animals and it was not uncommon for them to launch an attack, without threat, on an unfamiliar person. Such attacks were always extremely swift and the bites inflicted were severe.

In general, we may expect to find aggressive motivation virtually undiluted by any opposing tendency in two sorts of situation; firstly, where the opposition is negligible and secondly where some specific influence is in operation to raise aggression to an unusually high level. The Kunkels' (1964) observations on guinea pigs provide an example falling into the first category. The unresisting cowering foreign animal is attacked without hesitation and without threat but the rival who is ready to fight is threatened before hostilities are begun. In mammals, the presence of young is a factor which may increase parental aggressive tendencies far beyond their normal level and behaviour in these circumstances provides numerous examples corresponding to the second type of situation mentioned. In females, either pregnant or with young, this effect is so common as to be almost a general rule but in species where the male plays a part in caring for the young, he too may become unusually aggressive when a litter is being reared. An example of unhesitating, threatless attack against apparently superior odds by the male meerkat in defence of his young is described in detail in Chapter 10. Examples of similar behaviour on the part of a female are legion. The increased aggressiveness of a pregnant female rat is well known; indeed a vicious bite, without any warning, may be one's first intimation that a rat is pregnant. The most impressive example of this type of behaviour that I have ever seen, however, was an attack made on a full grown Alsatian dog by two female cats, both with litters. The cat with kittens is ready to spring to the defence of her young but in the domestic situation there is rarely any need to do so. As a result aggressive behaviour is denied its normal outlet and the cat becomes not merely ready to defend her kittens if necessary; she actually appears eager to do so. On the occasion in question, the two females were in this condition of eagerness to attack. The Alsatian happened to wander into the garden and made some sound which betrayed his presence. As though actuated by a single switch, the two cats leapt up from their litters, neck and neck they raced for the door, neck and neck they jumped a small fence separating them from their objective and hurled themselves upon what was to them a superb releaser for their pent up fury rather than a danger that had to be faced. The wretched dog, taken completely by surprise and utterly demoralised by this unprovoked assault, turned tail and fled, yelping like a hurt puppy, with one cat still on his back, raking him with her claws; like a jockey flogging the last ounce out of his steed. The same cats, meeting the same dog in other circumstances, when their aggressive motivation was not raised to such an unusually high level, would not attack but instead take up the familar arched-back defensive threat posture.

It thus appears that in mammals, as in other groups, there is no evidence not in accord with the idea that threat is characteristic of situations where there is conflict. The next point requiring study is the details of the behaviour involved in threat. Are the actions, so to speak, new inventions—movements evolved *ab initio* with no other function than to act as signals—or can they be derived from actions used more directly by the animal in the process of making its living?

In other words, do threat patterns have the characteristics to be expected if they have been evolved out of intention movements and displacement behaviour? In this enquiry, we must bear in mind that ritualisation will tend to have exaggerated the movements, both by increasing their amplitude and by making them more obvious by repetition. It may therefore sometimes be difficult to deduce what the original form of a movement may have been and it may be necessary to study a series of species before we can do so—as we have already seen was required in the case of tail-rattling threat in porcupines. Here it was possible to trace the ritualised movement back to tail trembling arising as an incidental accompaniment of a high general level of excitation. Much more commonly, however, the movements used in threat are related to those used in actual fighting and are clearly derived from intention attacking movements. This type of threat bears a direct relationship to fighting techniques, being at its simplest where the latter are simple but more complex where the fighting is stylised: moreover, it automatically displays the weapons which will be used if the attack is actually pressed home and therefore has an intimidatory quality.

In its most elementary form, threat derived from intention attack consists merely of thrusting the head at the opponent with the mouth open, ready to bite, thus displaying the teeth. The effectiveness of this display is frequently increased either by adding a sound component, by enhancing the optical effects or a combination of the two.

In cases where visual signalling cannot be very effective, the usual way of accentuating open-mouthed threat is by means of an accompanying vocalisation. This method is characteristic of the predominantly nocturnal marsupials, but the hairy-nosed wombat is unusual in that attention is drawn to the bared incisors by a tossing movement of the head. In the Dasyuridae, threatening is normally accompanied by a panting or hissing sound at low intensities rising at high intensity to a shrill squeak which, in the smaller species, sounds almost like an insect's stridulation. In shrews too, with their poor eyesight, the value of a visual threat is slight and here, as we have seen, a similar auditory component has been elaborated into the characteristic screaming duel.

In the Carnivora, threat is normally accompanied by growling, snarling or screeching sounds and in some species the opening of the mouth is modified in such a way as to accentuate the display of the canine teeth which constitute the main weapons. The lip is drawn up so as to bare the canine and expose the whole of its length, in the manner familiar to us in a snarling dog. According to Kleiman (1966b), this gesture is not universal in the Canidae. Apart from the domestic dog it occurs in the wolf, coyote (*Canis latrans*) and golden jackal. Many other canids merely open the mouth widely without the special lift of the lip and in some species there is no open mouthed threat. In this connection, the behaviour of the hyaenid *Proteles* is of some interest. Von Ketelhodt (1966) describes intense piloerection but makes no mention of snarling or opening the mouth and in three photographs of defensive threat, including the one used in Plate I, the mouth is firmly closed. Since the dentition is extremely reduced in this species, it is tempting to associate with this the absence of the usual carnivore tooth display, which in the present case would not serve to deter an opponent by a show of weapons but would merely disclose their inadequacies.

Display of the canines in Equidae and in some of the Cervidae has already been mentioned but it is in the hippopotamus that display of the teeth in threat reaches its most impressive form (Verheyen, 1954). The head is raised, the mouth opened widely and the head then moved around with a circular motion. The movement is more like a yawn than an attacking movement and it may indeed be derived from displacement yawning. This certainly seems to be the origin of the yawning sometimes used as threat in primates. In the chacma baboon yawning is an expression of uncertainty; in *Papio anubis*, while yawning may have the same significance it is also used as a form of threat by adult males (Hall and DeVore, 1965). A possible parallel to this type of yawning tooth display occurs in the Tasmanian devil. In this species, yawning occurs in circumstances very similar to those described for *P. anubis* and I interpret it as a rather ill defined low intensity threat.

In rodents, the highly specialised incisor teeth are capable of producing a noise when grated against each other. Tooth chattering or gnashing, which may be regarded as a form of intention biting, ritualised in such a manner as to provide its own auditory component, is very common as a form of threat. It occurs in primitive forms such as *Aplodontia* (Wandeler and Pilleri, 1965) and sciurids (Eibl-Eibesfeldt, 1951b; Balph and Stokes, 1963); in murids and cricetids and in New World hystricomorphs such as the guinea pig and acouchi. Eibl-Eibesfeldt (1958) suggests that all rodents may use this form of threat but I cannot recollect having heard it used by either *Hystrix* or *Atherurus*. In these species which rarely attack by biting, it may have been replaced by tail rattling. The American porcupine, *Erethizon*, which does not have sound-producing tail quills, does gnash its teeth in threat (Shadle, 1950). *Thryonomys* is another rodent in which I have never heard tooth gnashing. The noise made by these animals as they eat tough grass stems is so similar to tooth gnashing that confusion would be very likely to occur if the latter sound were used in threat. Here it seems to have been functionally replaced by growling, a sound I have heard used to warn off a would-be food thief.

Camels and dromedaries, as well as rodents, gnash their teeth when threatening and pigs too champ their jaws and grind their teeth both before and during a fight. They also produce an abundant flow of saliva, which the champing converts to a froth, so that they do literally "foam with rage".

If, however, the teeth are not the main weapons, then the intention movements of attack will not be related to biting and any ritualised version which serves as a signal will therefore differ from the forms of threat that have just been described. In both Old and New World Camelidae, where neck wrestling is an important part of fighting, the animals threaten with the head held high, so that the neck is displayed to the opponent. In the dromedary the throat beard adds to the effect (Gauthier-Pilters, 1959). This display may be regarded as basically a "weapon threat", but it has become ritualised so that it is now an intimidation display, serving to show size and strength to the opponent. This aspect of the threat is illustrated by the fact that the vicuña will stand on any convenient mound or tussock of grass to display, thus increasing his apparent height (Koford, 1957). Furthermore, when they actually move in to the attack, camelids lower the head.

In kangaroos too, the main form of threat combines intention attack with intimidation. As already mentioned, the animals fight reared up on their hind feet. A fight is normally preceded by threat, in which the opponents face each other reared up to their fullest height (Sharman and Calaby, 1964). The licking and biting at the presternal gland which accompanies this posture and its possible reassurance function, have already been discussed. A second threat posture, whose significance is obscure, is also briefly described by Sharman and Calaby. In this, the animals stand quadrupedally with the legs stiffly extended. These authors make a very interesting incidental observation; to wit, that hand reared male kangaroos, when they become adult, are extremely aggressive towards people. Presumably this is because, having become imprinted on humans, a man is accepted as a sort of conspecific but his behaviour is peculiar; he shows himself to be implacably hostile, continually threatening by his habit of approaching bipedally. The kangaroo is not vicious, he is merely responding to the threat made against him.

In the horned and antlered artiodactyls, the basic weapon threat is to present the horns towards the opponent with the head held low in the attacking position, often combined with a butting or tossing movement. Such threat occurs in Cervidae (Packard, 1955; Geist, 1964), in various species of antelope (Walther, 1958; Buechner and Schloeth, 1965), in the mountain goat (Geist, 1964) and in cattle (Schloeth, 1961). Many species of antelope, however, also have an intimidatory threat with the head held high and the horns not directed towards the opponent. We have already had an example of this in Grant's gazelle, but it is also characteristic of the Tragelaphines. Similarly, in wildebeest (Talbot and Talbot, 1963) holding the head high is a challenge, signifying readiness to fight. In view of the similarity to the camelid head high threat, it seems reasonable to assume that this posture is derived from an ancestral neck wrestling movement. Although neck fighting has been abandoned, the threat derived from it has been retained in its ritualised form and in some cases has even undergone further elaboration, as in the white throat display of Grant's gazelle. Some antelope, for example Thomson's gazelle, have a threat posture in which the neck is held up almost vertically but the nose is directed down, so that the horns are pointed towards the opponent. This appears to be a compromise between the head high intimidatory display and presenting the horns in straightforward intention attack. Walther (1964b), however, points out that in antelopes such definitive threat postures constitute only a part of the full expressive vocabulary. The positions of head, neck and horns and the orientation of the body can all reflect the current mood. To understand the full significance of all the variations shown in any particular species requires familiarity with its whole behavioural repertoire, but in general terms, the principles governing the relations between mood and posture are simple. The head held high expresses readiness to fight and may be the equivalent of a challenge; to direct the horns towards the opponent denotes readiness to attack, while turning the horns away, or even looking away, may indicate the offer of a truce—or signifies at least that if the opponent withdraws he will be permitted to do so unmolested. Since the horns are the most recently evolved weapons, one may regard a threat vocabulary in which the horns are of predominant importance as a highly evolved language. In species retaining more primitive methods of fighting, utilising other weapons, one will usually find that there is a type of threat corresponding to the intention attack movement for each weapon. Thus Geist (1964) points out that the moose, which kicks as well as using its antlers, may threaten not only by lowering these weapons towards its foe, but may also raise a hind leg, or toss its forequarters up as though in preparation for a kick with a fore leg.

Geist (1966) makes a further important point about the use of the horns in threat. In some species their display may constitute the major component of threat and its efficacy is therefore related to the size of the horns. In at least some of the species where the males are not territorial but establish a hierarchy, rank is determined mainly by horn size. The horns thus take on a new function as status symbols and it is in such cases that their positive allometric growth is most striking. Geist points out that in considering hierarchies, we must distinguish between species, such as cattle, in which the herd members grow up together and remain associated from those in which bands of adult males of a more temporary nature with a more shifting population are formed, like the bighorn and Stone's sheep (*Ovis canadensis* and *O. dalli*). In the former case, horn size is of less importance since judgement of the rank of a fellow member of the herd may depend largely on learnt individual recognition. In the latter, however, a newcomer joining a group must find his place in the hierarchy amongst animals that are unknown to him. It is in this situation that the status-indicating function of the horns is maximally important, for by responding appropriately to these symbols, the newcomer can find his place without having to challenge any rival except those who are his own equals.

In reindeer too, rank is related to antler size and their shedding is followed by a decline in status. Espmark (1964) notes that, in correlation with this, there is an adaptive significance in the fact that while males shed their antlers in early winter, females retain theirs much longer, until after the calving period. During the winter, the animals must scrape away the snow to reach the underlying lichens that form their main food and there may be competition for the cleared "feeding craters". The antlered female, at the critical period when food is scarce and she has a calf to care for, has become dominant to her erstwhile superiors, the now unarmoured and socially degraded males. She and her calf therefore have priority at feeding craters: not only is a male unable to dispossess her of a crater she has herself dug, she can displace him and take over his.

Threats derived from intention movements of attack are, in general, predominantly offensive threats. If, however, the animal is slightly on the defensive, the combination of movements of attack with those of withdrawal can create further types of threat behaviour. In its simplest form this can be seen in broadside threat, where the animal turns himself sideways on to his opponent. The conflicting urges to attack and to keep out of danger are simultaneously manifested along the length of the body: advance of the head, nearest the foe, is brought to a halt while the less exposed hind quarters are still moving forward, and the result is the broadside orientation. Broadside threat has already been mentioned in the mountain goat but it is also common in rodents. It is, however, most familiar to us in the arched back threat of the domestic cat. Here, the tendency for the forequarters to be withdrawn while the rear is still advanced produces not only a sideways orientation to the source of danger but also the arching of the back. This is a purely defensive threat and the cat often behaves in this way if cornered by a dog. She will not launch an attack unless he advances further and, if given the chance to escape, will do so. A similar arched back threat is also shown by a number of viverrids. Since they display the maximum surface area to the opponent, broadside threats have intimidatory value. In species where the hair is erectile, the adoption of the threat posture is accompanied by maximal piloerection. This accentuates the sudden apparent increase in size produced by the change in orientation and adds to the intimidatory effect. If combined with arching the back, there is also a sudden change in shape and the net effect is disconcerting.

In the black rhinoceros Goddard (1966) describes a form of broadside threat display in which, turning sideways to his opponent, the animal moves forward with a stiff-legged shuffling gait. This display is followed by defaecation and the animal also lowers his head and sweeps it from side to side, with the upper lip and the anterior horn scraping against the ground. The display was performed by a male towards another from whom he had fled a short while before. The situation appeared to be one in which the animal was strongly motivated to attack but too much intimidated to do so.

Pawing, thumping or drumming on the ground with the feet are movements used in threat by a number of species. Movements made with the fore and the hind feet have a different significance. Pawing with the fore feet is related to advance towards the enemy and forms part of offensive threat in many ungulates. The red squirrel also "bark rakes" with its fore paws when in an excited state (Eibl-Eibesfeldt, 1951b). Although the significance of this action is not always apparent it is a normal part of the male's "intimidation run" towards the female when he is entering her territory to mate, which suggests that it is aggressive rather than defensive. Hind foot drumming or thumping, on the other hand, is related to retreat rather than to advance. Species as diverse as kangaroos and rabbits use such an action as an alarm signal. In kangaroos the foot thump is actually made during the first few hops away from the cause of alarm and its linkage with retreat is unequivocal. In the rabbit the movement is ritualised and successive foot thumps are made before take off, or even without actual flight. Where hind foot thumping is used in threat, we would therefore expect it to be defensive rather than offensive, in view of its relation to flight. In the agouti, Roth-Kolar (1957) regards foot thumping as defensive. In the acouchi, D. Morris (1962) describes two patterns—the hind foot stamp, in which both feet are simultaneously thwacked on the ground in a slow rhythm and foot rattling, in which the feet beat a rapid tatoo and may move alternately. The former is used as an alarm signal and may also be performed by a defeated animal at the end of a fight. Foot-rattling is a threat action made by a losing animal at the stage in the encounter when he is ceasing the more offensive tooth gnashing and is making a call intermediate between aggressive scolding and the drawn out "mewing" which signifies defeat. In other rodents, foot drumming occurs in situations where there is clearly some excitement, but its exact significance is not certain. This is the case, for instance, in *Meriones persicus* (Eibl-Eibesfeldt, 1951a) and similar behaviour is mentioned by Getz (1962) in two species of *Microtus*.

In porcupines, threat, as befits such well armed creatures, is often used as a warning to other species. In both *Hystrix* and *Atherurus*, hind foot stamping is a more intense threat than is tail rattling and signifies a higher probability of attack. It is thus an exception to the rule that hind foot thumping is related to retreat, but it is the sort of exception that proves the rule. In these species, with their backwardly directed spines, the attack is an obliquely backward movement: if the threat is "back to front", so is the animal's method of advancing on its foe. The threat, despite being made by the hind foot, is thus a ritualised movement of attack, not of retreat.

The tendency for marking to be associated with hostile encounters has already been noted and there are a few cases where patterns related to marking constitute definite threat displays. The possible relation of the guinea pig's testis display to marking has been mentioned but the prairie dog and the hippopotamus provide clearer examples. In the former, (King, 1955) territorial disputes between males often take the form of an olfactory contest using the marking glands. The two animals rush towards each other but stop short before they actually meet. One then turns round, raises and spreads his tail, exposes his anal glands and waits for the other to approach and smell them. Having sniffed, the opponent then takes up the same posture and waits to be smelt in return and so, alternately presenting their anal glands to each other, the two may continue for some time. The whole may be regarded as a sort of competition in olfactory intimidation. It frequently ends in a draw; the two abandon the contest, start to feed and move apart without a fight. The hippopotamus' method of scattering his droppings at a marking place has already been described. When two rivals face each other, threatening with the yawning tooth display may be accompanied by defaecation with vibration of the tail so as to scatter the droppings, exactly as in marking. The social rank of a male is directly correlated with how effectively he can perform this display. It appears to act as a challenge and as a signal of readiness to fight; a male who does not respond in like manner is acknowledging inferiority and must yield place. Since a female responds to the challenge merely by slight tail movements without defaecation, the display may also serve as a means of sex recognition (Verheyen, 1954).

It is, however, in the skunks, where the secretion of the anal glands has been specialised as a defensive weapon, that a threat related to marking reaches its highest elaboration. His distinctive black and white colouration, easily visible in the twilight, marks the skunk as dangerous. If this alone is not sufficient to deter a potential attacker, the skunk threatens but only discharges his secretion at the foe as a last resort. In most extreme threat, the animal charges at the enemy, stops short and suddenly throws up his hind quarters so that he stands on his hands with the body vertical, the tail drooping down over the back or slightly to one side and the everted anal glands directed forward, ready to expel their noxious secretion straight at the enemy. At lower intensities the animal may merely jerk the hind-quarters upwards slightly. The hand-stand posture has been described in both *Spilogale* (Howell, 1920; Johnson, 1921) and *Mephitis* (Seton, 1920). Bourlière (1955) says that the threat attitudes of skunks "vary widely according to the species", but Johnson saw a single animal adopt different postures according to the intensity of the threat and it seems likely that some of the differences reported may be variation of this sort, rather than species differences. Johnson was also of the opinion that the hand-stand position was adapted to discharging the secretion upwards at an enemy much taller than the skunk and doubted if it would be used against a species of commensurate size. Both these points require further elucidation. Regardless of what the answer may be, the hand-stand threat posture is so extraordinary that its origin would be extremely difficult to imagine, did we not have, in the viverrids, species showing anal gland marking postures forming a complete series from the simple anal drag to the full hand-stand position. The skunk's "firing position" is virtually identical with that of *Atilax* when marking an overhanging branch.

One curious form of threat, whose origin is not clear, is shown by species possessing cheek pouches. The pouches are rapidly inflated with air, so that the head suddenly appears much bigger. *Cricetomys* at the same time rears on its hind legs and makes a puffing sound: the total effect is disconcerting and daunting. Hamsters also inflate the pouches in threat, but whether all rodents that possess pouches use them in this way is not known.

Piloerection, having the effect of increasing apparent size, is, as already mentioned, a common accompaniment of threat. It is particularly well developed in Carnivora, where special long hair on the shoulders, along the back or on the tail, emphasises the effect. It is also found in other groups such as ungulates, rodents and primates. Little information is available about piloerection in marsupials, but *Dasycercus* has a bright orangey-buff tail with a black tip of hair forming erectile dorsal and ventral crests and in *Sminthopsis* the fur on the shoulders is erected during a fight. In porcupines, where the hair is modified to form spines, piloerection constitutes a weapon threat which is unusual in not being derived from an intention attacking movement. The primary function of piloerection is in relation to thermo-regulation. Since this involves adrenergic control, it is natural that piloerection should occur in agonistic situations where there is adrenaline secretion. Direct nervous control has, however, been evolved as an accompaniment to structural specialisation: if there are areas with specially long hair functioning as optical signals, there is normally direct nervous control. A squirrel's tail, the hairs rising and falling with every change in tension, mirrors the animal's mood as sensitively as the changes in its pupil do the changes in incident illumination. In a porcupine, the control is extremely obvious. I have seen a *Hystrix* as he walked past a cat whom he rather disliked, erect the spines only on the side of his body next to the cat as he came level with him and lay them flat again as soon as he had gone by.

Optical emphasis is also often added by special coloration. The porcupine's white tail quills and the white throat patch of Grant's gazelle are examples of this. In *Atilax*, the rhinarium is dark dorsally but bright pink anteriorly. The pink patch at first looks rather absurd—like a small area accidentally missed when applying a coat of paint. Its significance only became apparent to me when I had seen threat behaviour. The snout is then wrinkled more than is common in other species, so that the pink patch is turned upwards and directed towards the foe. The pink lips and lining of the mouth are also displayed and the net effect of the sudden appearance of this light patch, contrasting sharply with the almost black fur, is highly alarming and adds a visual element to the loud threat screech. Patches or strips of contrasting colour on face, fore limbs or sides are very common in the Bovoidea. In the majority of cases their function has not yet been elucidated but it is likely that many of them serve to draw attention to particular signalling movements and postures.

Emphasis of threat by the addition of an auditory component is also common and has already been briefly mentioned. Since sudden loud noises are alarming to most animals, it is not surprising to find that even otherwise silent species may make some type of noise when threatening. The various types of hissing or squeaking made by marsupials are a case in point. Growling and snarling are also characteristic threat sounds and are made not only by carnivores but also by some rodents and by primates. The similarity of the threat sounds used by different species may reflect two different things. On the one hand, the general tension of the agonistic situation will make for abrupt explosive sounds related to sudden inspiration and expiration. On the other, there may be something to be gained from a universally understood code—Leyhausen's "esperanto of expression"—since warning off other species may be just as important as driving away conspecific rivals.

The higher primates, with their mobile grasping fore limbs, are unusual in making use of extraneous objects to produce noise. This may take two forms—slapping or beating with the hand against the ground or some convenient object such as a tree trunk and shaking or tearing off of branches. Displays involving these actions occur in situations where there is a high level of excitement. They may be performed on seeing a human observer or a predator or when two groups are meeting. The function seems to be a generalized intimidation rather than a specific warning of imminent attack. Such displays occur in both monkeys and apes and, indeed, I have seen something closely resembling the chimpanzee's branch shaking display performed by a small boy to impress a group of smaller children in whose play activities he was taking the role of initiator. Displays of this type have been described in baboons (Hall and DeVore, 1965) and in the rhesus (Hinde, 1966) and patas monkeys (Hall, 1966) and also in chimpanzees (Goodall, 1965) and gorillas (Schaller, 1963). In the latter the display is extremely complex and includes a series of different methods of producing noise: the animal first hoots loudly, then drums on his own chest with his hands and may later shake and tear down branches and finally thump the ground with his hands. In captivity apes are quick to discover and exploit the resonant properties of such objects as iron cage doors.

Although there are many species for which no information is available, what is known accords with the view that mammalian threat behaviour, like that of other groups, has evolved from the types of movements that animals are prone to make in conflict situations. It is instructive to pause for a moment and consider the differences in the behaviour of a carnivore towards its prey and towards a conspecific rival. The former situation, particularly if the prey is large or unfamiliar, can be one of conflict, just as the latter is. In fact, I have seen a young caracal when faced with his first live prey move hesitantly towards it, pause and make a displacement face-washing movement. And yet, while the conspecific duel is characterised by intensive threat, the prey is not threatened, even if the predator is quite clearly in some awe of it. The adaptive reason is, of course, obvious: the prey must not be scared away, whereas with the rival, nothing is more desirable than that he should make himself scarce. In the one case selection will have operated to suppress every movement liable to drive away the second partner in the interaction; in the other, to elaborate it. But one can go further than this. One of the most striking things that emerges from a study of mammalian threat behaviour is that almost all of it can be traced to origins in intention movements of attack, defence or retreat and very little to displacement behaviour. Moreover, the movements used follow a simple rule: in offensive threat intention movements of attack predominate, whereas in defensive threat they relate more to withdrawal and retreat. The animal virtually mimes its probable next course of action. Selection clearly has not been operating in a random manner on the principle of "take any movement the animal happens to make in the conflict situation and elaborate it into an efficient signal". If it had, then surely threats originating from displacement behaviour would be much more common. The Felidae, for example, are very prone to displacement face washing: why then is some ritualised form of this action, some paw waving movement for instance, not used in threat? Yawning, the only clear displacement activity that has been ritualised into threat, is not "irrelevant": it is an action which, because it displays the teeth, in fact produces a weapon threat. Cheek pouch inflation, whatever its origin, is also not "meaningless", but is intimidatory by virtue of the sudden increase in size it produces.

Walther was primarily concerned with threat when he expressed the opinion that thorough familiarity with the whole behavioural repertoire was necessary for a proper understanding of the expressive movements of any particular species. The converse proposition, that if you *are* familiar with the rest of the behaviour you can interpret the language of the signals, carries a very definite implication: to wit, that the sign language is neither abstract nor arbitrary but symbolic. I believe that this characteristic of mammalian threat behaviour is not accidental but is the result of selection operating on a system in which (as in all signalling) two parties are involved—the sender and the receiver. The efficient signal is the one which produces the appropriate response and the selective forces involved in the evolution of the signals are therefore shaped by the behaviour of the receiver. Some signals, for instance sudden loud sounds or sudden increase in size, depend on responses which are so generally present in vertebrates that they must have been evolved long before the mammalian stage was reached. For other, more specific signals one cannot invoke a pre-existing innate response, only waiting to be released. In such cases the significance of the signals must at first have been learnt. The ease with which this learning could occur would therefore be an important factor in the evolution of the signal. It is surely not over-anthropomorphic to suggest that it is easier to learn to associate with attack a movement that itself suggests attack than one which is quite unrelated. Mammalian threat behaviour does in fact fall into the two categories which these considerations suggest: either it produces effects which are very generally daunting or it is based on intention movements of the actions most likely to follow. Arbitrary or "meaningless" signals, if they occur at all, must be very rare.

Learning on the part of the receiver may have been important in the early stages of evolution of highly specific signals but it is only to be expected that a built-in response should in the end be evolved to replace the learnt one. Once this has happened, ease of learning no longer enters into the matter and the threat may remain operative even if the mode of attack alters. We have seen examples of this in the way *Rusa* bares non-existent canines and in the retention of head high threat in species that have abandoned neck wrestling as a fighting technique. In fact, Lorenz (1966a) regards some human threat postures as being related to the erection of neck and shoulder manes that we no longer possess.

We have so far spoken as though offensive and defensive threat were two sharply contrasted alternatives, but this is an oversimplification. All gradations between readiness for instant attack and readiness for instant flight exist and may be reflected in the animal's behaviour. The transition from attack to defence can be shown piecemeal, by a decline in frequency of purely offensive actions or by the bringing in successively of the various components of defence. For example, when two *Cricetomys* are reared up boxing and biting at each other, the first sign that one is getting the worse of it is that he makes fewer bites at his opponent and concentrates on boxing him off. A sign of his being still nearer to defeat is the replacing of tooth gnashing by the high-pitched protest squeak that signifies "Stop it", rather than "Go away or I will bite you." A transition from attack to defence is sometimes shown at a single instant along the length of the animal's body. This is particularly clearly shown when kittens are indulging in fighting play. One approaches the other to attack, but the nearer he gets, the more defensive tendencies are aroused. His head, closest to the "foe", is the region where he first expresses them and the head thus starts to assume the defensive "belly-up" posture before the hindquarters do, so that the kitten's head goes down and twists around, bringing his shoulders to the ground. His hindquarters continue to advance so that he pushes himself towards his opponent almost standing on his head before he finally rolls over into the full defensive posture, lying on his back with teeth bared and striking out with all four paws.

This belly-up defensive posture is characteristic of carnivores but it also occurs in other Orders, in species where the limbs are sufficiently unspecialised to be effectively used in the manner described. Its significance is unmistakable: on its back, the animal gives not the least hint of intention to advance in attack but, claws and teeth at the ready, it is obviously prepared to sell its life dearly.

From what has been said, it follows that a defensive threat, as well as signifying that further attack will bring retaliation, is also an admission of inferiority. As such, it could come to divert further attack by the winning animal, not only because the latter knows that to do so may be costly but also because he knows it is unnecessary: if a chance to escape is given him, the defensive threatener will take it. This certainly appears to be the case in the lemming (Arvola et al., 1962), where if a defeated animal turns and threatens, the pursuing victor frequently stops and allows him to escape. This can hardly be because he has suddenly been intimidated by the individual he has just succeeded in trouncing. As described, the threat appears to be an admission of defeat and to be accepted by the opponent as a sufficient victory.

In some species there is a further stage—surrender. Special submissive or appeasement postures have been evolved, signifying complete absence of hostility and having the effect of preventing attack by the victor. Such a posture adopted by a beaten animal at the end of a fight serves to prevent the winner from pressing home his advantage. Similarly submissive behaviour shown to an approaching superior may prevent any attack being made and an inferior, if he adopts a submissive pose, may be permitted to approach a superior more closely than would otherwise be possible. In solitary species appeasement is important primarily in the first type of situation, but in social species, where it is necessary for an inferior to remain associated with a superior without provoking attack, its main use may be in the other two sorts of encounter.

The value of an appeasement posture, provided the response to it is appropriate, is obvious, but how this type of behaviour ever came to be evolved in the first place requires some consideration. At first sight it might seem that the animal that refused to defend itself would just get killed. A consideration of the details of the appeasement postures used by different species and their relation to the rest of the behavioural repertoire shows that the problem is really not very difficult.

The commonest of all appeasement postures is to lie on the back, as in the belly-up defence posture, but with teeth and claws not at the ready. The vulnerable throat and belly are thus exposed and the appeaser is completely at the mercy of his opponent. This type of appeasement posture is found in dogs and rats—both social species—but also in the solitary hamster, in the vole *Microtus agrestis* (Clarke, 1956), and I have seen it in *Sminthopsis crassicaudata*. In the agouti, a defeated animal at first simply crouches motionless and this is often sufficient to prevent further attack. If the attack is continued, the loser may lie down prostrate and roll over on its side with its legs stretched out and its eyes almost closed. Even a persistent attacker now desists, examines the motionless body carefully and finally moves away (Roth-Kolar, 1957).

In view of the point previously made, namely that a defensive threat is also an admission of defeat, it is relatively simple to see how belly up defence could evolve into belly up appeasement. All that is necessary for this to be possible is that there should be a stage where forcing the opponent over to pure defence should be accepted by the superior as sufficient victory and the loser be given the opportunity to withdraw. Arvola's observations, which to me suggest that this is the case in the lemming, have already been mentioned. In this connection, it is interesting to note that some workers, making their observations in very unnatural situations, have actually interpreted the reared-up defence posture of the mouse as one of appeasement (Grant and Macintosh, 1963). This error could never have been made, had the behaviour of the mouse in face of predators been taken into account, and Crowcroft (1966), who studied his mice in approximately natural conditions, has no doubt that this posture is not a submissive one. Although this reared-up defence posture is not relevant to the evolution of belly-up appeasement, the fact that submission and defence could be thus confused is a striking testimony to their close relationship. From the condition where defence is accepted as an admission of defeat, further evolution could lead to the removal from the defence posture of anything liable to encourage attack, and this I believe to be the origin of belly up appeasement. It is an extension of the ultimate defence posture; stripped of the last elements of fight, it offers nothing likely to precipitate the discharge of the opponent's aggressive tendencies. In the hamster, the principle of removing all aggression-stimulating factors shows itself particularly clearly. In this species there are dark patches on the chest which are displayed in the reared up attacking posture but in appeasement the paws are placed so as to conceal them (Grant and Mackintosh, 1963).

The case of *Sminthopsis* is particularly interesting because the full appeasement posture of lying on the back is not always shown. Sometimes a male approaching a female and clearly anticipating trouble will almost take up the posture but not go quite all the way; sometimes he will get as far as laying down his head and shoulders, then right himself and continue his approach, but much more often he will only make the initial move of turning his head sideways as if about to lay himself down. This is always done so that the white throat is turned towards the partner.

If one had seen only this throat exposure, its evolutionary origin would be hard to guess. The intermediate stages, however, make it quite clear that throat exposure is not an independent piece of behaviour, evolved *ab initio* in its present form, but a movement derived from something else. Belly-up appeasement we have already seen can be derived directly from the corresponding defence posture and curtailment of the action to produce throat exposure only is simply a further step, one which could not have been taken until after the full posture had come to be an efficient inhibitor of attack.

It cannot, however, be assumed that a belly up posture is always one of defence or of submission. In shrews, for instance, Crowcroft (1957) interprets such a posture as a high intensity threat and found that its adoption frequently resulted in the opponent fleeing. Bunn (1966) confirms that the victor in a fight is the one who is left lying on his back as the other runs away. It thus seems that here the belly-up posture is aggressive, rather than defensive, and Bunn suggests that this is because the shrew fights most effectively on its back, so that the posture is genuinely as well as symbolically aggressive.

It is, of course, tempting to see death feigning as simply one stage on from appeasement—for what could be more inoffensive, less likely to provoke attack than a dead animal—and from the sort of behaviour described in the agouti, death feigning seems but a short step. Death feigning, however, is characteristically evoked by interspecific enemies, not by conspecifics. To decide whether it is likely to have evolved from an appeasement posture requires more knowledge of the behaviour of species which show it and of related species than we at present possess. Death feigning is best known in the American opossums but it is not peculiar to them; it is also practised by a number of placentals, the most familiar example being the fox. The habit is also shown by *Xerus erythropus*. I have witnessed this behaviour only once. The occasion was when a captured juvenile *Xerus* was brought in to the laboratory. As I opened the box containing the squirrel and put in my hand to pick it up, it heeled over and lay apparently dead. The body was motionless and no respiratory movements were visible; the paws were held close to the face with the digits rigidly flexed, the eyes were shut and the mouth gaped open. In fact the animal not only looked dead; it looked as though it had been dead for some time. This should have made clear to me what was happening, but in fact I thought, "If only they had got him to me a little sooner I could have saved him". I tried placing a little warm milk in his mouth but there was no swallowing response. I was still sadly contemplating the corpse when, after what I judge to have been about two minutes, it came to life and drew a few deep gasping breaths. The squirrel adapted to being handled so readily that I never succeeded in evoking the response again. The circumstances in which death feigning occurs in nature and what protection it may afford the animal are not known, but I am told that when dug out of their burrows, the squirrels will sometimes behave in this way. Edwards (1946) mentions what may be death feigning in an American ground squirrel—*Citellus mexicanus*—but his description is not very precise.

In dogs, the belly up submissive posture, although it may have originated in relation to defeat in a fight, is used in other contexts to forestall aggression. Puppies are prone to roll over in the submissive posture not only in face of a strange adult of their own species but show the same behaviour towards people; adult females may also behave in this way. *Lycaon* uses a similar appeasement posture (Kühme, 1965a).

Kaufmann (1962) has described in the coati, *Nasua narica*,

a posture which he believes may be one of appeasement. This species threatens with the head stretched out in front in line with the body and the long snout turned up to bare the teeth. In the "appeasement" position, in contrast, the head is lowered and the back curved. Kaufmann's observations, however, do not give much ground for ascribing an appeasement role to this posture: certainly it is a "weapons away" gesture, but its significance seems to be no more than the offer of a truce, "I am willing to stop, if you are", exactly like turning the head aside in antelopes.

In Artiodactyls also, lying down may be used as a submissive gesture towards a superior. Walther (1966) records it in various species of gazelle, in kudu (*Tragelaphus strepsiceros*), sitatunga and wildebeest and it also occurs in red deer and fallow deer, *Dama dama* (Burckhardt, 1958). Here its origin is probably not related to a belly up defence posture, which does not occur in these species. Walther suggests that it is derived from the common concealment posture of lying down as flat as possible with the head stretched out on the ground, which is adopted by the young in response to danger. This interpretation is strengthened by the fact that in some species, for example oryx and wildebeest, the submissive posture is accompanied by a vocalisation resembling the infantile distress call. If the situation is not particularly tense, the full submissive posture may not be adopted and lowering the head in the intention movement of lying down may suffice. This movement, which might otherwise resemble threat and so evoke attack instead of preventing it, is usually accompanied by turning aside, so that the horns are directed away from the superior. In black wildebeest, *Connochaetes gnou*, since fighting involves going down on the carpal joints, the danger of confusion with threat is greater and here the turning away of the head is very marked and the face at the same time is directed downwards: only if the subordinate goes as far as lying down completely, does he then extend the neck and stretch it out forwards. The uttering of a characteristic call, which accompanies the submissive behaviour, must also help to make clear that there is no aggressive intent and so prevent "misinterpretation". Wildebeest are exceptional in that the submissive behaviour can go one stage further. After he has lain down, if the superior still does not withdraw, the inferior may actually roll over on his side, belly towards the partner and with the side of the face flat on the ground. From this position no attack can be launched; the animal cannot even spring rapidly to its feet to defend itself and its vulnerable side and belly are unprotected.

In species such as Grant's gazelle, where threatening is done with the head held high, there is no danger of the head lowering of intention lying down being misinterpreted: here the submissive posture consists of lowering the head to the level of the carpal joint with the snout directed horizontally so that the horns point back and at the same time, the ears are laid back (Figure 10). This is the exact opposite of head

Figure 10. Grant's gazelle: appeasement by a juvenile male (on left) in response to high horn threat by an adult male. The adult is shown displaying full scale threat, so as to make clear the antithesis between this and appeasement. In actual encounters of the sort shown, the threat is usually somewhat less intense. (After Walther, 1965.)

high threat, in which the neck is stretched up, the snout directed downwards so that the horns slope slightly forwards and the ears are turned forwards. Burckhardt (1958) believes that similar appeasement postures in the chamois, *Rupicapra rupicapra*, and fallow deer are derived from the attitude adopted by the young when approaching the mother to suckle. Against this and in favour of the view that these postures are intention movements of lying down, is the fact that in the fallow deer, the full pattern of lying with outstretched neck is also used in appeasement. In red deer, the latter is the normal appeasement posture.

Zeeb (1959) describes a submissive facial expression shown by young horses in response to threat by an older animal. When the youngster assumes this expression, the adult ceases threatening. The submissive animal opens the mouth slightly and makes nibbling movements of the jaws without actually bringing the upper and lower incisors into contact. The corners of the mouth are not drawn up as they are in threat and the ears take up a position not seen in any other context. The pinnae are turned so as to project laterally from the sides of the head and the openings are directed downwards towards the ground. Zeeb interprets the whole as a ritualised form of the intention movements of social grooming. This expression has not been reported in any other equid but in Burchell's zebra* Backhaus (1960) describes a different form of submissive behaviour which occurs at the end of a fight. The winning animal lays his head on the loser's rump: if the latter permits him to do so, this acts as a token of surrender and the victor ceases hostilities. This case is interesting, since as a preliminary to mounting the male lays his head on the female's rump. Permitting this is therefore a close parallel to presentation of the hindquarters in a copulation invitation, which is a gesture of appeasement widespread amongst the higher primates. In this latter case the action may be derived originally from turning away as a gesture indicating absence of aggressive tendencies but in its developed form of full presenting there may be a double effect on a potential aggressor. Not only is he given the "soft answer that turneth away wrath", but the posture may also have the effect of changing the context of the encounter from an agonistic to a sexual one. In much the same way the liberation of a little urine by an appeasing puppy may convert the

* The taxonomy of this species is rather confused. German authors generally refer to it as *Equus quagga*, whereas English workers call it *E. burchelli*. The former name is based on the assumption that Burchell's zebra should be regarded as a sub-species of the now extinct quagga, the latter that it belongs to a distinct species. Since the evidence suggests that before the killing off of the quagga, it and Burchell's zebra lived in the same area without interbreeding, I prefer to regard them as distinct species and call the living form *Equus burchelli*.

situation to a parent-child relationship and the older animal usually responds by licking up the pup's urine. The same principle is shown even more clearly in a piece of behaviour seen in the titi monkey by Moynihan (1966). A subadult male, caged together with an adult male, was worsted in a number of fairly serious disputes. When he found escape by flight was impossible, the subadult would leap upon the other animal's back; just as a juvenile when alarmed jumps on its parent's back. Moyniham reports that this action by the younger animal "always inhibited further hostility by its opponent, probably because it stimulated parental tendencies in the latter".

Let us return for a moment and consider further the use of presenting as an appeasement gesture. The normal response is a brief "symbolic" mounting. This differs in detail from a true sexual mounting and is, in effect, a signal that the submission is accepted and hostilities are at an end. On the basis of this type of interchange, a number of further complexities have developed, anatomical as well as behavioural. In the females of a number of primates the sexual skin on the buttocks becomes turgid and swollen when the female is in oestrus: it could therefore serve as a means of transmitting to her companions information about her sexual condition. Rowell's (1967) work, principally with baboons and to a lesser extent with rhesus monkeys, however, indicates that, at least in these species, the sexual skin is of very little importance in this respect and some other facts suggest that its function may be rather more complex.* In a number of mammals, including monkeys, it has been found that while males reared without experience of females are capable of performing the normal copulatory patterns, their orientation is often at first imperfect (see Chapter 12). The primary function of the female's sexual skin may therefore have been to assist the male in achieving the correct orientation in mating, rather than to inform him that the female is in oestrus: certainly it makes her hindquarters very catching to the human eye as

* It may be noted in passing that the enormous swellings characteristic of caged female baboons are not seen in the wild. They seem to be a sort of physiological captivity artefact, presumably the result of the animals undergoing repeated cycling without breeding (Rowell, 1967).

she presents. Presenting, however, has come to be a gesture used in situations other than mating and it is not restricted to females. It is therefore not surprising to find that colouring which accentuates this gesture is also not restricted to oestrous females. In some species the female's skin patch is pink and quite visible even when she is not on heat and in some the male also has a pink patch of bare skin on his rump (see Wickler, 1967 for details). Such facts have led Wickler (1963) to suggest that the rump patch has the function of adding to the efficiency of the appeasement gesture and that where the male also has a patch this is a secondary condition; a sort of mimicking of the condition originally evolved only in the female. More surprising is the case of the gelada baboon, *Theropithecus gelada*, where the breasts and nipples of the female are curiously modified. The skin is naked and red with a border of white papillae and the whole mammary area bears a close resemblance to the female's own genital region. Wickler suggests that indeed the breasts are mimicking the genital sexual skin. The gelada is a species that spends a great deal of time sitting down and one can see that it might therefore be convenient to have one's behind before and so be able to transmit the appropriate signals without having to get up. The breast skin shows the same relation to oestrus as does the genital sexual skin: it is not, however, clear whether the function of the signals given by the breasts is to transmit information about the female's sexual condition or is related to appeasement, or possibly a combination of the two.

Related to presenting may be a curious form of threat described in the squirrel monkey, *Saimiri sciureus*, by Ploog and Maclean (1963). A dominant male may threaten a subordinate by spreading his legs and thrusting his erect penis towards his opponent. This threat is unique, since it cannot be derived from any behaviour directly concerned with fighting or with territorial defence by marking. Its origin can lie only in the appeasement ceremony and it is most easily explained as a ritualised form of the normal mounting response to appeasement presenting: the dominant animal is signalling to his opponent that he expects submission. The story can be taken one step further. The significance of the red and blue facial colours of the mandrill has long remained a puzzle. Wickler originally thought that, like the gelada's breasts, the mandrill's face might be mimicking his own buttocks. On investigation, however, the buttocks were found to be disappointingly normal in colouration. Morris and Morris however, thinking of the squirrel monkey's genital threat, examined not the buttocks, but the genitalia of the mandrill—and found that the penis is bright red and on either side of it the scrotal sacs make a blue patch. They suggest that the case of the male mandrill is indeed comparable with the female gelada: his face mimicks his genital region as her breast does and thus transfers the threat signal to a more obvious and easily visible area. Exactly the same conclusion was independently reached by Wickler (1967).

All this has led us away from the main theme of primate appeasement behaviour, to which we must now return. According to Hall (1966) the patas monkey is unusual in that submissive presenting does not occur and indeed, no appeasement posture is known in this species. Possibly the absence of presenting is, like the "adaptive silence" of the patas, a reflection of the fact that their method of coping with the dangers inherent in life in the open savannah involves remaining inconspicuous and ready for swift flight at an instant's notice. In the gorilla averting of the eyes or rapid shaking to and fro of the head are used to disclaim any aggressive intent (Schaller, 1963). Female chimpanzees present in appeasement but this species also has other forms of behaviour serving the same end (Goodall, 1965). These include a special facial expression, the appeasement grin, in which the lower lip is drawn down to show the teeth and gums; a subordinate may also placate a superior by reaching out to touch him on the scrotum or the lips.

The appeasement grin has probably originated from an expression denoting anxiety or fear. Hinde and Rowell (1962) describe a "frightened grin" as one of the characteristic facial expressions of the rhesus monkey. They note that the grin appears in situations where there is an inhibited tendency to flee and also that it is often a response to the approach of a dominant animal. From an expression of fear to one of submission or appeasement is no great step, nor is the further change to a simple greeting, in which the submissive element is reduced or negligible. Indeed, Hinde and Rowell quote a number of authors who have considered the grin in apes to be a simple greeting, rather than, as Goodall does, an appeasement signal. In ourselves, while a smile can certainly be evoked by purely pleasurable stimuli and is used as a simple greeting, it is also true that it frequently occurs in situations of slight tension and is used in a placatory manner.

The origins of the other appeasement signals mentioned by Goodall—touching the superior on the scrotum or lips—remain obscure. The latter, however, inevitably calls to mind the human suppliant gesture of touching the beard, to which reference is commonly made in classical Greek literature.

From this survey it can be seen that two major principles have been in operation in the evolution of appeasement postures. Firstly, most of them rely for their efficacy primarily on the fact that stimuli tending to evoke attack are minimised. The operation of this principle leads to elimination of stimuli which are characteristic of threat. Secondly, there is the principle of diversion: the signals of the appeaser may serve not only to prevent attack, but to turn the opponent's behaviour into some other, socially positive, channel—sexual and parental being the most usual.

This account is, however, unsatisfactory because it deals only with one side of the interaction. If appeasement behaviour is a product of selection, it is to be expected that, as in other communication systems, the behaviour of the responder should also have undergone evolutionary modification. In other words, we would expect specific responsiveness to the appeaser's signals to have been evolved, so that these no longer have a merely negative effect, in the sense of failing to evoke attack, but a positive one, actually inhibiting it: the appeasement, in fact, becomes a signal in its own right. As such, we may expect some degree of ritualisation to have occurred with the usual function of making the signals as clear and unambiguous as possible. That the situation is as just outlined seems indicated by three different lines of evidence. Firstly, an animal will sometimes actually "put on the brakes", stop short and fail to deliver an attack that had actually started, if the opponent appeases at the crucial moment. Lorenz (1963) quotes examples of this and I have seen it happen very dramatically in *Sminthopsis*. Secondly, there is the fact that a superior will sometimes signal to an inferior that his approach is unaggressive, which must reflect an emancipation of the signal from the original agonistic motivation. Altmann (1962) reports such behaviour in free-living rhesus monkeys. Lastly, if it were merely a question of removing aggression releasing stimuli, it is not easy to understand why appeasement gestures should often have come to be so accurately the antithesis of threat, for simple removal of the offending stimuli does not necessitate assuming the opposite pose. Furthermore, on this basis alone, it is very difficult to see how an intention movement could come to have any signal value. On the other hand, if we are dealing with a communicatory *exchange* system, in which the appeasement is a positive signal, evoking a positive response, then as in other signalling systems, we may expect ritualisation to occur. There is then no difficulty in understanding the characteristics of the signalling and why it has so often taken the form of threat antithesis. It also seems more reasonable to interpret as the end products of highly evolved ritualised communicatory exchanges those cases where exposure to the opponent of some particularly vulnerable part apparently inhibits attack; rather than to invoke some pre-existing, preadaptive inability to attack the defenceless. That such exposure should often occur in appeasement is an almost inevitable consequence of the principle of antithesis. The fighting animal must protect his weak points and if appeasement takes the antithetical form, it follows that they must be exposed. Lorenz (1963) has expressed exactly the same view.

Appeasement behaviour is not peculiar to mammals but occurs in other vertebrate groups as well. In mammals, however, the characteristic motivational complexity which has been mentioned in Chapter 2 lends a corresponding complexity to this type of behaviour and it is gross over-simplification to suppose that in mammals appeasement has a single motivation or fulfils a single function.

In the simplest cases the context is purely agonistic and

the function of the appeasement is to protect an inferior against imminent attack by a superior. In other cases, however, the situation may be more complex, motivations other than purely agonistic ones are involved and the function of the appeasement is correspondingly altered. A few examples will make this clear. In highly social Canidae, for instance, appeasing or submissive behaviour is frequently shown by an inferior animal when approaching a superior who is not in any way threatening him or giving any signs of intention to attack. Here the agonistic defensive motivation is combined with motivation to seek social contact, even though this necessitates approaching a potentially dangerous superior. The submission then has the function not of warding off an attack already in preparation but of permitting the establishment of social contact without thereby inviting attack (Schenkel, 1967).

In *Sminthopsis* the same combination of motivations is involved but the situation is slightly different. The male appeases when approaching the female during courtship and is particularly likely to do so if she has previously attacked him, so that, once again, the function of the appeasement is to facilitate the establishment of a social contact. The male, however, is not inferior to the female and might well be able to defeat her if they fought in earnest. It is his absence of aggressive motivation towards her, with consequent inability to fight back, that throws him on the defensive; the submissive throat-turning gesture has therefore taken on something of the function of a non-aggression signal. In the cane rat, this is even clearer. In this species the young wag their tails if approached from the rear by an adult and in this context the gesture is an appeasement signal in the simple and direct sense. Although tail wagging may not be sufficient to inhibit an attack that has already been launched, a youngster who wags his tail before an attack has started is not molested. The same tail wagging signal is used by the male in his courtship advances to the female. The situation, however, is not exactly as in *Sminthopsis*. A male is always superior to a female, even if she is bigger than he is, and a female never attacks her suitor; if she is not yet prepared to accept him, she merely flees. The appeasement gesture here is thus acting purely as a non-aggression signal; its function is to facilitate a social contact not by inhibiting attack but by reducing the probability of flight. In human terms, it is a gesture signifying friendship, not fear.

Other cases where an appeasement gesture is used as a friendly or non-aggression signal by a superior towards an inferior have already been mentioned and Pfeffer (1967) describes in the mouflon, *Ovis ammon*, behaviour which he regards as an appeasement ceremony but which could also be interpreted as a signal of friendship. After two males have fought and one has established dominance, the inferior is not subjected to further attacks. The two may graze amicably side by side and now and then perform the appeasement ceremony. This consists of the inferior licking the superior on the neck and shoulders, the latter often going on his carpal joints to facilitate the process (Figure 11). The only factor in

Figure 11. Appeasement ceremony in mouflon (*Ovis ammon*): the superior is kneeling to be licked by the inferior. (After Pfeffer, 1967.)

the situation showing that there is a submissive element in this ceremony is the fact that the exchange is not reciprocal: it is the inferior who licks the superior, but the reverse does not occur.

The examples which have been given are sufficient to indicate that in mammals submissive behaviour occurs in a wide spectrum of contexts. Starting from the purely agonistic, these grade over into the purely friendly behaviour which, in social species, may be shown reciprocally without evidence of any tension in either partner and which form the subject of the next chapter.

CHAPTER 8

Amicable behaviour

The social patterns dealt with so far have been mainly of a negative kind; patterns concerned with keeping away strangers or with diverting attack. The stability of a social group, however, does not depend solely on the fact that outsiders are attacked but within the group aggression is limited and controlled. There exist also types of socially positive, friendly or amicable behaviour. In addition to postures, expressions and sometimes vocalisations which denote that an approach is friendly not hostile, some type of special amicable behaviour exists in every social group, whether it is a permanent association of adults, a temporary one of parent and offspring or even a very transitory one of male and female lasting only for a short mating period.

Lorenz (1966a) considers that friendly behaviour should not be regarded merely as the expression of an independently existing bond between the animals concerned. In his view, once an innate amicable pattern has been evolved, this will show the accumulation properties characteristic of built-in forms of behaviour with an increasing need to be performed the longer it is denied outlet. A social companion provides the only adequate means of discharging the pattern: thus the situation is reached where it is not the existence of the companion that makes it necessary to show friendliness but the existence of the friendly behaviour that makes it necessary to have a companion. The behaviour has then become something which creates a link between the animals. Lorenz expresses this by saying that the pattern is not merely the expression of a bond; it is the bond itself. While there is a great deal of truth in this, even as applied to our own behaviour, it is not the whole truth. As just outlined, this view would seem to imply that any conspecific (or possibly any of opposite sex) should be equally suitable as a social partner. This, however, is not necessarily the case. Although we know very little about the factors determining them, individual preferences do exist in many species and are reflected in choice of a particular companion. For example, one of our tomcats once induced a strange young female to join the household and he became very devoted to her. He did this despite the fact that a female conspecific was already present in the home, reproductively fully competent (as her subsequent history showed) and of about the same age as the stranger. Although he mated regularly with each of them, his friendly behaviour was directed almost exclusively to the female of his choice; the other was no more to him than an adequate sexual object when she was on heat.

Furthermore, as Lorenz points out, once a bond has been formed between the partners, there are many species in which the attachment to a particular individual becomes very strong. Indeed, a situation with some of the properties of a positive feedback loop may be established, in which the friendly behaviour increases the closeness of the bond and the mutual attachment of the partners facilitates the performance of the amicable actions. In some cases the link with the special partner may become so enduring that if separation occurs, there may be prolonged or even permanent inability to accept a substitute.

These complications, however, do not invalidate the main point, namely that patterns of amicable behaviour play an important part in maintaining social cohesion, whether between a single pair of animals or within a larger group. Amongst mammals the commonest form of friendly behaviour is social grooming. I know of no species in which the mother does not do some grooming of her offspring* and grooming of the partner often forms a part of courtship behaviour. These special cases will be dealt with later and here we will treat only generalised friendly behaviour which may occur between any members of the group. Care of the fur is extremely important in mammals, both in general cleanliness and in keeping down ectoparasites; grooming is such a common and

* In *Tupaia* (Martin, 1966) this form of maternal care is minimal and is possibly restricted to the initial licking associated with removal of the embryonic membranes at birth.

frequent occupation that it is difficult to believe that its performance does not give some satisfaction to the animal. The obvious enjoyment of being stroked or petted shown by many tame mammals suggests that part, at least, of this satisfaction comes from the skin stimulation it provides. It is thus not surprising that to groom or be groomed by a fellow has become such a widespread form of mammalian amicable behaviour.

Social grooming, in addition to its friendly significance, has a more directly utilitarian aspect. It is principally directed to those areas of the body which are difficult for the animal itself to reach. The commonest of these are behind the ears, under the chin and, in species where the spine has not got a great deal of lateral mobility, at the root of the tail. In many cases some sensory specialisation appears to have been evolved so that titillation of these areas is particularly gratifying, as may be seen in the responses of tame animals to stroking on different parts of the body; it is, for instance, common knowledge that a cat enjoys being stroked behind the ears and under the chin. In some species this special sensitivity is so highly developed that certain areas constitute veritable pleasure spots. In *Thryonomys*, the rather thinly haired white patch under the chin is specialised in this way and a tame animal will turn its head and remain in a trance-like state of apparent ecstasy for as long as you are prepared to go on rubbing its chin. When two friendly animals meet, one will frequently turn its head and be briefly groomed under the chin by the other, so that the procedure forms a sort of greeting ceremony. I never actually timed how long the tame cane rat would lie still to be groomed but King (1955) reports that his tame prairie dog would do so for an hour. How important a social role stimulation of this sort can play is illustrated by an incident concerning an onager, *Equus hemionus*, in the Schönbrunn zoo, which is related by Antonius (1955). In this species, as in other Equidae, the root of the tail is a pleasure spot. The ass in question had taken a violent dislike to Antonius: whenever he approached its enclosure, the animal would hurl itself against the fence, biting and screaming in a frenzy of rage as though possessed by a devil and clearly ready to rend him to pieces, could it but reach him. One day it happened that the onager, distracted as it drew near by one of the other animals in the enclosure, turned so that its hindquarters were against the wire fence. Antonius promptly scratched the base of its tail with a heavy iron key that he was carrying. The effect was astounding. The animal stood quiet as a lamb and from then on no longer threatened Antonius as an enemy but approached him in what he describes as a state of "happy excitement" and presented its rump to be scratched.

In most primates grooming of each other is an important socially cohesive action but the gorilla is exceptional in that social grooming is almost restricted to mothers grooming infants and juveniles. Schaller (1963), in his extensive study, never saw adult males or females grooming each other. In many other species, including New World as well as Old World monkeys, it is common for one animal to solicit grooming by lying down in front of another. A chimpanzee may also solicit by standing with the head bowed, either facing or with its back towards the animal by whom it wishes to be groomed (Goodall, 1965). In baboons smacking the lips is an action which usually accompanies grooming: it may be a ritualised form of the movements made when eating the debris removed from the skin during the grooming. It is also used when a female approaches a mother with an infant, wishing to groom the latter and, in other contexts, as a generalised friendly signal (Hall, 1962b; Hall and DeVore, 1965).

In many social species, grooming may be mutual. Sometimes a grooming session may consist of the partners taking it in turn to groom and be groomed. In *Thryonomys* one individual may approach another, groom it for a while about the ears and throat and then, placing its head in front of the other's mouth, solicit grooming. The prairie-dog solicits rather similarly, by crawling under the nose of the partner (King, 1955). Sometimes, however, grooming is reciprocal and simultaneous. Horses provide one of the most familiar examples of this. The two stand facing in opposite directions and groom each other at the root of the tail or the base of the neck. A pair of peccaries will similarly groom each other round the region of the dorsal gland but may also take turns at attending to each others' ears and necks. In *Suricata* simultaneous mutual grooming is practised and here a response related to the reciprocity is easily demonstrated. If you scratch a *Suricata*, particularly at the root of the tail, it will at once start to groom whatever part of your skin (or clothes) happens to be nearest its mouth. In the natural situation, the same thing happens and when one individual grooms another, he is automatically groomed in return. Amongst the Canidae, mutual grooming is an important social activity in the raccoon dog and the bat-eared fox, *Otocyon megalotis*. Kleiman (1967) suggests that the conspicuous facial markings in these two species may be related to the fact that the face is the area most often groomed.

When two adults of the same species meet, it is necessary for them at least to be able to classify each other, even if individual recognition is not involved—male or female, foreign or familiar, friend or foe? The relevant information is provided by olfactory cues and smelling, either at the mouth or the ano-genital region, is a normal feature of encounters. It may be no more than a quick "recognition sniff" at the appropriate region, and in some species it is performed only when circumstances make it necessary. In *Sminthopsis crassicaudata*, for instance, I have found that a group of animals kept together in an enclosure do not indulge in recognition sniffing. They are familiar with each other and have learnt that only known and innocuous individuals are about; but add a strange animal and the picture changes at once. Agonistic responses to the sight of a conspecific reappear: in effect, the old security is gone and it can no longer be assumed that anyone met with is a "friend". For a while every encounter is accompanied by a tense interchange of mouth sniffing and only when the olfactory answer "familiar" has been given do the two relax and continue on their several ways. In other species simple recognition sniffing may be a feature of every encounter. I found this to be the case in *Xerus erythropus*: a pair kept together in an enclosure without the introduction of strangers did not abandon the habit of nose to nose sniffing. In this latter case the sniffing is something very closely approaching a greeting ceremony and in other species recognition sniffing has been elaborated into a ceremonial which certainly does act as a friendly greeting. In the prairie-dog, what King (1955) describes as the "identification kiss" plays this role. Whenever two members of the same coterie meet, each turns the head slightly sideways, opens the mouth and contacts the other. His tame animal would always kiss King's hand before making any further contact. When familiar members pass each other, the kissing may be done in a rather perfunctory manner but if there is any doubt as to identity—for example, if the meeting is near the boundary of the territory—the two, on sighting each other, lie down and wag their tails, then slowly creep forwards, tail wagging at intervals, until finally they make contact and kiss. King considers that the open mouth reflects a threat element in the encounter but this seems a trifle improbable in view of the fact that in rodents biting threat normally takes the form of tooth gnashing, not of opening the mouth. He does not consider the alternative possibility, which appears to me more probable, that the open mouth is a ritualised intention grooming movement.

In the yellow-bellied marmot (Armitage, 1962) recognition sniffing is directed to the partner's cheek gland. In this species, the sniffing closely approaches being a greeting ceremony but is less elaborated than in the prairie dog and more restricted in its occurrence: it is usually seen in encounters between young or between an adult and a youngster but rarely between two adults.

In *Lycaon*, members of a group greet each other by nudging with the nose and licking at the corners of the mouth (Kühme, 1965a). This, however, is not simply an elaboration of recognition sniffing, but is derived from the infantile method of soliciting an adult to disgorge food. In this species, all the adults who have participated in a successful hunt cooperate in regurgitating food not only for the young, but also for the adults who have remained behind to guard them. Having fed, the guards may in turn regurgitate if solicited, so that a piece of flesh from the kill may traverse several stomachs before finding its ultimate resting place (Kühme, 1964, 1965b, c). This communal food sharing and the greeting derived from it are thus forms of behaviour which belong basically to the parent/offspring relationship, but have acquired a wider context and therewith a new role in promoting group coherence.

In the spotted hyaena, smelling of the ano-genital region

has been ritualised into a greeting ceremony which is regularly performed when members of a group meet. Each raises a hind leg displaying the genital region and permits the partner to smell. The greeting is performed in the same way by both sexes and Wickler (1964), who describes the behaviour, suggests that the great enlargement of the clitoris of the female is related to its function as an optical signal in this ceremony.

In Primates, both a mouth/mouth and a mouth/genital contact may be used in greeting. Baboons, for instance, may greet each other by lip-smacking followed either by the two making a mouth to mouth contact or by one embracing the hindquarters of the other and nuzzling or smelling the genital region. An adult may also "kiss" the posterior of an infant (Hall and DeVore, 1965). Although olfactory stimuli no doubt play a role in these greetings, in view of the relative unimportance of the sense of smell in higher primates, it is likely that the tactile element is equally involved. In chimpanzees, tactile stimuli predominate in greeting behaviour and, correlated with their much greater manual skill, contact of the hand with the partner's body is much more important than it is in baboons. One partner may approach the other and touch him with the flat of the hand or the back of the slightly flexed fingers on head or shoulder, groin or genital region. The hand touching may be mutual, or they may fling their arms round each other in an embrace (Goodall, 1965). The human habit of holding hands as a token of affection presumably reflects the same tendency to contact the partner with the highly sensitive fingers. Indeed, although the hand shake may have been ritualised as a demonstration that the right hand carries no weapon, its roots may well go back to this same tendency.

The various forms of behaviour which contribute to sharing of smells within a group have already been mentioned. These may also be included in the category of friendly behaviour. Possibly allied to smell sharing may be the use of a common dunging place, which occurs in some species, for instance the vicuña (Koford, 1957). Communal dunging is also practised by some of the social viverrids. Roberts (1951) says that in the yellow mongoose, *Cynictis penicillata*, if the colony is a large one, a number of communal dunging places will be found nearby. In my meerkats, communal dunging occurred only as a transitory phenomenon when the youngsters were two or three months old and the adults did not add their droppings to those of the young. In non-social species, as we have already seen in Chapter 5, communal dunging, far from playing any socially cohesive role, may have an opposite significance and serve as an aid in avoiding encounters.

In most social species, social grooming combined with some form of greeting suffices to maintain the stability of the group. In the social Canidae, however, where there is cooperative hunting, there are special forms of behaviour which serve not only to create a bond between the members but may also synchronise mood between them. The most familiar of these is choral howling, which is characteristic of wolves, jackals and coyotes. Crisler (1959) relates how she and her husband cemented their friendship with their tame wolves by joining them in bouts of community howling. Your dog is doing exactly the same thing when he comes and accompanies your piano practice by companionable howling. Lorenz (1963) quotes Allen and Mech as having observed that a pack of wolves in one of the Canadian National Parks, when about to set upon and kill a moose, used first to indulge in a bout of communal greeting, nuzzling each other and tail wagging together. This behaviour was interpreted as having the function of ensuring that all were simultaneously in the mood to make the cooperative kill. Kühme (1965a) found that in *Lycaon* also an outburst of communal greeting was part of the routine of setting forth to hunt. Another canid social habit is yawning together. This constitutes a ceremony which normally occurs before leaving the dens after emerging from rest and has the function of ensuring that individuals do not start to move off until all are roused and ready. Once again, our dogs show signs of this habit. Human yawning is contagious to dogs and most of them will yawn if you merely pretend to yawn in front of them. The solitary cat, in whose life social yawning plays no part, will show no such response.

Although the biological function of amicable behaviour is clear enough, it has certain characteristics which, at first sight, seem a trifle surprising. The first of these is the almost insatiable lust for being groomed that many social species show. With a human companion who is prepared to indulge them, they often develop a sort of addiction and accept, or even demand, an amount of attention which would be out of the question in the natural situation. My tame cane rat and King's prairie dog provide examples of this and meerkats also become addicted to being petted. Even non-social species may become surprisingly eager to be caressed, although in their normal existence grooming occurs only between parent and child or between sexual partners. This is true not only for domestic species, like the cat, but also for wild ones, for example *Atilax*. The reason for this is probably quite simple. The readiness of the human companion to provide almost unlimited grooming is something that has no counterpart in nature. There has therefore been no need to evolve any strict limiting mechanism. This is the same explanation as Lorenz (1966b) has put forward for our maladaptive tendency to become addicted to sweet things. In natural circumstances, to take the sweetest is a perfectly safe guide to food choice and no upper limit is required. It is only when we have provided ourselves with artificial foods that the absence of such a limit can lead to harmful eating habits.

The second characteristic of amicable behaviour which seems surprising is the combination of this extraordinary capacity for "affection" with extreme ferocity towards unfamiliar conspecifics; a combination typical of most social species. A moment's reflection, however, shows that it is only the fact that strangers are excluded that makes group cohesion possible. This, however, merely leads to the further question: what is it about an alien that makes him evoke an aggressive instead of a friendly response; wherein does he differ from a group member? Such evidence as there is indicates that smell is one of the main factors involved. This was clearly shown in the case of the sugar-glider, where a female marked by a strange male was promptly attacked by her erstwhile companions. The behaviour of meerkats too shows the importance of olfactory stimuli, both in arousing aggression against strangers and in counteracting it within the group. Our tame females were always very prone to attack an unfamiliar woman, although a man was usually tolerated. If a female visitor had been in the house, the meerkat would smell the place where she had been sitting and make the growling noise that normally greets the discovery of dog's droppings. On one occasion one of the animals launched a vicious attack on the arms of a man with whom she was well acquainted and normally friendly. It was later discovered that a dog had previously slept on the jersey he was wearing at the time. On another occasion when a female member of the family had returned home after a long absence, the female meerkat at first attempted to bite her. Not being permitted to do so, she would bide her time and then bite and worry the cushion of any chair just vacated by the object of her hostility. She continued to behave in this way for about a fortnight. The opposite effect of a familiar smell in checking aggression was shown in the behaviour of the male towards me when I returned home after an absence of a little over six months. As I entered the house, he rushed to the attack but as his nose approached my legs he checked suddenly, sniffed and failed to deliver the intended bite. His subsequent behaviour illustrated the importance of another factor in the recognition of group membership—to wit behaviour according to a known routine. The meerkat remained somewhat hesitant and suspicious towards me but since lunch was ready, we went at once to the dining room. I took up my accustomed place at the table and he too was back on the instant, in his accustomed place on my shoulder and the familiar routine of our previous relationship was re-established, like a set of displaced gear wheels slipping back into register.

In normally non-social species, the tolerance of known individuals which is often shown by captive animals has, in some cases, been shown to depend on familiarity with their olfactory characteristics. In *Sminthopsis*, for instance (Ewer, 1968) I have found that mouth smell is the most important factor in recognition and visual cues are not involved.

Crowcroft's (1966) observations on mice also demonstrate the paramount importance of smell in this species as an indicator of group membership or otherwise, but they also show something further. The effect of the foreign mouse's smell is not merely to stimulate attack on its author: it arouses aggressive behaviour in a more generalised and far-reaching manner. Crowcroft found that once alerted to the presence of a stranger in his territory, a mouse does not merely attack the foreigner when he meets him. If contact is lost, the territory owner will search systematically for the intruder; moreover his readiness to fight is such that brief "mistaken" attacks may be made when a known individual is encountered, to be corrected rapidly in response to the familiar smell.

In carnivores, the visual stimuli provided by a strange conspecific are not normally alone sufficient to evoke aggressive responses. This is true, at least, for cats, dogs, meerkats and the kusimanse, as shown by their responses to mirrors. Since they do not normally have any means of becoming familiar with their own appearance, their first reactions to a mirror constitute a test of the response to the visual cues provided by an unfamiliar animal. In all the species mentioned, the normal response is to show interest and to approach without threat. This is often followed by attempts to locate the conspecific behind the mirror and when this fails, interest usually wanes quite rapidly. If, however, the stranger should show any threat behaviour, this will at once evoke an answering threat. This was very clearly demonstrated in the behaviour of a black-footed cat, *Felis nigripes*. The animal was exploring a bench in the laboratory one evening. The lights were on and it was dark outside, so that the glass of the windows provided a good reflecting surface. Presently the cat noticed the window ledge, a few inches above his head, and jumped up. He was startled by the sudden appearance of his reflection and instantly bristled in defence. The reflection did likewise and the animal found himself suddenly launched into a threat duel which ran through the whole gamut of the expressive vocabulary, ending in the pose characteristic of maximal aggressive threat (See Figure 3, page 19). With two so exactly matched rivals, there was only one possible end to the contest. When maximal threat received an exactly equal answer, there was no attempt to launch an attack: having clearly met his match, the animal retreated and jumped down again onto the bench. He then moved along to a second window, hesitated a moment, bristled slightly and jumped up on the ledge. There, of course, his rival again threatened back and the performance was repeated but this time more briefly and the retreat to the bench was made sooner. The cat then returned to the first window, hesitated for a while and then, very cautiously, raised his forequarters and peeped over the ledge—where he was just in time to catch his rival doing exactly the same thing. He hastily withdrew and made no further attempt to renew the acquaintance.

This encounter illustrates an important point. A stranger, away from home ground, is very likely to react with some defensive movement to the sudden appearance of another individual. By so doing, he is providing stimuli which, if not counteracted by some recognition signal such as a familiar odour, are alone enough to start hostilities. In the event, his foreign smell will act in the same direction and serve to precipitate attack.

There are, however, cases in which even a familiar smell can fail to stave off attack. I have seen an incident of this sort involving three domestic cats who normally lived in harmony; but whether the same thing would happen with a truly social species I do not know. The cats in question were two females and a male. When treating one of the females for an infected sore, I put a white bandage round her lumbar region. She went out into the garden, where her unorthodox appearance promptly evoked first threat and then violent attack from her companions so that it was necessary to rescue her and restore her normal visual characteristics.

Another incident in which attack was evoked by a familiar individual showing behavioural instead of visual abnormality, relates to a pair of *Cricetomys*. The two, a male and female, had lived together in peace for a long time. One morning the male was found dead, with his lower jaw wedged between the bars of the cage and his throat torn out. The throat wounds could have been inflicted only by his mate. Since the animals will often gnaw at the bars of the cage, particularly if not provided with enough hard food, the most likely interpretation of the situation is that in gnawing at the bars, the male's teeth had slipped so that he fell forwards and his jaw was thrust between the bars so that he could not free himself. His struggles, constituting quite unprecedented behaviour, must have so upset the female that she attacked him, and, powerless as he was to resist, killed him.

These examples all serve to illustrate a point of some

importance. We may speak of "individual recognition" in other species but this does not necessarily have exactly the same connotation as it does when applied to our own intraspecific relationships. To us an individual is exactly what the term implies—an indivisible integrated personality. It is, however, not legitimate to assume that other species are necessarily capable of the degree of abstraction which such a concept implies. For them the "individual" may be, to a greater or lesser extent, a bundle of discrete releasers, some more important than others, their relative valencies changing according to the situation. In Chapter 10, the responses of parents to their young will provide further examples of the dissociability of the releasers provided by a single "individual". One cannot therefore assume that in every species where it is shown, individual recognition must have exactly the same subjective basis. The higher primates may have something approaching our own sort of concept of each other as true individuals. In other species this capacity is probably much less developed and their relationships are therefore much more at the mercy of specific responses to discrete sets of stimuli provided by their fellows.

CHAPTER 9

Courtship and mating

Courtship consists of special forms of behaviour performed by male and female as preliminaries to mating and has the function of ensuring that the latter results in fertilisation or successful establishment of pregnancy in the female. Exactly how courtship achieves this end varies from group to group. In many invertebrates and probably lower vertebrates also, one of its roles is related to recognition of the species of the partner. In such cases courtship is a reciprocal affair and comprises a series of actions which release appropriate responses in the partner. If the latter is not a conspecific of opposite sex, then failure to give the correct response will terminate the exchange and so prevent the occurrence of a biologically useless mating. The responses of the female are normally linked with her physiological condition, so that if courted when she is not capable of a fertile mating she is unresponsive, and again a useless mating is avoided. The courtship of the hymenopteran *Mormoniella*, described by Barrass (1959, 1961), functions in this way to ensure that the male copulates only with a receptive female of his own species.

In mammals, there is no need for special behavioural signals to transmit to the male the necessary information as to the species, sex and sexual condition of the partner; her smell alone is sufficient to provide it and the function of courtship must therefore be sought elsewhere. In certain other groups courtship is known to have an effect on hormone production, bringing one or both of the partners into a fully sexually competent condition. This has been very clearly demonstrated, for example, in the case of the ring dove, where Lehrman (1958, 1959) has shown that the sight alone of a courting male causes secretion of oestrogen and progesterone in the female, as a result of which she becomes ready to lay and to incubate. While some stimulatory effects of this type cannot be ruled out in mammals, they are certainly of minor importance. Artificial insemination of an uncourted female is successful and in some species there is virtually no courtship. This is true, for instance, of the African elephant. Here the male merely approaches the female and lays his tusks on her back as a signal of intention to mount. If she is receptive she permits him to do so and if not, she walks or runs away (Buss and Smith, 1966). Furthermore, it is a common finding that if a female is first put with a male when she is already at the peak of her heat, successful copulation may occur almost at once with little or no courtship. The latter is therefore not a physiologically necessary part of the mating process.

In mammals, the main function of courtship seems to be simply to ensure that when the female has reached the point in her sexual cycle when copulation is most likely to result in pregnancy, a sexually excited male is present, ready to pair with her; in other words, to ensure correct timing of mating. The female's period of heat and sexual receptivity corresponds with this stage in her cycle but, in general, she becomes attractive to the male considerably earlier. The duration of the female's heat varies from species to species but is sometimes very short—a matter of a few hours. If the female were attractive to the male only during this short receptive period, there could be danger of her failing to find a mate in time. The fact that she becomes sexually attractive a little in advance introduces a factor of safety: the male is generally to hand before he is required and must be kept in attendance on the female until the final stage is reached and she is ready to accept him. His courting activities ensure that he does remain present and eager until the right moment comes: they may possibly accelerate its arrival. The sexual selection which results from competition between courting males has already been mentioned. It seems quite possible that in some species this has led to the evolution of a secondarily prolonged "attractive-but-not-receptive" stage in the female.

Exactly what happens during a courtship depends on the relative states of sexual readiness of the partners when they meet. Typically the male will have found the female before she is receptive and the courtship may then be prolonged. It may also be a rather one-sided affair, with the male making all the advances and at first being evaded or repulsed by the female. As previously mentioned, if the female is already receptive when the male encounters her, courtship may be much abbreviated and mating occurs almost at once. The situation where a receptive female meets with a male who is not sufficiently sexually motivated to mate is less common, but by no means exceptional. In most species the female will then show some type of behaviour which increases the male's sexual excitement and induces him to mate. The courtship procedures within a single species may thus show a considerable range of variation in different circumstances and a single mating may not display the complete behavioural repertoire. While this lability may add to the difficulties of studying and describing courtship, its functional significance in relation to ensuring that mating occurs at the correct stage of the female cycle is obvious; all of which underlines the fact that such is the fundamental role of mammalian courtship.

In view of these considerations, mammalian courtship may be expected to show certain characteristics. Should the not-yet-receptive female merely tend to evade the male, this should favour the evolution of some type of courtship pursuit; should she defend herself more actively, the evolution of a courtship with similarities to fighting or threat is more probable. At the same time, there are likely to be signals made by the male to indicate that his approach is not aggressive. These would most simply be derived from ordinary friendly behaviour or from intention movements of mating. When the female takes the initiative, we may expect her courting to take the form either of a direct invitation to mate by presenting her hindquarters and adopting the mating posture or of some behaviour likely to raise the male's general level of excitation. The latter may very much resemble the tactics most commonly used by youngsters to persuade a littermate to join in play; to wit, inviting pursuit by running up to the partner and away again or dashing by close in front of him or else initiating some sort of semi-playful fighting. Lastly, just as an intimidation display enhances the effect of an associated threat, so a display of the male's size, strength or special adornments may have an effect on the female's readiness to accept him and may form a part of courtship.

Unfortunately it is not a simple matter to find out how far these expectations are borne out in practice. Mammalian courtship behaviour is by no means fully documented and its characteristics make it peculiarly difficult to study. It is not a rigid sequence, every action of which must be performed in a determined order before mating occurs. According to the circumstances, as already explained, a mating may be rapid, with virtually no courtship, or be preceded by long periods in which elements of courtship are performed, broken off and begun again, before the consummation is reached. Single observations may therefore give a rather misleading picture and prolonged study may be required before the full repertoire of courtship behaviour is elucidated. Consideration will therefore be limited to a few major orders in which a reasonable number of species have been investigated.

In marsupials, courtship has been described only in macropods and dasyurids. The most thorough study that has been made is Sharman and Calaby's (1964) on the red kangaroo. A male normally smells the cloaca and pouch of any female he encounters as a matter of routine and may also smell her urine, even placing his nose in the stream of fluid as she micturates. This interest in female urine is by no means a marsupial peculiarity.

In many placentals males show a curious response to odours produced by a female, particularly to the urine of an oestrous female. Typically the head is raised, the lips turned back and the nose wrinkled and breathing is stopped for a moment. To a human observer, the facial expression is rather suggestive of disgust. The German name for this procedure is *Flehmen* and, since there is no English equivalent, it is convenient to use the same term. Flehmen is apparently universal in Bovidae and Camelidae, where it is not necessarily restricted to males, although it is always more commonly performed by them than by females. It also occurs in the black rhinoceros (Goddard, 1966), in a number of carnivores and is reported in a bat (Mann, 1961). Flehmen can often be evoked by strong odours which are quite unnatural; for instance, my male *Suricata* would regularly flehm in response to the smell of sherry; the normal context, however, is sexual. Although the expression is quite striking, it does not appear to function as a signal. Knappe (1964) points out that all species known to flehm possess well developed Jacobson's organs and he believes that flehmen is in fact an action which serves to open the orifices of the *ductus incisivus* and so bring Jacobson's organ into contact with the smell; a procedure which may possibly have a sexually stimulating effect. Although marsupials possess a Jacobson's organ and although the male kangaroo shows such an interest in female urine, flehmen has never been described in a marsupial and Sharman (personal communication) tells me that he never saw his kangaroos do so. Although he does not show this overt response, the male kangaroo's olfactory testing of the female serves to inform him about her sexual condition. If she is not coming into oestrus, he takes no further interest but if her heat is starting, he attempts to stay close to her; she moves off and he follows. This may take the form of a vigorous pursuit or a leisurely following, during which the male often grasps the female's tail in his paws, as though attempting to restrain her. He also often emits a soft clucking sound, an unvoiced noise made by movements of the tongue and lower jaw. This appears to reflect a state of high excitement and, although common during sexual following, clucking is not restricted to this context but may also occur in other circumstances. The pursuit may last up to two hours, but if the female is already fully on heat when the male arrives, it may be curtailed to as little as a couple of minutes. The female finally stands for the male and takes up the mating posture, with her fore paws on the ground; he at once grasps her round the lumbar region and mounts. Copulation lasts some 15–20 minutes.

If a male and female *Sminthopsis crassicaudata* are kept together, the male starts to pay his attentions to the female some days before mating occurs. In the first stage his interest in smelling her cloaca increases and a brief routine check up is replaced by prolonged smelling. If the female is sitting down, the male may pull at her hindquarters with his paws to gain access to her cloaca. The female, however, usually runs off, pursued by the male, looking as though his nose were tied to her hinder end. Sometimes he takes a grip with one paw on the fur on her flanks and holds her back—a procedure reminiscent of the kangaroo's tail holding. In the next stage the male' interest is no longer restricted to the female's cloaca. After a less prolonged sniff, he also smells her mouth and, placing his paws on her shoulders, may groom her about the face and ears, but usually he does not yet attempt to mount. Interchanges of this sort may be punctuated by short pursuits. When the female has come fully into oestrus, mounting may occur virtually the instant the pair have emerged for the night's activity period. After a quick smell at cloaca and mouth, or even at the mouth alone, the male lays his head on the female's so that the tip of his snout lies between her ears but does not grip with his jaws, clasps her round the body with his fore limbs and places his feet just outside hers and intromission follows. Successful mounting may be preceded by a few scuffles, in which the female resists and threatens but a serious fight does not develop. During a pursuit I have heard one of the animals make a sound which I have not heard in any other context; a very low intensity unvoiced "da-da-da-da". It was probably made by the male and seems to correspond to the kangaroo's "clucking".

It is characteristic of Dasyurids that copulation is extremely prolonged. In *Dasycercus* and *Dasyurops* (Fleay, 1965) it is said to last for 24 hours; Woolley (1966) reports a maximum of 12 hours in *Antechinus stuartii* and I have watched a pair of *Sminthopsis* in continuous copulation for 11 hours. On the following night, the same pair again copulated over a period of 9 hours, but this time in a series of briefer sessions, the longest lasting approximately two hours. In the initial stages, there are jerking movements of the forequarters and the male's snout can be seen to move about convulsively. Because the animals are small and, being nocturnal, must be observed in dim light, it is not easy to make out exactly what is happening and the relatively long soft fur adds another difficulty. The impression that could easily be gained is that the male has hold of the female by the back of the neck and is shaking her about unmercifully. This is quite erroneous: the jerking is caused by movements of the male's fore paws, not unlike the rubbing movements made by many placental mammals. His jaws are not gripping the female's fur at all but he is moving his head about vigorously so that his chin is rubbed on the back of her head. The paw movements are seen only in the early stages and their function may therefore be to induce the female to take up a position facilitating full intromission. The chin rubbing, however, recurs at intervals whenever the female makes an attempt to break away and appears to have the effect of calming her down. Several times during the course of the copulation, the female ran to and fro, apparently trying to escape and with remarkable skill the male maintained his position, his chin pressed firmly on the back of her head and his paws clutched round her body. When she attempted to run down a burrow, he succeeded in diverting her by jerking his body backwards and sideways, still clutching her round the middle, so that her paws were lifted off the ground and at the same time her body was swung round, so that the direction of her flight was forcibly altered.

In *Antechinus*, according to Marlow (1961), courtship consists of the male gripping the female in his jaws by the scruff of the neck and following her about the cage for some half an hour or so before copulating. In view of what happens in *Sminthopsis*, I am inclined to doubt whether the male is in fact actually gripping with his jaws and suspect that he may, like the latter, merely be pressing his chin on her head. Marlow's animals were placed together only when the female was already on heat and full courtship was therefore not seen. In a single pair of *A. flavipes* which I kept, there was considerable preliminary chasing, but the male appeared to be insufficiently excited to mate and copulation was not seen. As the female began to come on heat, the male would smell her cloaca and I several times saw him lay his head on hers, very much as *Sminthopsis* does. She always made off at once and, as her heat developed, this grew into prolonged chasing. *Antechinus* is an extremely agile climber and the two dashed hither and thither amongst the branches with which their cage was furnished at such speed that it was impossible to follow their movements. Whenever the male managed to corner her, the female resisted and threatened loudly and there was a short scuffle. During intense pursuit, a little "cha-cha-cha-. . ." noise was sometimes audible which, like the similar call in *Sminthopsis*, was probably made by the male. The female was more agile than the male and he finally seemed to lose heart and gave up chasing. She then showed the ambivalent type of behaviour, sometimes called "coquettishness", which is so commonly seen in female placentals. Simultaneously attracted by the male and ready to flee from him, the female will approach him if he moves away, but will take to flight if he approaches her. The female *Antechinus*, when the male ceased chasing and sat still, at once oriented her activity towards him and would direct her course so as to run past close beside him and thus stimulate him to a fresh pursuit, only to threaten if he caught her up. After this chasing had gone on for a couple of hours, the female suddenly retired into the nest box and although the male followed, the ensuing silence suggested that no mating took place and certainly the female did not produce a litter.

Marsupial courtships, so far as they are at present known, thus fall into the category of a ritualised pursuit with the addition, at least in *Sminthopsis*, of special procedures which appear to pacify the female. One of these, grooming, is a generalised type of friendly behaviour; the other, head rubbing, appears to be a special pattern not seen in other contexts and may possibly be a ritualised form of the intention movement of mounting. Sharman and Calaby do not express any opinion as to whether the handling of her tail by the male has any special effect on the female kangaroo. There is no evidence in any of the species mentioned that the "mating call" of the male has any effect on the female: it may be a purely incidental expression of the male's excited condition.

The significance of the extremely prolonged copulation in these animals is a problem that has exercised the minds of all who have studied them and one to which no definitive answer has yet been given. I believe that its significance is a purely negative one. In placentals, although individual copulations may be quite brief, they may be repeated frequently during the period of the female's heat. Kühme (1966), for instance, records a pair of lions copulating repeatedly over a period of 6 hours; Hall and DeVore (1965) observed a female baboon to copulate with three males a total of 93 times during the 5 days of her oestrus. According to Reed (1946), golden hamsters may copulate as many as 170 times while Steiniger (1950) estimates that during the 6–10 hours of her oestrus a wild *Rattus norvegicus* may copulate 200–400 times with 6–8 males.

It seems that, in general, as long as she is on heat, a female is willing to accept a male and a sexually competent male is ready to copulate with her at any time during this period. The dasyurid characteristic is not that sexual relations last particularly long but that the pair remain *in copulo* continuously for so long, the male not dismounting after an ejaculation. It may therefore be that in regarding the dasyurid copulations as peculiar we are taking an unduly eutheriocentric attitude and possibly we should rather consider that separation once an ejaculation has been achieved is the phenomenon requiring an explanation. In view of what has been said above, it seems only natural to expect that, in the absence of any contrary factors, the partners should simply remain paired until either the male is sexually exhausted, the female's heat has come to an end or dawn sends the animals back to their burrows. There is no more need to assume that the extremely prolonged concourse is *necessary* to achieve fertilisation than that the female baboon requires ninety-three copulations to produce the same result.

Two different sorts of factor may have been important in producing a selective pressure leading to the evolution of briefer mountings in other groups—predator pressure and anatomical characteristics. Prolonged copulation would be extremely dangerous in an environment where predators abound, for the animals are particularly vulnerable when paired. It may be significant that in the placentals it is not amongst species likely to be preyed upon, but in the Carnivora and what is more, in primitive representatives of the order, that the most prolonged individual copulations occur. Vosseler (1929) records that in *Cryptoprocta* copulation lasts up to 2¾ hours and amongst mustelids a similar duration is not uncommon: the ferret, *Mustela furo*, for instance, may remain *in copulo* for almost three hours (Hammond and Marshall, 1930). Reed (1946) quotes an account of mating in the sable, *Martes zibellina*, according to which, in up to thirty copulations, the pair may be united for a total of as much as 18 hours. It may therefore be that brevity of mounting has often arisen as an adaptation to predator pressure, an adaptation which was not necessary for the dasyurids in an environment where they were themselves the only effective predators.

It may also be significant that amongst the large herbivores, it is the rhinoceros, in which attack from predators can hardly be a very serious risk, that has the longest copulation—about half an hour (Goddard, 1966).

In the evolution of the highly specialised penis structure and corresponding extremely brief mountings characteristic of advanced artiodactyls, it seems almost certain that cursorial adaptations which render a prolonged concourse mechanically inconvenient have played an important role. Anatomical considerations too may have played a part in determining the relatively short mounting period (up to 20 minutes) found in the red kangaroo (Sharman and Calaby, 1964). In *Perameles nasuta* the mountings described by Stodart (1966) were extremely brief, lasting only a few seconds. The adaptive significance of this and whether or not it is characteristic of Peramelidae as a whole remains to be elucidated.

This digression has carried us away from the topic of courtship behaviour. Amongst the Rodentia, a large and diversified order, it is not surprising to find a considerable variety of procedures. In the Sciuridae, courtship is of the pursuit type. In the ground squirrel, *Citellus armatus*, Balph and Stokes (1963) describe two types of male approach to the female. Either of these leads to a pursuit which ends with the pair retiring to the burrow, where copulation was presumed to follow. If the male approaches rapidly, he may strike and bite at the female; she retaliates and the two may then chase each other, the roles of pursuer and pursued being interchangeable. The male, however, may approach more slowly and less aggressively and the two may nuzzle each others' noses but, again, a pursuit follows. A courtship chase is also characteristic of the palm squirrel, *Funambulus pennanti* (Purohit et al., 1966). In the grey squirrel, *Sciurus carolinensis*, the procedure is more ritualised (Bakken, 1959). The male approaches the female, flicking his tail in a fore-and-aft direction. The female moves off and the male follows, waving his tail slowly. When she finally stands, the male displays for a moment, waving his tail in a circle, before making the final approach. In the European red squirrel, the procedure is further elaborated (Eibl-Eibesfeldt, 1951b). The pursuit may be a genuine chase or a gentle "symbolic" following. The female accelerates if the male approaches too rapidly, but if he shows signs of giving up, she waits for him or even moves towards him. As he approaches, the male makes a special call which closely resembles the contact-keeping cry of the young. Eibl-Eibesfeldt believes that this has an inhibiting effect on the female's aggressive tendencies. A comparable "baby" call is also made by the male field mouse, *Microtus arvalis*, during his courtship following of the female (Eibl-Eibesfeldt, 1958). Before the final approach and mounting, the male squirrel performs an elaborate tail display. Stopping some foot or so away from the female, he turns broadside on to her, waves his tail horizontally a few times, then round in a wide circle and finally, with the hairs maximally erected, he brings his tail down over his back in a slow, impressive movement and stands thus displaying to the female for up to a full minute. This, the most complex sciurid courtship so far described, includes, in addition to ritual pursuit, both a placatory signal (the "baby" call) and an optical display.

Amongst the South American rodents, courtship has been described in the guinea pig (Kunkel and Kunkel, 1964), the agouti (Roth-Kolar, 1957) and the acouchi (D. Morris, 1962). Although the latter two are closely related, their courtships are dissimilar. In the acouchi there is simple following of the female by the male, who carries his tail high and wags it from side to side, apparently as a signal that his approach is not aggressive. The female stops at intervals and then moves on again but finally she remains still, adopts the mating posture and the male then mounts. In the agouti, the courtship is more violent and abounds in threat actions. The male approaches the female slowly; he makes a variety of vocalisations, squeaking, twittering and gnashing his teeth; he marks frequently with his anal gland, his hair bristles and he may pause and drum on the ground with his fore paws. The female too is aggressive and if he comes near will at first drive the male off, rising on her hind feet and biting at him. After up to an hour of this, the female permits the male to come close to her; at last he makes contact and grooms her around the nose. Finally, she allows him to smell her genital region and to mount. In the guinea pig smelling of the female's nose, body and genital region occurs early on in the courtship.

The male then slowly circles round the female giving a prolonged trilling call. This is not a specific mating call, as both sexes may make it in other contexts which suggest that it signifies some degree of anxiety. If the female remains sitting, the male will circle her more closely and may extrude his testes briefly now and then. As already mentioned, a testis display also occurs in threat. The postures accompanying the display are not identical in the two cases and the motivations may therefore be slightly different. Ultimately the female takes up the mating position, with her back legs extended and the male mounts. The male mara, *Dolichotis patagona*, also displays his genital region to the female during courtship (Roth-Kolar, 1957).

A peculiarity which is sometimes shown in the courtships of the guinea pig, acouchi and mara is the liberation by the male of a jet of urine at the female. Kunkel and Kunkel (1964) also report this habit in the chinchilla and the hare-mouse. This is not a regular part of the courtship sequence but is done if the female fails to adopt the mating position. It appears to reflect a sort of frustration and the female's response to it is negative rather than positive. In the guinea pig a female may also enurinate a male if she is being pestered by him and is not yet ready to mate, while in the mara, young may behave in this way if they are eager to get some food but deterred from approaching by the presence of adults. The response in all these cases reflects some state of irritation or frustration and there is nothing to suggest that the urine spraying has a sexually stimulating effect. Enurination also occurs in the American porcupine, *Erethizon dorsatum* (Shadle, Smelzer and Metz 1946; Dathe, 1963), apparently with a repellant rather than a stimulating effect. In Lagomorpha too, enurination of the female is a relatively common accompaniment of courtship (Southern, 1948). Myers and Poole (1961) found that the females usually showed no response, other than shaking the urine off their fur, and consider the enurination as an incidental product of the male's highly excited condition. Enurination is also mentioned by Dathe (1963) as occurring in the coypu and in the Indian porcupine, *Hystrix leucura*: he considers that the habit may play some role in the establishment of dominance.

In the Muridae, courtship is usually very simple, consisting mainly of the male following the female and attempting to lick and nuzzle her genitalia as a preliminary to mounting; there may also be a little grooming by the male of the female's face and ears. In the house mouse and in the Persian gerbil (Eibl-Eibesfeldt, 1951a) courtship is of this simple type. In wild rats, according to Steiniger (1950), what happens depends on the circumstances. If only a single pair is present, there may be no following but after some mutual crawling over and crawling under, the female adopts the mating posture and the male at once mounts. Within a colony, however, all the sexually active males pursue an oestrous female and there may be prolonged chasing before she stands and a male is able to place his paws round her flanks and make the vibratory movements to which she responds by adopting the mating posture. Copulation is very brief, a matter of only two or three seconds and as the male dismounts the female turns round and bites at his head. He normally avoids this attack by leaping backwards as he dismounts. Barnett (1963) mentions the male's backward leap, but not the female's attempt to bite.

In the giant rat, *Cricetomys gambianus*, the procedure is a little more complex. The male attempts to smell the female's genitalia but at first she repulses him, rising on her hind feet and boxing him off with her paws. He responds by rearing up too. Although they have thus assumed the typical fighting posture, no hostile action follows; the female is purely defensive and the male is attempting to get close to the female to groom and pacify her, not to bite her. Reared up thus, with paws on each others' shoulders, they may move over a distance of several metres, the heavier male usually advancing and the female giving way. Finally she allows him to start grooming her about the face, ears and neck. The grooming is usually accompanied by rubbing with the cheek glands and shortly after this stage is reached, the female permits the male to mount, usually first retiring to a protected corner.

In the hamster, *Cricetus* (Eibl-Eibesfeldt, 1958), the male, on entering the female's territory, first marks extensively before he starts to court her. The courtship consists of the male following the female and, as in the squirrel and field-mouse, making a call resembling the distress cry of the young. Following continues until the female signifies her readiness to mate by raising her tail and the two then retire into the burrow to copulate.

The most complex rodent courtship that has to date been described is that of the little grasshopper mouse, *Onychomys leucogaster* (Ruffer, 1966). The complex procedure may be broken off and resumed again several times and can last anything up to three hours. In the preliminary stage, after the two have circled round each other a few times, the female takes the initiative, following the male and sniffing his genital region. After again circling each other, the two then rise on their hind legs and bring their noses into contact. The female may interrupt such a naso-nasal contact by dropping to the quadrupedal posture and smelling the male's genitalia before resuming the nasal contact. At this stage, the courtship may break off, presumably because the female's heat is not yet sufficiently advanced for the male to become sexually excited. At any rate, when the courtship is resumed, the male shows more interest in the female and he now follows her and, as before, the following is now and then interrupted by a "nose kiss". Finally, after a kiss, the female remains sitting on her haunches while the male sniffs her genitalia and then parades back and forth in front of her a few times. He then again smells her genitalia, rises on his hind feet and grooms her face. This final sniffing and grooming may be preceded by the male's moving under the female and rubbing his back against her belly. At this point the courtship may again be broken off for up to half an hour. When it is resumed, the final phase has been reached: the partners are both ready to mate and it is only a matter of the male's final stimulation of the female to induce her to adopt the mating posture. She now lies on her side and the male grooms her neck and along her side. This may or may not be preceded by some circling and a sort of embrace, the two grasping each other with their fore limbs. The female then adopts the mating posture, tail to one side and hindquarters slightly elevated and the male mounts. During copulation the pair roll over on their sides—a relatively common happening during the mating of small short legged mammals as diverse as mice and weasels.

In this courtship the most striking element is the nose kiss. Two possibilities exist as to its origin. Firstly, it might be an elaborated form of greeting sniffing but the upright posture adopted speaks against this interpretation, as does the fact that *Onychomys* is not social. It seems more likely that it is a ritualised form of a fighting posture, as the rather similar rearing up of *Cricetomys* most clearly is. Its significance is then that of showing that no attack will in fact be made. Despite the adoption of the boxing stance, biting is completely inhibited, so that its performance by the pair constitutes a sort of mutual non-aggression pact.

The courtship of *Thryonomys* is of some interest, in that it combines a number of elements seen elsewhere in different groups. When making his initial approaches to the female, the male wags his tail from side to side which, like the up and down wagging in the acouchi, appears to signal the non-aggressive nature of his intentions. At first the female flees precipitately and the male chases but gradually she becomes quieter and allows him to draw near. He then places himself in front of her and, still tail wagging, treads alternately with his hind feet in what may be a ritualised intention copulation movement. The female may again move off and the procedure can be repeated a number of times. At some stage a nose kiss exactly comparable with that of *Onychomys* may occur, the two rearing up on their haunches and remaining motionless for several seconds with their noses in contact. In the final stage, the male, once the female has come to tolerate his presence close in front of her, breaks off his treading and grooms her round the ears and neck and is finally permitted to mount.

In the porcupines there can be little question of a male forcing his attentions on an unwilling female. In both the American *Erethizon* (Shadle, 1946) and the African *Atherurus* the female becomes sexually importunate when she is on heat and stands before the male inviting copulation with her hindquarters raised, her tail turned aside and her spines relaxed. The penis is sufficiently long for the male to achieve intromission standing behind her on his hind feet, his paws not clasping her but merely resting lightly on her back or sides. Little is known about courtship. In *Erethizon*, Shadle, Smelzer and Metz (1946) report that before mating, the male and female, reared on their hind feet, may make a naso-nasal contact, much as *Thryonomys* and *Onychomys* do and they also found that the male sometimes sprayed the female with urine. As in the South American rodents, this had no sexually stimulating effect on the female and seemed merely to reflect male frustration. In *Atherurus* the single pair I have watched were a young female and an old male. As she came on heat, the female became very active and would dash hither and thither with surprising speed and agility. Finally she would stand in front of the male and invite copulation by adopting the mating position; he would then mount as already described. I never saw the male take the initiative, which may possibly have been because he was rather old. He was, however, still sexually competent, as an offspring was produced. The information is still very meagre, but in both Shadle's *Erethizon* and my *Atherurus*, when mating was actually observed, it was the result of invitation by the female, not of male demand. This suggests that in these dangerously armoured species it may be normal for the female, at least in the final stage, to take the lead—in fact, in the porcupine world it may be always Leap Year!

In the spiny anteater and in hedgehogs the male is also faced with the problem of making his approaches to an armoured partner. The courtship procedures are basically similar in that there is prolonged following of the female by the male, who attempts to smell her genital region but is at first repulsed and a true copulation attempt is not made until the female invites it. In *Tachyglossus* copulation occurs ventro-ventrally with the pair lying on their sides (Dobroruka, 1960). During courtship the male, sniffing behind the female, attempts to get his snout under her body and push her over into the mating position. This he fails to achieve but finally the female turns on her side for him and permits him to copulate.

In the hedgehog (Herter, 1957) there is prolonged following. The female is at first aggressive and attacks the male by butting and biting at him. He retreats but continually renews his efforts to approach her, threatening by hissing as he does so. In one pair whose mating was timed this went on for 20 minutes. The female then suddenly ran a short distance, lowered her spines and lay down with her hind legs stretched out backwards so that the vaginal opening was exposed. The male immediately came up, reared on his hind feet and, with his paws lightly resting on her back, copulated. The penis, like that of the porcupines, is sufficiently long to permit intromission without it being necessary for the bodies of the partners to be closely pressed together.

In all these spined species the same principle appears to operate. The male is attracted to the female but his approaches obey a strict rule of "thus far and no further", and as long as the female is unwilling he does not make any direct copulation attempts. The female does not show a gradual increase in receptivity: instead her behaviour switches over rapidly from the stage of active resistance to active soliciting. It is the suddenness of this reversal in the female from non-receptive to demanding, rather than merely receptive, that constitutes the main behavioural characteristic adapted to circumventing the difficulties raised by the nature of the defensive armouring.

In the Carnivora, courtship, although it may be prolonged, is not elaborate. The heat of the female is generally rather long lasting and her smell attracts the male before she is ready to accept him. The male may therefore be in attendance on the female for some time before mating occurs. It is common amongst carnivores for the male to have a strong inhibition against biting a female. She is therefore able to repulse him in the initial stages, since he does not retaliate when she snaps or claws at him but merely draws back, and after a slight pause, renews his efforts to approach her.

In dogs the courtship consists of little more than persistent following of the female by the male, who attempts to smell her genital region. She bites at him and he retires temporarily. As she comes more fully into heat, the female ceases to repel the male and may even invite copulation by standing in front of him in the mating posture. In wild Canidae, as far as is known, the procedure is similar and courtship is often accompanied by specific vocalisations, whose function seems usually to be that of assisting in bringing the partners together. In the fennec, Gauthier-Pilters (1962) reports a mating call made by both sexes. In the raccoon dog (Seitz, 1955) the male has a mating cry, the "yearning call", which closely resembles the distress whine of the young, but the female rarely calls.

Seitz notes that a similar "symbolic whine" also occurs in foxes. Van der Merwe (1953) describes a special mating call made by the female black-backed jackal and replied to by the male and Seitz (1959) mentions an "entreaty call" made by the male golden jackal, *Canis aureus*, to the female.

Very little appears to be known about courtship in bears. Meyer-Holzapfel (1957) merely records the male showing increasing tendency to smell the female's genital region as she comes to heat. The pair of grizzlies, *Ursus arctos*, seen mating by Mundy and Flook (1964) were already copulating when first seen. The male was clasping the female in the lumbar region and resting his head on her back, but not gripping her neck, and he nibbled her ears during pauses between successive bouts of pelvic thrusting. Meyer-Holzapfel gives a figure of polar bears copulating, showing the male gripping the female in his jaws by the back of the neck.

The Poglayen-Neuwalls (1966) have described mating in the olingo, *Bassaricyon*, but no details of courtship are given beyond the fact that there is a characteristic mating call, given by both sexes. The male grips the female by the neck and may drag her about thus before mounting, but the neck grip is not used during the actual copulation. If, during copulation, the female attempts to raise her head, the male presses it down again with his snout but does not grip her.

In the viverrid, *Suricata*, the neck grip may be used in an exactly similar way as a preliminary means of inducing passivity in a recalcitrant female. The first mating I observed in this species was between an adult female who, to judge from her behaviour, had been on heat for some time before a male was obtained for her. Since he was a stranger, she at first attempted to drive him away. He, however, at once seized her by the neck and dragged her about and she then became passive and permitted him to mount. The procedure appeared so violent that I feared for the female's safety and at first separated them. The male maintained his grip with such tenacity that considerable force was required to open his jaws but I found that, despite this, the bite was inhibited: the jaws were locked in position short of full closure holding only a fold of the female's skin and there was no wound or mark on her. On later occasions, when the pair knew each other, mating was preceded by bouts of more or less playful fighting in the course of which the degree of excitement necessary for mating was reached. One curious feature of this fighting was that frequently one of the animals would grip the partner's nose but without closing the jaws sufficiently to cause any hurt. The significance of this nose bite is obscure but it is not peculiar to *Suricata*, for Schenkel (1947) mentions nose gripping as occurring in the mating behaviour of wolves and Fisher (1939) saw the male sea otter grip the female by the nose or chin. In later matings between a young female and the same male *Suricata*, by then older and less sexually excitable, the female frequently took the initiative and would incite the male to courtship fighting by nipping at him, biting particularly at his cheeks.

In *Helogale undulata*, Zannier (1965) does not mention the neck grip but it is possible that, as in *Suricata*, it is used only if it is necessary for the male to subjugate a hostile female. In Zannier's animals courtship began with mutual smelling and licking of the partner's genitalia, which might be initiated by either the male or the female. The male then starts to follow the female, who moves away from him, but stops periodically and the male attempts to mount. During this following a mating call is made. At first the female moves off when the male tries to mount, wriggling out of his grip if he has managed to clasp her. Finally she stands for him and, placing his paws on her slightly arched back, the male presses her down and clasps her flanks; she turns her tail aside and intromission follows. During copulation the male bites the fur of the female's side and neck in what appears to be a modified form of the neck bite.

In Mustelidae, the neck grip is commonly used in the preliminary stages of mating in much the same way as in *Suricata*. The courtship of the skunk, *Mephitis mephitis*, studied by Wight (1931) showed an interesting peculiarity. The initial stages were characterised by non-serious fighting and by the male dragging the female about by the neck. As he mounted and clasped the female with his fore-limbs, however, the male performed an action which does not appear to have been recorded in other species. This consisted of a rapid scratching of the hind foot directed towards the female's vulva. This happened not just on one occasion but as a regular accompaniment of mounting and Wight concluded that it constituted "a normal titillating act preliminary to copulation" and was of the opinion that it induced the female to adopt the mating posture. In a number of small rodents and also in *Sminthopsis*, the male's foot may vibrate during copulatory thrusting but this appears to be purely incidental and usually the foot does not touch the female's body.

Amongst the Felidae courtship behaviour is best known in the domestic cat. When she comes into heat, the female may make a mating call. There is considerable individual variation in readiness to do this, Siamese cats being particularly vocal at this time. The call is not essential in summoning the male, as the smell of the oestrous female is sufficient to attract males—at least in normal urban conditions. Usually several assemble and courting of the female is accompanied by fighting amongst the males. At first the female refuses to allow a male to approach her: she evades him by fleeing and also repulses him by spitting and striking out with her claws. As he approaches the male makes a low call which, to the human ear, has something appealing about it and is very different from the loud "caterwaul" with which he challenges a rival. Gradually the female allows the male to get nearer and her increasing receptiveness is shown by her rubbing her head against any convenient object and rolling on the ground in front of him. He is still repulsed, however, when he attempts to make contact with her. Normally his aggression towards her is completely inhibited; he merely draws back with half-closed eyes and ears laid back as she strikes at him, in most dramatic contrast to the ferocity he shows in threatening and beating off a rival. There are, of course, individual differences in this respect and exceptional cases of attempted rape do sometimes occur. Finally the female allows the male to come close enough to touch her and he at once grips her by the back of the neck. If she is fully receptive, she remains passive and the male mounts, clasping her flanks with his paws and treading with his hind feet. The female responds by assuming the mating position and intromission follows. The copulation is brief and is usually accompanied by a screeching cry from the female and a growl from the male. As the male dismounts, the female frequently turns and strikes at him with her claws; he avoids the blow, as the male rat does, by a quick backward leap. This attack by the female does not invariably occur.

Although the tomcat is a byword for promiscuity, a genuine personal attachment can develop between a male and a particular female and in such cases the female may simply remain quiet for a moment and then go over into post-copulatory rolling and licking of her genitalia. Copulation is repeated a number of times with the female at first becoming increasingly eager and the male gradually less so, so that the initiative gradually passes over from male to female. Normally the female will accept several males in turn before her heat has come to an end. In the large Felidae the male normally does not grip the female's neck in his jaws until the moment of ejaculation. Antonius (1943) says that this vestigeal use of the neck bite is characteristic of lion, tiger and jaguar and is usual in pumas: he also saw it in the one North American lynx mating that he observed.

In the Pinnipedia courtship is abbreviated in the species where the social organisation is a harem system. The male has been defending his females from rivals and herding them together to prevent them straying for some time before they give birth to their young and experience the post-partum oestrus during which mating occurs. Preliminaries inducing the female to permit the male to approach are therefore unnecessary. In *Arctocephalus gazella*, for example, the preliminaries to copulation are brief and occupied only four minutes in one case timed by Paulian (1964). The male and female circle round each other, grooming each others' heads and necks, the female then extends her body, the male gives a characteristic cry, snorts and mounts. He grips the female by the neck with his jaws and she diverges her hind flippers to facilitate intromission. Once this is achieved the male releases his grip on the female's neck and moves his head about as he thrusts. Paulian noted that during copulation the female moved her hind flippers so that they stroked the male's scrotum but was not certain that this was invariably done.

Amongst the monogamous species males and females do not come together until after the female has left her young, and elaborate courtship displays are known in some species. In the harp seal, *Phoca groenlandica*, (Mohr, 1956) the males assemble in the water beside the ice floes on which the females are lying. Here they perform a display which seems designed to attract the female's attention, bobbing up and down in the water with the body held vertically. This is followed by a sort of demonstration of swimming virtuosity, the animals circling about, sometimes on their backs, and turning somersaults in the water. Although frequently performed communally and then giving the impression of a sort of competition in which each male is attempting to catch the attention of a female, the display is not omitted even if only a single male and female are involved. The male sea leopard, *Hydrurga leptonyx*, also performs a swimming display in front of the female before approaching her to mate.

One of the most interesting characteristics of carnivore sexual behaviour is the widespread use of the neck grip, either during copulation or as a preliminary to it. The neck grip also occurs in the closely related Pinnipedia and amongst Carnivora is so widespread that it is legitimate to regard its absence as atypical and probably secondary. In the Carnivora an oriented neck grip is also very commonly used in two other contexts—prey killing and transport of the young. In prey killing the bite is carried home with the full strength of the jaw muscles. In the other two contexts, as Leyhausen (1956) points out, it is not merely a case of biting more gently: the jaw closing muscles are strongly contracted but full closure is inhibited by simultaneous activation of the antagonistic muscles. The occasional cases of wounding or death of mate or young that do occur are the result of failure of this inhibitory mechanism.

In other Orders too the same three contexts may call forth the oriented neck bite. In rodents, for instance, those species that do kill other small mammals use the neck bite—but there are few species that behave in this manner. Similarly, a neck grip is occasionally seen in copulation but usually it is used only when things do not go quite smoothly and the male is having difficulties in achieving intromission. This is the case in the golden hamster (Dieterlen, 1959) and in the guinea pig. The dormouse is exceptional in that a copulation neck bite is normal (Kunkel and Kunkel, 1964). In transporting the young, although a neck grip is sometimes used by rodents, a grip near the centre of the body is much commoner. These data suggest that in this Order, the neck grip is something vestigeal, appearing in rather atypical circumstances, rather than the result of an independent evolution in isolated cases, paralleling the typical carnivore patterns. The sporadic occurrence of the neck grip in other Orders—for instance in prey killing in *Dasycercus*, during mounting in shrews (Crowcroft, 1957)—points in the same direction: that the neck grip is not something evolved by the Carnivora but something of much greater antiquity, merely elaborated in that Order. Indeed, it is more than likely that the neck grip goes back in vertebrate history beyond the origin of the mammals. Many modern lizards use a neck grip in copulation and here it functions not only to restrain the female but also to ensure that the male is correctly oriented to her. I do not know of any review of killing techniques of living reptiles but the neck armouring characteristic of the ceratopsian dinosaurs is difficult to interpret otherwise than as a defence evolved in relation to a predator using an oriented neck bite. It would, in fact, be surprising if this were not so. The neck grip is superior to any other hold in dealing with a long bodied creature armed with teeth, whether the aim be temporary restraint or killing. It would therefore speak very poorly for the efficacy of natural selection if this had not been "discovered" at an early stage of tetrapod evolution. Since teeth and jaws capable in one bite of inflicting a lethal wound on an adversary of equal size are found only in advanced predators, it would have been perfectly possible for the death bite and the mating grip to have had a common origin and have subsequently become differentiated. Only after such differentiation had occurred and an inhibited neck bite had been incorporated in the behavioural repertoire would it have been "safe" to utilise the same pattern in the transport of the young. Of the uses to which the neck grip is put, the latter is the only one whose evolution cannot have antedated that of a mammalian type of reproduction.

The more effective the killing bite becomes, the more essential is its inhibitory control in mating and transport of the young. This implies that the mate or offspring must do nothing which might switch off the inhibitory control and so trigger off a genuine killing bite; in other words, selection will operate not only on the gripping pattern, but on the response to it. Carnivora do, of course, show an appropriate response. When picked up by the neck, the young hang limp and passive and, making no struggle or resistance, do nothing to precipitate a genuine attack. The response persists in the adult and here selection is operating in the same way in relation to the mating grip. In *Suricata*, the adult passivity response is very striking. A violently struggling animal, lying on its back and defending itself with teeth and claws against being picked up, scolding at the top of its voice, becomes instantly limp and quiescent if one succeeds in picking it up by the scruff of the neck. The correlation with the way in which the neck grip is used in mating is clear enough. We should thus not regard the preliminary dragging about of the female by the scruff of her neck as the primitive form of behaviour and the neck bite during copulation as more advanced. Rather, they are two ways in which a neck grip, originally used to enable the male to mount, has become specialised subsequent to the evolution of passivity in response to being carried. The fact that the dragging about may be omitted and that, in any case, the grip is released before mounting, makes it clear that the preliminary drag is anything but primitive. If the female did not have a specific response to the grip, she would merely take the opportunity to make off as soon as released. All of this leads to the superficially paradoxical conclusion that the reason why it is the Carnivora that show the maximal utilisation of the neck grip in mating and in transport of the young is precisely because in them it is most potentially dangerous.

Before turning to consider artiodactyl courtship, it is necessary to define what we are including in this category of behaviour, for within this Order, reproductive and territorial activities become so interwoven that drawing a sharp line between them is impossible. In species where the breeding herd consists of a single male with a group of females, the male may defend a definite area. The females may stay within the territory of their own accord or it may be necessary for the male to prevent them from straying beyond its boundaries. A final stage is reached when the defence of a specific territory is abandoned and the male simply defends his group of females and herds them together to prevent their becoming dispersed. All the females are defended and herded and as they come into oestrus, the male mates with them.

Driving away of rival males obviously belongs in the context of territorial behaviour even though "territory" may have become translated from "within my domain" to "near my females". The herding together of the females is a new type of behaviour evolved in relation to the particular type of social organisation, as a means of keeping the females within a territory and paradoxically, making possible the subsequent abandonment of spatial territorialism. Its function is to ensure, on the one hand, that when they come on heat the females are available as mates to the herd owner and, on the other, that they are not available to any rival male. It is not very meaningful to argue as to whether herding should be regarded as belonging in the context of territorial or of reproductive behaviour, since in such cases, the two have virtually coalesced. It seems reasonably clear, however, that herding of the females is in fact derived from the preliminary "driving" of the female which will be described later and which forms a normal part of courtship behaviour. However, since the present aim is comparative, it seems desirable, as has already been done in the comparable case of the Pinnipedia, to limit the term "courtship" to the activities related to the male's inducing the oestrus female to mate—in fact to what Schenkel (1966b) calls pre-mating behaviour.

In the hornless Camelidae, unlike other artiodactyls, copulation occurs with the female lying down. Courtship is not elaborate and consists mainly of a preliminary pursuit or driving of the female by the male, who then attempts to make her adopt the mating posture, often by rather violent means. The guanaco, *Lama guanicoe*, for instance, bites at the female's legs and neck wrestles with her to push her down. The dromedary also uses his neck in the same way. A characteristic grunting call is made by the male in many species (Pilters, 1956).

This type of courtship, in which the male is using ordinary fighting techniques to subjugate the female, is not found either in the Suidae or the Bovidae, where the possession of potentially lethal weapons makes it essential for the female to be treated with more restraint. In these families, courtship is more complex and includes special procedures which reduce the female's readiness to flee or to fight. In warthog (Frädrich, 1965) there is a long preliminary following of the female, during which the male emits a characteristic vocalisation and attempts to bring his snout into contact with her genital region. Finally she pauses and the male then massages her flanks and hindquarters with rapid movements of his snout. Although these movements are often quite rough, they appear to have a calming effect on the female and induce her to remain quiet while the male mounts. In domestic pigs the procedure is very similar. The male makes a characteristic grunting call as he follows the female and rubs his snout against her sides and her genital region, until she stands and permits mounting.

In the horned artiodactyls, the courtship procedures are further elaborated and show a considerable range of diversification but in all species studied the courtship begins with following of the female by the male. In some cases the male actually seems to be encouraging the female to move and the procedure is termed *driving* but in others it has more the characteristics of a ceremonial parade. The commonness of flehmen in the artiodactyls as a response to the smell of the female's urine has already been mentioned. In the Camelidae, flehmen is not particularly linked with mating but in many of the Bovidae, it forms a regular part of courtship. At an early stage in the procedure the courting male stimulates the female to micturate, usually by licking or nuzzling at her genital region, and when she responds he smells the urine and flehms. This urine testing appears to be primarily a method whereby the male assesses the condition of the female but where it has been incorporated as a regular part of courtship it probably serves to increase his readiness to mate.

Although many species remain to be investigated, the studies that have been made to date, particularly those of Walther (1958, 1964a, b, 1965), make it clear that there are two quite distinctive types of courtship which are characteristic one of the Tragelaphinae, the other of the Antilopinae. The different courtship patterns are correlated with different mounting postures. In the tragelaphines, when the male mounts, his head is laid down in contact with the female's body. In the antilopines, contact with the female is minimal; the male holds his head high and in some species he does not even touch her flanks with his fore limbs. As examples of the two types, we may take the courtship of the kudu, *Tragelaphus strepsiceros*, and of Thomson's gazelle, *Gazella thomsoni*, described by Walther (1964a, b).

In the kudu (see Figure 12), Walther divides courtship into three phases. The initial stage is characterised by mutual

Figure 12. Kudu (*Tragelaphus strepsiceros*): mating. (a) Driving; male follows female. (b) Coming alongside. (c) male lays his head across female's neck. (d) Mounting position, with head pressed on female's body. (After Walther, 1958.)

sniffing, flehmen by the male in response to urination by the female, flight or aggressive behaviour on the part of the female if the male approaches too closely or too fast, and some threatening on the part of the male. Gradually the female becomes more tolerant of the presence of the male and the second phase is reached when she permits him to stand close beside her with his head level with her rump. This may be interrupted by brief bouts of defensive nose jabbing by the female towards the male, who replies by threatening with the head held high and he also turns his head to one side. Walther regards this "sideways facing" also as a form of threat, but it seems to me more likely that it is an indication of unwillingness to carry out the threat—in fact, the typical truce offer. In any event, once the female will permit the male to stand close beside her, the third phase begins with the male driving the female. As he follows her, he periodically comes up alongside her, often with his cheek in contact with her side; once he draws level with her shoulders, he pushes his neck over hers for a moment, then falls back a little: the whole process may be repeated several times. At first the female still shows some defensive responses and may snap or nose jab back towards him but gradually this ceases. Once she slows down sufficiently or stands for him, the male lays his head on the female's back and attempts to mount. She may move on again but finally adopts the mating position with her head low and her tail raised. The male mounts with his chin pressed down on her shoulders and his neck lying along her back.

Walther regards the male's placing his neck over that of the female as derived from neck wrestling. Although with the evolution of horns, this primitive form of fighting has been abandoned in serious contests, it persists in this ritualised form as a part of courtship. His subsequent laying of his head along her back is, like the similar behaviour of the elephant, a ritualised intention mounting movement. By signalling thus his intention to mount, the male is, in effect, asking the female if she is yet ready to accept him and when she responds by assuming the mating posture, she is giving the affirmative answer.

Although there are differences in the details of the courtship procedures, all the Tragelaphinae studied show the characteristic coming alongside the female during driving, the laying of the head on the female's back and the copulatory position with the neck in contact with the female's back.

In Thomson's gazelle, the procedure shows a basic similarity in that there is a preliminary phase during which the male follows the female, who gradually permits him to approach more closely, leading to a final stage during which he signifies his readiness to mount by a special signal and she responds by indicating her readiness or otherwise to accept him. The type of signalling involved, however, is different. Visual signals predominate, contact is reduced and the whole procedure is more elaborate. When the male first approaches the female, she may flee and wild chasing may occur. Presently, however, the female slows down and the courtship proper may be said to have started. The male now follows the female, performing an impressive display with his nose alternately lowered so that the neck is held horizontally and then raised to its fullest extent, so that his horns point downwards and backwards along his neck. As he approaches her, the female urinates and the male smells the urine and flehms. He then again follows her, continuing the alternation of head-high and head-horizontal postures. As his excitation grows, he stops for a moment with the head held high and beats a rapid tattoo on the ground with the hoofs of his fore feet, producing what Walther calls the "drum roll". Following now passes over into a stylised mating parade during which the male follows close behind the female, virtually nose to tail. The movements of the two may be so perfectly synchronised that Walther declares they look like a single creature with eight legs. The point has now been reached where the male must find out if the female will permit him to mount. This is done by a characteristic movement. He raises one fore foot and taps it against the female's hind leg. German authors call this movement the "Laufschlag", which is commonly translated as the leg beat. Its function is, like the tragelaphine's neck contact,

Figure 13. Uganda kob (*Adenota kob*): mating. (a) The leg beat. (b) Mating position with head held high and minimal contact with the female's body. (After Buechner and Schloeth, 1965.)

to ask the final question "Are you ready?" Silence gives consent and if the female does not move off, the male mounts (see Figure 13). It is typical of gazelles that the male maintains the reared up mounting position without supporting his weight on the female's body and the two may actually continue to move forward at a slow walk during the copulation.

The two most striking characteristics of this courtship are firstly, the alternating head-horizontal/head-high display and secondly, the leg beat. In view of the mounting posture, with the head held high, Walther concludes that the head-high movement is a ritualised intention mounting movement. The head-horizontal closely resembles the tragelaphine neck wrestling movement and may, like the latter, be a highly stylised remnant of a long abandoned fighting technique. In other species the leg beat is not necessarily as formalised as it is in Thomson's gazelle and may, in fact, be a quite vigorous kick. This suggests derivation from a genuine kick used in fighting.

A courtship including the use of the leg beat and culminating in copulation with the head held high occurs in all Antilopinae and also in the Hippotraginae (Walther, 1958), but there are, of course, detailed variations from species to species. In the Uganda kob (Buechner and Schloeth, 1965), in correlation with the extremely small mating territories, the preliminary chasing and the mating parade are omitted and the male approaches the female in a characteristic prancing gait with the head held high, displaying the black patches on his fore limbs and his white throat patch. This species is also characterised by the fact that after copulation has occurred, there may be a repetition of a number of the movements used in courtship, constituting a sort of post-copulatory display. This also may be related to the small size of the mating territory and may function to keep the animals within its boundaries. The leg beat occurs also in the courtship of Ovinae (Geist, 1964) and of the okapi, *Okapia johnstoni* (Walther, 1960), and giraffe (Innis, 1958). Further comparative details may be found in the papers of Buechner and Schloeth (1965) and of Walther (1958, 1965).

The most interesting general point that emerges from these studies of courtship in the artiodactyls is the close parallel between the evolution of fighting techniques and of courtship. In the case of fighting we have already seen how, with the acquisition of horns, the primitive all-in camelid type of encounter was replaced by more stylised contests. The same process has occurred in the evolution of courtship ceremonies. In the camelids, courtship is primitive and the female is treated with considerable violence. In the bovids, stylised movements, many of them derived from fighting, have replaced direct attack. Furthermore, these stylised signal movements are often derived from a technique now obsolete in its original context of genuine fighting. In much the same way, touching the forehead in salute—a ritualised intention movement of doffing the hat to a superior—can serve us as a greeting even in the absence of the hat.

In the Perissodactyla, the same principles can be seen to operate. The rhinoceroses are armed with a potentially lethal weapon and, as we have already seen, their fighting is stylised. Courtship too shows restraint and although it includes a number of elements relating to fighting and threat, these do not lead to any serious consequences. In the unarmoured Equidae, on the other hand, where fighting techniques are not highly stylised, courtship may be correspondingly violent.

In the black rhinoceros (Goddard, 1966) courtship begins with the male following the female. During the initial stages the female micturates frequently, the male smells the urine and then flehms. He may also perform the pattern of rubbing his horn on the ground with a side to side swing of the head and adopt the stiff-legged gait which has been described in the broadside threat display and the pair may indulge in some sparring with the horns. In one of the six matings observed, the female charged the male when he attempted to approach. He did not retaliate, but withdrew and performed the stiff-legged horn rubbing display. Presumably, as in the carnivores, his tendency to respond aggressively is inhibited by the fact that the attacker is an oestrous female, not a rival. The motivation for the display is thus essentially the same as when Goddard saw it performed by an inferior male to a superior; in both cases there is aggressive motivation which is held in check by a conflicting tendency. Before mounting occurs the male lays his head on the female's rump and if she stands for him, he then mounts. There are usually a number of preliminary mountings before intromission is achieved and between these the pair may walk about or feed. Copulation when it finally takes place, is prolonged by comparison with other large herbivores and lasts for approximately half an hour.

In the wild ass, *Equus asinus*, where males and females remain separate outside the breeding season and the partners are thus strangers when the first approaches are made, the courtship consists of a violent subjugation of the female by the male. He pursues her, biting at her and kicking her, until she is prepared to stand for him (Antonius, 1937). Without this preliminary fighting, she is not prepared to mate. In the horse, on the other hand, where the male guards and herds his troop of females, the two are already familiar with each other and courtship is simple and peaceful with the male following the female until she is prepared to stand for him. Antonius (1937) points out that this difference in their courtships explains why the cross of male ass/mare is much commoner than that of stallion/she ass. To the male ass, the mare is merely a female who offers unusually little resistance whereas the stallion does not include in his courtship repertoire the preliminary fighting, without which the she ass is unwilling to accept him.

In the higher primates too, the same "familiarity principle" operates and courtships are not elaborate. In monkeys, when the female comes on heat it is common for her to form a consort relationship with an adult male. The two remain in each other's company and copulation occurs at intervals without significant preliminaries. In baboons (Hall and DeVore, 1965), langurs (Jay, 1965) and howler monkeys (Carpenter, 1965), the female, as she comes into oestrus, approaches the male and takes the lead in consort formation. Carpenter relates that in the howlers, as their sexual excitement increases, a special form of behaviour is often shown and copulation usually follows. According to Carpenter the male and female "exchange rhythmic tongue movements with their mouths partly open". This description is not very lucid and possibly it is simply a version of the friendly lip-smacking gesture already described in the Old World monkeys.

In the bonnet monkey, *Macaca radiata*, (Simonds, 1965) it

is the male who initiates mating. Approaching the female, he raises her tail and examines her genitalia: if she is receptive she presents and mounting follows. If she is not ready to receive the male, she runs away.

Of courtship in the apes, very little is known. Sexual behaviour, according to Schaller (1965c), plays a very minor role in the life of the mountain gorilla and he observed very few matings in the field. In captive animals mating was often preceded by prolonged wrestling, chasing and fondling but this was not seen in the wild.

Goodall (1965) observed thirty-two chimpanzee matings in four of which the female solicited by crouching in front of the male and inviting copulation. In nine cases the male simply approached the female, who promptly crouched and copulation followed at once. In the other nineteen cases the male performed a distinctive display in front of the female. Leaping into the tree in which the female was sitting he swung from branch to branch with his body held vertically and his hair erected, especially on head, shoulders and arms. Once the display was performed on the ground, and the male, instead of swinging from branch to branch, stood upright in front of the female and rocked from one foot to the other. Usually the female at once responded to the display by adopting the mating crouch but on three occasions the male touched her lightly on the back with one hand before she did so.

This brief survey is sufficient to show that all the types of behaviour predicted at the beginning of this discussion (p. 201) do, in fact, occur in mammalian courtships. The commonest pattern of all is pursuit; indeed, it is well-nigh universal and may take any form from a violent chase, through a more decorous following to the highly stylised mating parade of Thomson's gazelle. Courtship fighting is also common and may be violent, as in Camelidae and some Equidae, or relatively restrained and playful, as in *Suricata* and *Mephitis*. More stylised movements derived from fighting also appear, the most striking examples being found amongst the Tragelaphinae. Threat behaviour too is widespread and examples are found in most groups. The curious nose kiss of *Onychomys* and *Thryonomys* may represent a specific courtship pattern evolved from what was originally a threat posture. Special signals by the male indicating that his approach is non-aggressive have also been found, for example the tail-wagging of the acouchi and *Thryonomys*. Female invitation, both by presenting and by inciting the male by playful fighting or "coquettish" pseudo-flight are also recorded. Finally, male displays also occur, notably the red squirrel's tail display, the swimming display in Pinnipedia and the chimpanzee's branch swinging display: possibly the Uganda kob's prancing should be included in this category.

In addition to all these expected types of behaviour, one that was not predicted on general grounds has been found: the use by the male of a special call resembling, to the human ear at least, the distress call of the young. A mating call of this type is known in three rodents; squirrel, fieldmouse and hamster as well as in some of the Canidae. While the advantage to the male of this, as it were, cashing in on the female's protective responses towards her young may be obvious enough, how this behaviour came to be evolved is less clear. It is the initial stages that present a difficulty for, provided the male makes a call which bears sufficient resemblance to the baby call to evoke some appropriate response in the female, there is no difficulty in understanding that it may then be improved upon and made more closely to resemble the genuine baby call. The problem therefore is—why should the courting male ever come to make a call bearing any resemblance to that of the youngster in distress? In the case of the canid yearning call, the answer is probably very simple. The call is made by the male when he is seeking a female but has not yet found her (Seitz, 1955). His situation is therefore not unlike that of the youngster when he finds himself separated from his fellows—he is alone and requires the presence of a conspecific and the two calls may therefore have basically the same motivation. When he has actually found his mate, a different sort of soft "youff-youff-youff" call is made. The rodent calls may have begun in the same way. Although now made in the presence of the female, they too may have originally been uttered while she was being sought. In the course of evolution, the calls may have become deferred to a later stage in the mating sequence with a consequent change in function from that of summoning the female to that of inhibiting her aggressive tendencies.

CHAPTER 10

Parent and child

All young mammals are fed on milk secreted by the mammary glands. This feeding process involves special behaviour on the part of both mother and offspring: the young must be able to locate the milk supply and to suck once they have found it, while the mother must behave in such a way as to facilitate their doing so. Maternal care, however, usually goes much further than the mere provision of the infant's first food. Commonly the mother not only feeds her young but protects them, cleans them and keeps them warm and in some cases she plays an active part in their first introduction to eating solid food. Frequently the care of the young is the responsibility of the female alone but in some species the male also has a role to play. Responses of parent and young must be adjusted to each other and it is therefore not desirable to discuss maternal or parental behaviour without at the same time dealing with the responses of the young to their parents.

Although these considerations apply equally to all mammals, the modes of reproduction are so different in monotremes, marsupials and placental mammals that it is convenient to deal separately with the three groups.

MONOTREMATA

Not very much is known about parental behaviour in monotremes and little can be added to Burrell's (1927) account which, although dealing mainly with *Ornithorhynchus* also contains some information about *Tachyglossus*. The two differ in that the egg of the latter is incubated in the pouch, whereas the aquatic *Ornithorhynchus* is pouchless and remains in the nest during the incubation period, with her body curled round the eggs.

When she is about to lay, the female platypus excavates a special nesting burrow, usually 15–20 feet long and 1–1½ feet below the surface. Sometimes the same burrow is added to in successive years and may then ultimately reach the surprising length of 100 feet. The fore limbs are the main digging organs, but the snout assists to some extent. At the end of the burrow a chamber is excavated and this is lined with grass, leaves, rushes or whatever similar material is to hand, to form the nest. Once the nest is completed, the female blocks the burrow up, usually at several points along its length, and she remains inside until the eggs have hatched.

In *Tachyglossus*, since the egg is carried in the pouch, no special brood nest is required and the female's activities need not be curtailed during the incubation period. How the egg reaches the pouch is not known for certain, but Burrell's opinion, that the body is sufficiently flexible for the egg to be passed directly into the pouch from the cloaca, seems more plausible than the alternative that the female places it in the pouch with her paws. From his account, I have always taken it that Burrell visualised the animal as remaining in its normal quadrupedal posture during this process, with the posterior part of the back very much curved: the upside down position shown in an illustration in Bergamini (1965) seems most improbable and is certainly not the obvious interpretation of what Burrel actually says.

After hatching, the young *Tachyglossus* remains in the pouch until it has grown too large to be accommodated any longer. In the specimen studied by Griffiths (1965) this occurred when the youngster was a little over 20 cm long and weighed 400 g. The pouch thereafter regressed, although the mammary glands remained fully active. The mother subsequently fed the infant at irregular and surprisingly long intervals of $1\frac{1}{2}$–2 days. Between feeds, the baby remained sleeping, making no effort to crawl about. Presumably in natural conditions the youngster would be left concealed in some type of refuge, but no details are known. When the female returned to feed her youngster, she assisted him to find the mammary area; she nudged him with her beak until he lay between her fore limbs, then, arching her body up, she would push him under her with one paw. Suckling occupied roughly half an hour, the two mammary glands being drained in turn. During this period a large amount of milk was consumed—48 g on one occasion. Griffiths is of the opinion that there is a specific milk ejection mechanism, probably controlled by oxytocin. Burrel notes that in the platypus the caruncle on the end of the muzzle becomes hard only some time after hatching and suggests that its function is to stimulate the mammary area and induce milk let down. Mr. A. K. Minchin of the South Australian Museum (personal communication) found that an orphan *Tachyglossus* could most easily be induced to take milk from his cupped hand. He noticed that the youngster rubbed and shoved against his hand with the beak, presumably also an action whose function is to stimulate milk let down. The milk was not licked or lapped but taken with a munching-sucking action. Griffiths also, judging from the sounds made, says that his naturally suckled youngster also sucked. Burrell (1927) says of the platypus that the hair covering of the mammary areas "serves, apparently, in the place of a teat and the young pluck at this and suck the milk from it . . .". The often repeated statement that monotremes do not suck, but lick up the maternal milk, thus appears to be incorrect. It seems that even if there are no teats, the easiest way for the youngster to feed is not by licking, but by sucking. This is a satisfactory conclusion, for it simplifies the problem of the evolution of teats. To a youngster that licked, a teat would be only an embarrassment, but to one that munched or sucked, it would be an assistance, so surely sucking must have preceded the evolution of teats. Had the young of monotremes really been lickers, we would therefore have been faced with the problem of why the ancestors of the other mammals were different and what started them sucking in the first place.

In these oviparous animals, lactation cannot be initiated as a result of the processes involved in laying the egg, but must depend on stimuli received from the activities of the young after hatching. Burrell notes that in the platypus lactation is not established until two or three days after the young have hatched and he is of the opinion that the same is true of *Tachyglossus*.

MARSUPIALIA

In marsupials there is no need for initiation or intensification of nest building just before parturition since the pouch provides shelter for the newborn young. Shortly before giving birth, however, the female commonly spends more time than usual in licking her pouch (Plate IV). In *Megaleia rufa*, the lining of the pouch during mid-pregnancy is normally covered with a dark coloured scale. This is removed and the inside of the pouch thoroughly cleaned before parturition. The female may start to lick her pouch 24 hours before she gives birth and cleaning is greatly intensified in the three or four hours immediately preceding parturition (Sharman and Calaby, 1964). Hartman (1952) reports that in the American opossums also the female cleans the pouch thoroughly just before giving birth. In one instance, a female which he erroneously believed to be pregnant was injected with pituitrin in the hope of inducing parturition. No young appeared but the female at once responded by licking out her pouch.

Impending parturition is also characterised by the female taking up the birth position, which she may do repeatedly before the birth actually occurs. *Megaleia* sits with the body inclined forwards and most of the weight supported on the tail, which is turned forwards between the hind legs. The legs are extended forwards and, if possible, the back is rested against some convenient support. The opossum too gives birth in a sitting position with the legs stretched out forwards.

Gestation in marsupials is usually not prolonged beyond the luteal phase of ovarian development. Even in those species where pregnancy continues into the post-luteal period, the gestation periods are relatively short—*Potorous tridactylus*, the potoroo, with 38 days, being the longest known (Sharman, 1965). The marsupial placenta is not capable of supporting a rapid growth rate, the young are born in a very embryonic condition and are extremely small. Even in *Potorous* the birth weight is only a third of a gram and the maximum recorded (in the grey kangaroo,*) is less than a gram. The newborn young are nevertheless capable of finding the maternal teat

* The nomenclature of this species is in some confusion. It is variously referred to as *Macropus giganteus*, *M. canguru* and *M. major*. The position is complicated by the fact that Sharman and his co-workers have recently produced evidence that the western and eastern grey kangaroos constitute two distinct species (see Kirsch and Poole, 1967). The gestation period of the eastern greys is approximately 5 days longer than that of the western and, presumably, the birth weight of the young is also greater.

quite unaided. In some of the more primitive dasyurids, for example *Antechinus* and *Dasycercus*, this is not a particularly difficult task: the teats lie near to the cloaca and are not concealed in a closed pouch but merely protected by flaps of skin. These are no impediment and may even assist by directing the young towards the mammary area. In the more specialised Macropodidae, however, the difficulties are much greater. The pouch is large and it opens forwards. As an adaptation to a leaping gait, the advantage of this arrangement is obvious in terms of preventing young being accidentally lost from the pouch, particularly at the stage when suckling is intermittent and the grip on the teat is periodically released. The task that now faces the newborn young, however, is a formidable one: the distance to be traversed is considerable and the pouch must be entered before a teat can be found. Apart from licking it free of the embryonic membranes and subsequently rupturing the cord by further licking, in no species that has been studied does the female give any further assistance to the young. It makes its way to the pouch entirely by its own efforts (Plate V). At birth, the fore limbs are well developed and provided with claws capable of grasping the mother's fur but the hind limbs are mere buds. Locomotion is by alternate movements of the arms, rather reminiscent of the swimming stroke used in a slow crawl, and the head is turned to the contralateral side as each stroke is made. In *Megaleia rufa*, Sharman and Calaby (1964) found that it took the young about three minutes to reach the rim of the pouch and disappear inside. If replaced at the vulva, they were capable of performing the journey a second time but one, made to do the trip three times, although it reached the pouch, failed to attach to a teat. The route taken is normally very direct but it is still not certain what factors guide the young on its journey. At birth neither the eyes nor the internal ear are functional (McCrady, 1938) but the nostrils are open and the olfactory region of the brain well developed. Since only the fore limbs are functional, upward crawling is very easy; indeed, McCrady was of the opinion that the structure of the neonate is such that if placed on a slope, it must needs crawl upwards, its weight sufficing to ensure the orientation. The ability to find the pouch, he considered, was thus dependent on a response to gravity; but one of an unusual kind, since the receptors of the ear were not involved. Hartman (1952) was of the same opinion and reports that in the Virginia opossum, *Didelphis marsupialis*, "whenever we tilted the skin, the embryos invariably travelled upward, even away from the pouch". These observations, however, are not in accord with those of Reynolds (1952), who found that the young opossum could change course and enter the pouch from the side. Sharman and Calaby found that the same was true of the young *Megaleia*: with the mother anaesthetised and lying so that her ventral surface was horizontal, the young could still find the pouch. These authors also mention a newborn *Trichosurus* finding the pouch after working round in a semi-circle. It seems clear therefore that although an automatic orientation against gravity may be responsible for starting the young off in the right direction, there are other factors involved in completing the journey to the pouch.

The often quoted notion that the mother licks a moist track along her fur, which the youngster then follows, has been shown by Sharman and Calaby to be untrue. In view of the sensory equipment at its disposal, the most plausible hypothesis is that the youngster responds to olfactory stimuli emanating from the pouch, the side to side head movements suggesting a klinotactic mechanism. If this is correct, then it might be possible to attract the young to a swab thoroughly impregnated with pouch smell, but to date no definitive tests have been performed and Sharman and Calaby's conclusions are therefore purely negative: the youngster can reach the pouch "without necessarily travelling along a completely upward path, without a moist track on which to crawl and without assistance from the mother's paws or lips". From the time that it finds the pouch and starts to suck*

* The often repeated statement that the milk is actively ejected down their throats and that young marsupials do not suck has been shown to be incorrect in every case that has been investigated (Merchant and Sharman, 1966). It is equally untrue to say that the youngster becomes so firmly attached to the teat that it cannot be removed without damage. Merchant and Sharman have shown that in kangaroos, provided the "joey" is removed gently and assisted in reattaching itself, there is no particular difficulty and no damage results. I. Eberhardt (personal communication) says that the same is true of the koala.

until it is ready to make its first excursions out of the pouch, the only maternal care the young marsupial requires is attention to its toilet. The mother frequently licks inside her pouch. Presumably she stimulates the young to micturition and defaecation by licking the cloacal region, at the same time removing the products, so that the pouch remains clean. Certainly, like the young of altricial placental mammals, pouch young in species as diverse as macropods and dasyurids do respond appropriately to cloacal stimulation.

In some species, for example *Sminthopsis*, the female becomes exceedingly aggressive when carrying pouch young, just as many placentals do when pregnant. Indeed, the first intimation that a litter is born may be violent threat on the part of a previously peaceful female, when the lid of her nest box is raised. Beach (1939) also noted that a female *Marmosa* with young would rear up and threaten at any disturbance.

The stage at which the young first emerge from the pouch differs from species to species. In the kangaroos and wallabies, where a single joey occupies the large pouch, emergence does not occur until the youngster is able to eat solid food; indeed, eating usually starts a little earlier, the joey leaning out of the pouch to nibble at the vegetation as the mother stoops to graze. In *Megaleia rufa* the youngster is about 190 days old when it makes its first expeditions into the outside world. The first trips last only a few minutes, the youngster stays close to its mother and soon climbs back into the pouch. The mother crouches down to assist, but the youngster climbs in on its own. Approaching from the front, it feels with head and paws along her ventral surface, working gradually back from her shoulders, until the paws grasp the rim of the pouch and the head is thrust inside. The youngster then dives in, head first, and executes a somersault combined with a sideways roll, so that it is back in the usual orientation, right way up and with its head protruding. The periods spent outside the pouch become gradually longer and finally, at the age of 235 days, the point of no return is reached. This is largely determined by the mother, who may refuse to stand for the youngster to climb in and may also by a sudden relaxation of the pouch musculature drop out one that has succeeded in getting inside. Although the mother's behaviour is thus the deciding factor in terminating pouch life, she herself must be responsive to changes in her offspring. Merchant and Sharman (1966) have made inter-specific interchanges of pouch young and have found that when a red kangaroo female fostered a young swamp wallaby, *Wallabia bicolor*, the final exclusion from the pouch did not occur until 267 days, which is far beyond the usual 235 days for her own offspring but normal for a swamp wallaby.

In *Megaleia*, if the female mated successfully at her previous post partum oestrus, she will be carrying a delayed blastocyst. This resumes its development before the current joey leaves the pouch permanently and is born very shortly after this occurs—usually within 24 hours. It is tempting to assume that the behavioural changes inolved in rejecting the older youngster are brought about by the hormonal events associated with impending parturition. Sharman and Calaby (1964), however, have shown that this cannot be the case, for the youngster was refused admission to the pouch at exactly the same stage by females that did not carry a delayed blastocyst and were not about to give birth. Sixteen young whose mothers promptly gave birth to another young one had a mean pouch life of $235{\cdot}5 \pm 2{\cdot}13$ days; while for eight, whose mothers did not do so, the figure was $235{\cdot}4 \pm 1{\cdot}56$ days.

The young-at-heel, although not entering the pouch again, remains with the mother and is still permitted to put its head inside the pouch and suckle. It continues to do this until it is approximately a year old. If the youngster becomes separated from its mother, it makes a special call: any adult female responds to this cry by calling in return and approaching. A mother whose youngster is missing will also actively seek for it, running to and fro, calling and investigating other young. Although apparently able to recognise their own young, the response of the female to foreign young is variable; some will readily accept a foster young-at-heel, others will refuse to do so. Sharman and Calaby do not say whether the youngsters show any differences in readiness to accept a foster mother. In their experiments with inter-specific transfers of pouch young, Merchant and Sharman (1966) found that the female red kangaroo would respond to the distress call of a fostered grey kangaroo or swamp wallaby, although their calls are very different from that of its own young. Since foreign young are easily substituted at any time during the period when the joey remains permanently in the pouch, it seems likely that the bond between mother and young is formed only when the latter starts to emerge from the pouch. During this period the female may have the ability to learn the characteristics of her own offspring's voice, apparently even when it belongs to another species.

In polytocous species, the position is different. The litter outgrows the pouch at an earlier stage and there are now two possibilities—either the young must be left in a protected nest of some description or the mother must continue to carry them about with her until they are able to fend for themselves. In American opossums the young are carried about, clinging firmly to the mother's fur. Hartman (1952) says that "all opossums carry their young on their backs from the time that they become too large for the pouch". This may be true for the genus *Didelphis* but, according to Beach (1939), the young of *Marmosa cinerea* are left behind in the nest once they cease to remain permanently on the teats. The female Beach studied showed retrieving behaviour and when the young were removed from the nest and scattered about the cage, she collected them all and returned them to the nest. This she did, not by picking them up in the mouth, but by pushing them under her with paws and snout: they at once cling to her hair and then locate a teat and grip it and are thus carried back to the nest. *Dasycercus* retrieves scattered young in exactly the same manner. Beach's *Marmosa* showed no ability to discriminate between her own and foreign young and made fruitless efforts to collect two new born rats.

In the dasyurid, *Sminthopsis crassicaudata*, the young, once they have outgrown the pouch, are left in the nest while the mother goes out to find food. At this stage in their development, the mother carries in grass, leaves etc. for bedding much more frequently than at other times.

Even before their eyes are open, the young *Sminthopsis* show defensive behaviour. If disturbed they will first threaten, hissing with open mouths, looking like a family of nestling birds gaping for food. If the disturbing factor does not promptly disappear, they will then scatter, leaving the nest cavity and burrowing into the bedding in all directions. Normally they rest in a compact group in contact with each other. The young remain in the nest for about 3 weeks after leaving the pouch. By then they are completely furred and their eyes have been open for at least 10 days. The first journeys outside the nest are very brief and are made quite independently of the mother. The youngster emerges from the nest and moves about cautiously for not more than a minute or two; it does not go more than a few inches away and then retires again to the nest. How many of these preliminary voyages of exploration are made by each youngster is not clear, but within a couple of days of the first emergence the young leave the nest, run about actively and will seize and eat suitable moving live food. They are also able to perform all the normal toilet patterns and frequently indulge in the characteristic sand rolling.

Immediately after emerging for the evening's activity, the young show no response to·the mother; they may meet her repeatedly but pay no attention and merely continue their activities of exploration and food seeking. After they have been active for some time, however, there is a change. If the mother happens to come close to a youngster it now suddenly runs to her, clings to her fur with all four paws and is carried back to the nest: sometimes a few more passengers may be picked up en route. The young are simply deposited in the nest; the mother comes out again at once and does not remain in the nest to feed them. The youngster, at least in the first days after leaving the nest, has no understanding of the significance of its response: it is not clinging to the mother because it is lost or tired and consciously wishes to get back to the nest. I have frequently seen the response shown by one who had just gone in and out of the nest box on its own and was sitting within an inch or two of the doorway. Indeed, I have several times seen one, sitting thus within easy reach of home, chase its mother hither and thither all round the enclosure, expending many times the energy required to walk home by itself. The response to the mother is shown on any particular evening only after the youngsters have been out for some time, but the nature of the factors controlling its appearance are not known: physical exhaustion, hunger or waning of exploratory drive are the most obvious possibilities. The response gradually becomes weaker and so does the attraction of the original nest. After a few nights the young cease to return to it and, if other refuges are available, they will start to adopt new homes. There is little tendency for the young to stay together; apart from occasional food thefts they are almost completely indifferent to each other and I never saw them play together. It seems clear that in natural conditions complete break up of the family would follow very shortly after the first emergence of the young from the nest. If a male is present, the break up may be accelerated, for the female comes on heat within a day or two of the first emergence of her young from the nest. If a male is encountered, she will spend the entire night, and probably the following night too, in copulation and the young will be left to fend for themselves.

In *Dasycercus* the young emerge from the nest at a rather earlier stage, before they are capable of eating solid food. At first the mother retrieves them by pushing them under her, when they cling to her belly and quickly find and grip a teat. A little later, they themselves jump onto her and cling to her back, like young *Sminthopsis*. In neither *Sminthopsis* nor *Dasycercus* have I seen any evidence of the mother assisting the young to find food, but such assistance does occur in the long-nosed bandicoot, *Perameles nasuta* (Stodart, 1966). In this species, when they first emerge from the nest the young follow the mother and forage with her, nosing about as she digs for insects and thus finding their first food.

EUTHERIA

In placental mammals it is not, as a rule, possible for the mother to carry the newborn young about with her in continuous contact with their source of food and warmth, as the female marsupial does. There are therefore two possibilities—either they must be born sufficiently advanced to be capable of independent locomotion and of keeping warm, or they must at first be protected in some type of nest or refuge. Broadly speaking, one may therefore consider the young as being either precocial or altricial, although within each category there is a considerable range of variation in the stage at which the young are born and no sharp dividing line can be drawn between the two categories.

In general, if the adult normally occupies a protected lair or burrow, the young are altricial; if there is no such home and the limbs are not adapted to constructing one to cope with the special needs of the young, then they must needs be precocial. This correlation is shown very clearly in the rodents. The young of burrowing species are very immature at birth; hairless, with eyes and ears non-functional and with very poor powers of thermo-regulation. On the other hand, the young of obligate surface dwellers, such as the guinea pig and the agouti, are extremely precocial. Similarly, amongst lagomorphs, the young of the burrowing rabbit are altricial, those of the hare precocial. Dieterlen (1963) may be consulted for a detailed summary of the variations in the developmental stage at which the young are born in the different families of the Rodentia.

In the Carnivora the young are generally cared for in some type of refuge and they too are altricial. They are, however, born slightly more advanced than is typical of burrowing rodents: although blind and with very feeble powers of locomotion, the young are furred at birth. The spotted hyaena is exceptional in that the young are born in a very advanced stage and are virtually precocial (Grimpe, 1923).

Amongst the precocial species, although born furred and with eyes and ears functional, there is variation in the locomotory ability of the newborn young. Some follow the mother by their own efforts almost at once, for example cetaceans and a few ungulates, whereas others at first spend much of their time resting in some spot where immobility and cryptic colouration reduce their chances of being discovered by a predator. Hares and many of the Cervidae and Bovidae fall into the latter category.

There is, of course, a negative correlation between litter size and the stage at which the young are born: large litters are characteristic of altricial species but in the precocial forms, there are commonly only one or two young at a birth. The correlation is, however, by no means absolute, since the developmental stage of the young is not the only factor involved in determining the litter size. Some altricial species produce relatively few young. In the giant rat *Cricetomys*, for instance, the normal litter number is four and the young are no more advanced than those of *Rattus*. Polytocous pigs and monotocous primates both give birth to semi-altricial young. The Cervidae are not all monotocous; in the roe deer, *Capreolus capreolus*, two or three young are commonly produced (Bubenik, 1965) and Frechkop (1955) cites a report of a Chinese water deer, *Hydropotes inermis*, carrying seven foetuses. Whether such large numbers of young are actually born or reared is not, however, known.

Behaviour related to the care of the young normally begins during pregnancy. In altricial species the nest is prepared in advance and nesting activities are usually intensified or initiated at this stage. Many species prepare a special brood nest. This is true, for instance, of the rabbit, where a nesting burrow is dug, lined first with grass or other similar material and finally with hair which the female plucks from her own chest and belly. Ross et al. (1963) have studied the hormonal control of nesting behaviour in the rabbit and conclude that the most important factor is the change in the progesterone to oestrogen ratio that occurs towards the end of pregnancy. *Tupaia* is unique in that a breeding nest in which the female will give birth to her young, is constructed entirely by the male, over a period of 1–5 days before parturition occurs (Martin, 1966). In this case the effect of the hormonal changes that occur during pregnancy must be indirect: most probably the male responds to changes in the female's smell, but it is not impossible that changes in her behaviour provide the relevant stimulus. In social burrow-dwelling species, where a nest is already to hand, the pregnant female frequently becomes more aggressive and adopts a special part of the burrow system as a breeding nest, from which she now excludes other members of the group. This is the case in such species as the rat and the prairie dog. While the nest building habits of small mammals are well known it is possibly a little surprising to find that this activity also occurs in the Suidae, as a prelude to parturition. The warthog normally sleeps in a burrow, frequently an abandoned aardvark hole. When about to give birth, the female prepares her refuge by bringing in grass or similar bedding material (Frädrich, 1965). Other wild species that do not use a burrow scrape out a hollow in the ground and fill it with nesting material, and domestic pigs behave in the same way (Hafez et al., 1962). Although the hippopotamus sometimes gives birth in shallow water, Verheyen (1954) regards this as abnormal. According to him, if she is undisturbed, the female always gives birth on land and prepares the chosen site by tearing up the vegetation and trampling it down to form a sort of nest.

In species that do not build a nest but take advantage of natural refuges, the pregnant female becomes increasingly interested in such places. She spends considerable time investigating them and resting in them, becoming gradually more attached to a particular one. This behaviour is familar in domestic cats, where the factors that are most important in deciding the suitability of a refuge are that it should be enclosed and protected and not brightly lit: a soft substratum is also preferred. Dryness is, presumably, also a desideratum. The lioness normally withdraws from the pride some time before parturition and seeks a place which offers suitable cover and is within easy reach of water and favoured hunting grounds.

Although in the Felidae the natural behaviour of the parturient female is to seek solitude, really tame domestic cats show an apparently contradictory type of behaviour. In cases where the bond between cat and owner is close, the female will refuse to remain in her selected "birth place" unless attended by her human companion and will emerge, miaouing piteously if left alone. Indeed I have often heard cat owners say that they were firmly convinced that their cat could delay its parturition, at least for some hours, until an absent owner returned.* I do not believe that this behaviour reflects a seeking for social contact but rather that it is related to the increased security felt in the presence of the human companion: what the cat is really seeking is the situation where anxiety is

* Koch's (1951) observations show that mares can exercise some control over the timing of their parturition and choose to give birth when external disturbances are minimal and Estes (1966) says that the same is true of wildebeest.

minimal. Certainly this seemed to be the explanation of the fact that my female *Suricata* was at first very unwilling to leave her newborn young in her sleeping box and insisted on carrying them inside my shirt—her place of maximum safety.

The pregnant female may also show characteristic changes in her toilet behaviour. She devotes more time to cleaning herself and directs her attention particularly to the mammary area. This type of behaviour, although extremely obvious in the domestic cat, is not considered in any detail by Schnierla et al. (1963) in their account of maternal behaviour in this species. The laboratory rat appears to be the only animal in which a quantitative study has been made (Rosenblatt and Lehrman, 1963).

These authors found that during the last week of pregnancy the time spent licking the nipple line was more than doubled and there was also a striking increase in the attention devoted to the anogenital region. There was some decrease in the time spent licking the non critical regions of fore paws, back and sides, but this did not compensate for the extra time spent on the mammary and genital areas and the total time devoted to the toilet increased. I know of no investigation of the function of this behaviour but it seems very likely that the licking may have some stimulatory effect on the mammary glands. Certainly it begins earlier and is carried out more persistently than could be required by hygienic considerations alone.

Schnierla et al. (1963) mention that the lioness, towards the end of her pregnancy is said to pull out the hair round her nipples with her incisor teeth. Presumably this is a response to a loosening of the hair in this region, comparable with what occurs in the rabbit over a larger area of the ventral surface. Its function may be to facilitate finding of the teats by the newborn cubs. It would be extremely interesting to know whether the same behaviour occurs in any other species and whether there are any grounds for supposing that the rabbit's behaviour may have been derived from a response of this kind.

In precocial ungulates, pre-parturitional behaviour is simpler and consists merely of leaving the group and seeking out a secluded spot. This behaviour is shown by domestic sheep and goats (Hirscher et al., 1963) as well as by wild ungulates. Even in the large herds of migratory wildebeest the female moves to the edge of the herd, or a little distance from it when about to give birth (Talbot and Talbot, 1963). Altmann (1963) notes that in the moose, *Alces alces*, and elk, *Cervus canadensis*, the female is still accompanied by her young of the previous year and when she withdraws in preparation for parturition she drives the juvenile away. The same behaviour is shown by roe and red deer (Bubenik, 1965) and by many bovids (Walther 1964b, 1965).

As parturition becomes imminent, there is a change in the attitude of the female; conspecific young now become attractive to her. Indeed, female sheep and goats (Hirscher et al., 1963) may attempt to drive off the true mother and "steal" the young of others that have given birth a little earlier. In the cat I have seen at this stage an extremely dramatic reversal of behaviour towards a kitten of the previous litter. As her pregnancy went on, the female had become increasingly hostile to him and finally he was greeted with a spit and a clout with the paw every time he appeared. An incident of this type occurred just before the cat retired to her usual kittening place. Half an hour later, the youngster chanced to approach her and was now licked and caressed: in the brief space of half an hour he had metamorphosed from an undesirable, to be driven away at all costs, into an object of extreme maternal solicitude.

Not very much is known about pre-parturitional behaviour in precocial rodents. Kunkel and Kunkel (1964) merely mention that although normally tolerant of strange youngsters, a female guinea pig in advanced pregnancy will drive them away. The only detailed study is that of Dieterlen (1962) on the spiny mouse, *Acomys cahirensis*. This species is unique amongst the Muridae in being precocial. There are only two or three young in a litter and they are born fully furred and with their eyes open. Unlike most other mammals, the female shows no tendency to go apart from her fellows when about to give birth. Just before parturition she becomes very eager to groom the young and will often steal a neighbour's pups, hold them in her paws and groom them. She may even suckle them before her own young are born. The urge to care for a pup is so strong that if none is available, she may attempt to groom an adult instead—but she treats it as though it were a pup, attempting to hold it in her paws as she grooms, a pattern never shown in ordinary social grooming between adults.

Although there is, for each species, a characteristic birth position normally adopted by the parturient female, individual exceptions are not infrequent. The typical posture is related both to the anatomical characters of the female and to the size of the foetus. In altricial species, where the young are small, the female usually crouches or curls round slightly, so that the vaginal opening is directed anteriorly and the foetus is expelled forwards under her own body. Where the foetus is large, the curving of the birth canal which this entails is usually not possible; in precocial species the female therefore commonly stands during parturition and the foetus is generally expelled backwards. This is true not only of ungulates but also of *Acomys*. There are, however, some exceptions to this generalisation; for example, both the guinea pig and the spotted hyaena give birth in a more or less sitting position and the precocial young are expelled forwards (Dieterlen, 1963).

The female frequently licks her vulva before the expulsion of the foetus and removes any birth fluids which ooze from the vaginal opening. Her activities are thus oriented to her own hinder end before the birth occurs and once the foetus is expelled, it is normally licked free of the embryonic membranes; birth fluids are licked up and the vaginal opening licked clean. The neonate is dried and cleaned by further licking and when the placenta emerges, it is usually eaten. The umbilical cord may either be eaten down to where it joins the umbilicus, as in most rodents and carnivores, or, as in many ungulates, a considerable length may be left attached, to dry up and break away within a day or two. As might be expected, the processes of licking the young and eating the placenta are omitted in many aquatic species but even in terrestrial forms they are not universal. In the great apes the placenta is frequently left unconsumed. Naaktgeboren and van Wagtendonk (1966) give instances of this in captive gorillas, chimpanzees and orang-utans. They also quote Manley as relating how a primiparous chimpanzee in the London zoo did not eat the placenta but carried it around, together with her infant, for the first 24 hours, by which time it was fly blown and stinking. The cord ultimately broke but how this happened was not seen. Goodall (verbal communication, 9th Ethological Congress) has seen exactly the same behaviour in the wild. On the other hand, a chimpanzee that gave birth to her first young one in the Berlin zoo did eat the placenta (Heinroth-Berger, 1965).

In the Camelidae maternal attention immediately following birth is very poorly developed. The embryonic membranes rupture as the young is born and it frees itself from them by its own efforts. According to Pilters (1956), the mother subsequently bites through the cord and licks the foal dry, although the placenta is not eaten. Koford (1957), however, says that in the vicuña there is no general cleaning of the young, merely a little nuzzling around its muzzle. This appears to be a primitive state of affairs, for in the Cervidae and Bovidae the placenta is eaten, every trace of membranes and birth fluids meticulously cleaned up and the calf thoroughly licked.*

Eating of the placenta bears no relation to the normal diet and even the most exclusive herbivores show an avidity for it. This unusual appetite is normally of brief duration: for example, guinea pigs presented with a fresh placenta 12 hours after they had themselves given birth refused to eat it. This suggests that the hormonal changes associated with parturition are involved. This seems to be true also of carnivorous species, for hypophysectomised cats do not eat the placenta (Lehrman, 1961). In this respect, as in so many others, *Acomys* is exceptional. Any adult female, at any stage in her sexual cycle, treats a newborn pup exactly as though she had just given birth to it herself; she will clean off the membranes, bite through the cord, lick the pup clean and eat the placenta.

Since the placenta contains lactogogic hormones, its consumption by the female may assist in establishing lactation. Its direct nutritional value may be regarded as incidental but

* Estes (1966), in his observations on the blue wildebeest, found that jackals or vultures often snatched away the placenta and devoured it before the mother had time to eat it herself but there was no suggestion that, given the opportunity, she would not eat it in the normal way.

may not be without some importance, especially in predators. Certainly the amount of meat consumed is sufficient to make it unnecessary for the female to leave her young during the first 24 hours in order to hunt for food.

The licking and cleaning of the young that follows directly upon birth has a number of different functions. The most simple and obvious is that if the membranes do not rupture at birth, the head of the foetus must be freed from them to permit it to draw its first breath of air. As Barcroft (1946) has shown, mechanical stimulation of the muzzle region (in fact, of the receptive field of the trigeminal nerve) has a strong excitatory effect on the respiratory centre so that, as she frees the head from the membranes, the mother also administers stimuli which assist in initiating breathing. It can hardly be just a happy chance that this should be so; it is much more likely that the high sensitivity of the respiratory centre to this particular type of cutaneous afferent inflow is a product of selection.

Where the young are born in a nest, the sanitary advantages of a thorough post-partum clean up are obvious. Many of the artiodactyls that give birth in the open are, however, just as meticulous in this respect as any nest dweller. Possibly the smell of the birth fluids might attract predators and their removal may therefore assist in protecting the young.

Drying off the young is clearly advantageous in reducing the danger of chilling, particularly in species born in the open. The stimulation provided by the licking, however, seems to have further effects; possibly it assists in establishing the peripheral circulation. At any rate, lambs not licked have a higher infantile mortality than ones that receive the normal treatment (Alexander and Williams 1966; Hirscher et al., 1963).

Particular attention is always given to licking the anogenital region of the offspring and this stimulates the first defaecation and micturition. This is true even in the Camelidae (Pilters, 1956), where post partum licking is minimal. In those Bovidae and Cervidae where birth is followed by a lying out period (*vide infra*) micturition and defaecation have to be stimulated in this way for several days and the mother consumes the products. This, as Walther (1964b) suggests, is probably a defence against predators and prevents the calf advertising its hiding place by the smell of its urine and faeces. The guinea pig also deals with the faeces of the offspring in this way for the first few days (Kunkel and Kunkel, 1964). In altricial species anogenital stimulation plays a more important role and without it the young are incapable of micturition and defaecation for a longer period, in fact, until they reach the stage when they are ready to leave the nest. Here nest sanitation alone makes this arrangement desirable, quite apart from any question of the responses of predators, but in some of the Canidae, including the domestic dog, the female continues to consume the urine and faeces of the pups for some time outside the nest. This appears to have the function of concealment of the den by preventing the accumulation of juvenile droppings in its immediate vicinity.

In precocial species, the parturitional licking of the young has another function. By orienting the mother to her offspring, loss of contact between the two is prevented and the first steps towards the formation of the bond between mother and young are taken. The degree to which maternal responses become attached specifically to the female's own young varies from species to species. In general, precocial species will mother only their own offspring and reject foreign young, whereas altricial species are much more ready to adopt foster young. This is correlated with the degree of danger of confusion, which is very slight if the movements of the litter are restricted within the confines of the home nest. The female has no need to recognise her own young individually, so long as she responds to her own nest. In a social species with mobile young, however, total confusion would result if there did not exist some means of linking the mother specifically with her own offspring. Recognition of own young and rejection of strangers is characteristic of the large ungulates but it also occurs in Pinnipedia where the crowded conditions of the rookery make it easy for the young to stray from their own homes into a neighbour's domain. Bubenik (1965) believes that the number of young is also a factor of importance in determining how specific is the attachment of the female to her own offspring and notes that in the roe deer, where two and three young are normal, exchanges could easily be made during the first 3 weeks, but in the monotocous red deer an exchange at the age of 3 days was impossible.

The development of the female's attachment to her own offspring has been most intensively studied in domestic sheep and goats. Her own smell on the foetal membranes does not appear to play any essential role: if a substitution is made immediately following birth, the mother accepts the foreign youngster as though it were her own. A ewe remains ready to accept a lamb for a period of at least 8 hours after parturition and if her lamb is removed at birth and returned 8 hours later, it is still accepted (Smith et al., 1966). In goats the female's receptiveness is less lasting and kids may be rejected after a separation of as little as 1 hour (Klopfer et al., 1964). Using ewes that gave birth virtually simultaneously, Smith and his co-workers were able to study the development of the attachment of the mother to her young. If the female had been allowed to lick her lamb for a period of no more than 20 minutes, a substituted foreign lamb would still be accepted, but after longer periods the strange lamb was rejected. The first half hour or so of maternal care thus appears to be required for the attachment to be formed. In goats, Klopfer et al. found the process was much more rapid—5–10 minutes of nuzzling the kid sufficed.

The primary factor involved in subsequent recognition of the young appears to be its olfactory characteristics. This is reflected in the various expedients commonly adopted to induce a female to adopt a foreign youngster—rubbing the foster young with the mother's birth fluids, with the skin of her own lamb or even against her own body. Later on, recognition from a distance by vocal or visual signals may be added but these are confirmed by an olfactory check up once contact is made. This is true in Pinnipedia as well as in bovids and cervids.

It is, in general, true that recognition of own and rejection of foreign young has the function of ensuring that confusion of young does not occur and the ability to behave in this way is characteristic of species in which such confusion is a genuine hazard. There are, however, special circumstances which produce an exception to this rule. If the young are all very much of an age, if they remain close together and if the females' feeding times are approximately synchronised, then complete promiscuity would be just as effective as individual recognition—probably even more effective, as no time need be wasted by the mother in attempting to locate her own youngster in particularly difficult conditions. If, in addition, the mother's responsibility begins and ends with feeding her young and its safety does not depend on her vigilance, then all need for a special bond between the two vanishes.

This is exactly the situation in the Mexican guano bat, *Tadarida brasiliensis*. The colonies studied by Davis et al. (1962) had their summer roosts in caves, where the young were all born within a brief period around mid-June. Davis and his co-workers found that the females do not carry their young with them when they fly out in the evening to feed but deposit them in special nursery areas, where they remain clinging to the cave roof. On their return, the females fly at once to the nursery where they are promptly set upon by every youngster within reach and each feeds the first two that succeed in grasping her teats. The females thus form a sort of communal dairy herd. A few weaklings may get pushed aside and die but since each female has only one young but can suckle a pair, the majority get fed and the system apparently operates very successfully.

Our knowledge of the role played by hormones in maternal behaviour is still rather fragmentary but enough is known to make it clear that the degree to which they control the female's maternal responses varies considerably from species to species. In general, if the mother becomes free of the young of the previous litter only shortly before parturition, then her behaviour towards young shows a sharp alteration at or around the time of parturition. If, in addition, the female's responses become specifically linked to her own offspring then this carries with it the implication that her general responsiveness to young is of brief duration and is soon replaced by hostility to any but her own offspring. This has been shown to be the case in sheep and goats and is probably true of many other ungulates. It has been clearly established that the female's readiness to accept a lamb decreases rapidly shortly after parturition (Hirscher et al., 1963; Smith et al., 1966), but how much her responses to her own accepted youngster are hormonally controlled is not known. In order to find this out, it would be necessary to allow the attachment to be formed and then study the effects of altering the mother's hormonal condition.

No systematic studies have been made in carnivores but it is well known that cats and dogs can be induced to accept foreign youngsters, sometimes even those of a different species. In zoos, if their own mother's behaviour proves inadequate, cubs of the larger felids are sometimes fostered on a domestic bitch. A more surprising case is the rearing by a domestic cat, belonging to Dr. C. K. Brain, of a pair of young *Cricetomys*, together with her own kittens. Dr. Brain attributed the readiness with which the cat accepted the young rats to the fact that they were given to her immediately after the birth of her own litter (personal communication). In view of the dramatic reversal at parturition which has already been described in the response of a cat to an older kitten, it seems very likely that this is correct and that, as in the ewe, the readiness of the mother to accept young is maximal at this stage.

In species where maternal responses are not strictly limited to the animal's own young, they may also be largely independent of hormonal control. This applies both to solitary and to social species. In many of the primates, for example, both juvenile and adult females, regardless of their hormonal status, will show maternal responses to a newborn infant. Similarly in the solitary golden hamster, virgin females show some maternal responses to young aged 6 to 10 days (Richards, 1966a, b). An extreme case is shown by laboratory mice, where Noirot (1964) has found that virgin females show slightly more maternal responses to pups than do lactating females. It would be extremely interesting to know whether wild mice would behave in the same way, or whether this represents a behavioural anomaly produced by selection in the abnormal captivity situation.

Although it is convenient to speak of the female's responses to a pup as though they were something quite fixed and definite, Rowell (1961) points out that this is an oversimplification. In the golden hamster, she found that the behaviour of females was influenced not only by their own reproductive condition and immediate previous history but also by whether they had been kept singly or in groups and also by the circumstances in which the young were encountered. If the test pups were placed in the nest, they were more likely to evoke maternal responses than if the female encountered them elsewhere. With *Cricetomys* I have found the reverse effect in the response to the young of a foreign species (a striped mouse, *Lemniscomys striatus*): placed just outside the nest, the foreign pup was picked up and retrieved but, once inside the nest, it was promptly attacked.

This last incident illustrates a point that has already been mentioned in Chapter 8—that "individual recognition" need not, in other species, have the same conceptual basis as it has in ourselves. This is particularly obvious in the responses of the mother to the young. We regard the youngster as an entity; it is a baby rat, mouse, kitten or the like. For the mother, however, it may be much truer to say that it is a collection of releasers and it is therefore not surprising that there may be circumstances in which contradictory responses to different aspects of the youngster are possible. The young striped mouse presumably provided enough of the stimuli usually given by a displaced youngster to evoke the retrieving response, possibly by emitting an ultrasonic cry, but once inside the nest its strangeness became its most dominant characteristic and it now evoked attack. Rowell (1961) has shown that in the golden hamster even a conspecific pup can elicit from an adult not only responses which are appropriate to young but also behaviour appropriate to prey or even to a sex partner. Similarly I have seen a domestic cat show ambivalent behaviour towards a strange kitten, her response depending on which end of the youngster she contacted. She happened to approach it first from the rear, sniffed its ano-genital region and at once started toilet licking; the kitten turned round and, on smelling its face, the female hissed and struck at it with her paws. She alternated in this way two or three times before the kitten finally made off. Clearly for the cat, one end of the kitten said "youngster-to-be-cleaned", while the other bore the message "stranger-to-be-driven-away". A female *Cricetomys*, who had only a single pup of her own, showed very similar responses. When presented with a strange conspecific pup, she at first sniffed its face suspiciously and made intention movements of attack, but when the ano-genital region was presented to her, she at once started to lick and permitted the pup to suckle. Two days later, a second pup was successfully added to the litter in exactly the same way.

So far we have dealt with what is only one side of an interaction: the behaviour of the mother and her responses to her young. No less essential, however, are the special responses of the young towards the mother. Even the most altricial enter the outside world equipped with all the patterns necessary to ensure their establishment in their new environment. The most striking example of this has already been described—the ability of the "embryonic" newborn marsupial to find its way to the maternal milk supply. For the newborn placental too, this is the first problem that has to be solved and all are equipped with searching behaviour enabling them to find a teat, together with an appropriate response to it when they meet it. In altricial species, the teat-finding behaviour is in fact very similar to that of the newborn kangaroo, except that the path followed is less direct. Almost as soon as it is free of the membranes, the neonate moves about in contact with the surface of its mother's body, propelled mainly by its fore limbs. As it moves it makes side to side head movements, very like those of the newborn marsupial, until the muzzle comes in contact with a teat, or the hairless area round it, and almost at once the teat is taken in the mouth and sucking starts. McBride (1963) has shown that in pigs the search is not random: the piglets move with the lie of the mother's hair. On her belly the bristles are so arranged that this response keeps the young in the region of the udder and so facilitates teat location. The mother usually assists, at least to the extent of adopting an appropriate nursing position and in many species will pull towards her a youngster that has lost contact with her body.

In the larger ungulates, suckling occurs with both mother and young standing up. Once it has got to its feet, the youngster starts to nuzzle against the mother's body in search of a teat.

The first orientation is normally towards her axilla or neck, but gradually the youngster works its way back, until the udder is reached. Frequently, but not invariably, the female assists by nudging her offspring towards the udder and, indeed, her licking of its anogenital region usually has the effect of turning it in the right direction.

Apart from taking hold of the teat and sucking, the newborn mammal frequently has some type of behaviour which stimulates milk flow. In the species that suckle standing up, this takes the form of butting with the head against the udder and piglets use a massaging movement of the snout. In altricial species alternating movements of the fore paws are used—the "milk tread". According to Eibl-Eibesfeldt (1963) the milk tread is shown by newborn rats and hamsters but in kittens I have found that a coordinated tread usually takes a few days to develop. At first the kittens use the paws to steady themselves and to push back the mother's skin on either side of the teat, so that it is more conveniently accessible, but the full pattern of alternate treading does not at once appear. This does not necessarily mean that the full pattern has not yet matured or that it has to be learnt. It may merely be that the tread is not evoked so long as the milk comes as fast as the kitten can cope with it. In this connection, the guinea pig is of some interest. Born at an advanced stage, the youngster normally supports itself on all four limbs as it sucks and does not show the milk tread. Eibl-Eibesfeldt (1951a), however, reports that if a youngster is born prematurely, the milk tread is present and it can also be evoked from the foetus *in utero*.

Once it has been cleaned, has found the teat and has started its first feed, the newborn youngster has entered upon the first stage of its independent existence: the suckling period. The characteristics of this period are, of course, very different in altricial and precocial species. It is convenient to deal first with the latter, amongst which the large ungulates have received the most intensive study. In both the Bovidae and the Cervidae, there are many species in which the calf, although capable of locomotion, does not at once follow the mother. Instead there is a "lying out" period, during which the calf remains in some concealed spot and the mother periodically visits it to feed it. Some authors (Walther, 1965) have attempted to make a clear distinction between those species that do and those that do not have a lying out period. In practice such a division proves difficult and, as Hirscher et al. (1963) note, opinions differ as to how wild sheep and goats should be classified. Instead of expecting the animals to sort themselves neatly into one or other of two pigeon-holes, it is more profitable to adopt a different approach. The suckling young has two basic responses; to lie down and sleep, once its stomach has been filled, and to follow its mother. According to the circumstances, it has proved advantageous to emphasise one of these responses at the expense of the other. This has led, on the one hand, to the evolution of a lying out period and all the behaviour associated with it, the tendency of the youngster to follow its mother being suppressed at first. Following the other possible pathway has resulted in restriction of the resting tendency of the young; in some cases it is represented by no more than a sleep after taking the first feed, which may be regarded as a vestigeal lying out period, the youngster thereafter accompanying its mother.

The factors which favour the evolution of a long lying out period are firstly, that the habitat should provide suitable cover; small bushes or, at the very least long grass. Secondly, if the species is not solitary, the social organisation must be such that the mother will have no difficulty in rejoining her group with her calf at the end of the lying out period. The converse factors, a habitat deficient in cover and the absence of a stable home range or territory, work in the opposite direction and favour reduction of the lying out period and early appearance of following behaviour. In view of these considerations, it is not surprising to find that, in addition to species with a well-defined lying out period (like the tragelaphines and gazelles—Walther, 1964a, b, 1965) and others in which it is vestigeal (like wildebeest and domestic sheep), there should exist yet others where the behaviour is intermediate and which are therefore more difficult to categorise.

Although it is convenient to speak of a lying out period, this does not mean that there is an abrupt change over from lying out to accompanying the mother. Instead, the calf begins to spend a longer time with the mother after she has fed him, although he still rests on his own a little later. Gradually these periods spent apart become fewer and finally he is accompanying his mother all the time.

Grant's gazelle, studied by Walther (1965), will serve to exemplify the type of behaviour shown in species with a well defined lying out period. The females leave the male's herd to give birth and thereafter they form a group of their own, somewhat peripheral to their male's group, while they care for their calves during the first couple of months of their lives. After the first feed, the calf itself moves off and seeks a sheltered spot, under a bush or in long grass, and here it lies down and sleeps. One particular calf which Walther watched in detail always chose a spot on the upper part of a certain hillock. Although a single refuge might be used repeatedly on different occasions, the calf never returned after a feed to the place it had just vacated. The areas utilised by different calves overlapped. The choice of a lying out place is made by the calf itself and it is not a matter of the mother concealing her offspring. It is, however, the mother who determines the feeding times and she summons her calf when she is ready to suckle him. She moves over towards his lying out place and circles about, looking towards him and calling and finally stands slightly sideways on, not "pointing" directly towards him. The calf then rises to his feet and if he does not at once see her, the female attracts his attention by nodding her head up and down. The calf then approaches and after mutual nose sniffing, the female licks his anogenital region and the calf seeks the udder and sucks. The two stay together for some time; there are several bouts of suckling and grooming of the youngster and the two move about together for a while. If the calf is left behind he runs after the mother, overtakes and passes her and then waits to be overtaken in his turn. Finally, the youngster moves off and again lies down in one of his refuges. He is carefully watched by the mother as he takes up his position. Although she then moves off to graze, the female remains on the alert and is ready to defend the calf, if danger threatens.

Walther was able to see what happened on the appearance of two different predators—jackals and spotted hyaenas. When a jackal came to within about 100 metres of her calf's refuge, the female went straight towards the intruder and drove him away. The jackal always made off without putting up any resistance. A hyaena evoked a response at a much greater distance, 500 to 800 metres. On sighting one, the female at once hurried towards her calf and ran by close beside him. He responded by jumping up and following and was thus led round to the side of the hillock away from the hyaena. The female returned alone and kept a watchful eye on the predator as long as it remained in the vicinity. The opportunity to see what would happen if the hyaena actually located and attacked a calf did not arise.

As the calf grows older, it spends more time with its mother along with the other members of the female herd and finally, when it is about 2 months old, the female and offspring leave the female herd and return to the male's group.

Very similar lying out behaviour is shown in the Cervidae. In red deer, for instance, the female and calf remain apart for 3–4 weeks before rejoining the herd (Bubenik, 1965). The calf, after feeding, goes off and lies down and the mother remains nearby, taking up her position down wind from its refuge. If alarmed, the calf gives a distress call, to which the female at once responds. She will attack or threaten a human intruder and an experienced animal will attack a dog on sight, without waiting for her offspring's distress call. Her response is not specific to the voice of her own calf and she will alert to the cry of any youngster. It is, however, governed by her own reproductive condition and a female just before parturition is unresponsive. In the roe deer, twins or triplets avoid each other and each chooses a separate lying out place.

With this method of caring for the young, one may contrast the situation in the blue wildebeest—a species that is typically highly mobile and frequents very open areas lacking in cover suitable for concealment of the calf. According to Talbot and Talbot (1963) the calf rises to its feet within 5 minutes of birth, follows the mother within another 5 minutes and is able to move as fast as an adult within 24 hours. Here the separation of the mother and young from the herd is minimal, both in space and time. The parturient female merely moves to the edge of the herd or slightly away from it, and rejoins it with her calf within 24 hours of its birth. Even in groups of wildebeest living in the non-migratory territorial manner described by Estes (1966), there is very little change in the relationships of parent and offspring. Although the females with young calves tend to remain together and form a group of their own, there is no lying out period and the calf, very shortly after its birth, follows the mother to join the group to which she belongs. In domestic sheep, the lamb at first normally lies down and sleeps by itself after feeding, but its tendency to remain apart until the ewe returns is not very strong and the lamb may wander off on its own when it wakes (Hirscher et al., 1963). In sheep suckling occurs very frequently; about once every 10–15 minutes in the mouflon, according to Pfeffer (1967), and once every three quarters of an hour in domestic sheep (Ewbank, 1967). This constant renewal of contact between the pair has the effect of preventing the lamb from straying far from its mother and so minimises the danger of its getting lost. In the Camargue wild cattle, Schloeth (1961) found that the calves lie out for a period of 3 or 4 days and then join the herd with the mother.

The pronounced lying out period, protected not only by concealment but also by maternal vigilance, characteristic of Grant's gazelle, appears to be a very successful method of rearing the young and Walther never saw a calf taken by a predator. This, however, should not be taken to imply that lying out is a more efficient type of juvenile behaviour than following. The circumstances of habitat and social organisation already referred to must affect the issue and correlated with variations in these factors must go alterations in the relative efficiency of the two types of behaviour. Absence of lying out in the wildebeest is an adaptation to a migratory existence, in which lying out would be impossible and if in Ngorongoro the calves fall prey to crocutas more readily than do those of other species that lie out, this means only that here the wildebeest are existing in a habitat differing from the one to which they are most fully adapted.

In the precocial ungulates, the defence of the young is, as a rule, the mother's responsibility and the calf's safety depends not on its own strength but on its ability to respond to the mother's alarm signals. The same is true of the Pinnipedia.

Paulian (1964) describes how, if threatened, the female *Arctocephalus* will pick up her pup by the neck and carry it into a cave or rock cleft, where it remains until the danger is past. Standing at the entrance, the mother faces the enemy and will fight, to the death if need be, in defence of her young. In other precocial species, the youngster may play a more active part in its own defence and its defensive patterns may appear very early. This is true of both porcupines and hedgehogs. Porcupines (*Hystrix*) during their first few days of life already show the full adult threat repertoire. They will erect the quills, turn their backs towards an enemy, rattle the tail, stamp with the hind foot and grunt (Roth, 1964). Similarly, within 24 hours of birth, although their eyes are still shut, young hedgehogs will erect the spines, turn the back towards a disturber and snort (Dimelow, 1963).

A curious feature of the maturation of precocial ungulates is the length of the lactation period. Although the young start to eat solid food at an early age, they continue to suckle, so that there is a long period during which the calf is receiving milk, as well as eating on his own. In wildebeest, for instance, suckling continues until just before the birth of the next calf (i.e for almost a year) although the young start to eat grass when only a few days old. Complex though ruminant digestion may be, it is difficult to believe that it requires almost a year to become fully established. In this connection, it is of interest to note that young ungulates commonly eat earth at the stage when they are starting on solid foods (Bubenik, 1965), a habit which probably relates to the establishment of the normal gut flora. Information about the development of digestive efficiency would be extremely interesting, but whatever may be the situation, the behavioural consequences of the slow transition to fully independent feeding is that, despite the advanced stage at which they are born, the young remain associated with their mother over a long period. The primary function of this association may be to keep the youngster under the mother's protective vigilant care, rather than to supply it with food. There are, however, other important consequences. Most observers are of the opinion that the formation of large social groups is based on family relationships and the tendency of the young to remain associated with the mother, even after they have attained full physical independence. It may therefore be that the psychological effects of the prolonged lactation are just as important as the physiological ones, if not more so, and the prolonged suckling period may have been evolved under selective pressures which have little direct connection with nutrition.

Altricial young are not at first capable of maintaining their body temperature at the normal level and are dependent on their mother for warmth as well as for food. It is usual for the mother, during the first day, to spend most of her time curled round her young in the nest, so that even when they are not suckling, they are in contact with her body. Subsequently the time spent with the young gradually declines. This is true for rodents, including both laboratory rats and wild species such as *Peromyscus* (King, 1963), and also for carnivores such as cats (Rosenblatt and Lehrman, 1963) and dogs (Rheingold, 1963). I have also found the same in meerkats. Even during the early period, however, the mother must leave the nest now and then and the young have a response which serves to prevent their becoming chilled in her absence. This consists of a tendency to come to rest only when in contact with a littermate's body. The result is that the young huddle together and thus, by reducing the total surface exposed, they minimise heat loss. This behaviour also has the effect of reducing the likelihood of straying out of the nest. In meerkats the contact response of the young is very strong and each individual attempts to take up a position with his head laid over the body of a fellow. In a litter of three young this often resulted in an extremely symmetrical arrangement in which each kitten had its head over one littermate's neck, his own neck in turn being similarly used by the third. The contact response is dependent on temperature and is reversed if the ambient temperature rises. On a really hot day a litter of kittens will be found lying as far apart as their nest box allows and avoiding contact with each other, instead of forming the usual compact mass, one on top of another. Goethe (1940) reports the same behaviour in relation to temperature in a litter of polecats. In addition, in very hot weather the mother would wet her belly in her water dish, then curl round her young until she had damped them and thus she assisted them to keep cool.

The result of the normal contact behaviour is that even where the individual youngster may be incapable of maintaining its body temperature, the litter as a group can thermoregulate. This makes possible a surprising reversal of the usual maternal behaviour which occurs in some species. The mother's presence in the nest is necessary for the welfare of her young but her comings and goings, by making its location more obvious to predators, may be a source of danger to them. There may therefore be some advantage in her not visiting the nest frequently. We have already seen that once her youngster is out of her pouch, the female spiny anteater returns to feed it only at long intervals.

A similar pattern of nursing behaviour is known in two placental orders. The rabbit comes to the nest to feed and groom her young only once in 24 hours (Venge, 1963; Zarrow et al., 1965). Field observations by Ingles (1941) indicate that a similar "absentee system" is adopted by the cottontail. It is not surprising that in the draught free, fur lined nest, the naked young do not become chilled; but it is surprising that the stomachs of the young and the mammary glands of the mother should be capable of receiving and delivering a sufficient supply of milk to last so long. In *Tupaia* (Martin, 1966) the process has gone even further and the female goes to the brood nest to feed her young only once in 48 hours. At the first feed the newborn tree shrew increases its weight from approximately 9 to 15 grams. Its stomach is thus capable of accommodating two thirds of its body weight of food—a truly gargantuan feat. In this case the leaf nest is less protected than the rabbit's burrow and the young are highly peculiar in that, although they remain within the nest during the suckling period, they are precocial. They are born with their eyes open, furred and even a solitary youngster can maintain its body temperature. The young are also capable of grooming themselves from birth and Martin is of the opinion that in her short visits, which last only 5 minutes, the mother merely feeds the young but does not groom them. Apparently she does not even carry out toilet licking to remove urine and faeces for the nest becomes soiled with these products (Martin, personal communication).

The length of the intervals between successive sucklings in the rabbit and tree shrew is in striking contrast to their brevity in the mouflon. That frequency of suckling should be adaptively related to mode of life is not in itself surprising but the magnitude of the disparity shown—from 15 minutes at one extreme to 48 hours at the other—is impressive testimony alike to the powers of natural selection and to the adaptability of mammalian digestive and lactational physiology.

In addition to their feeding and thermoregulatory behaviour, altricial young usually have a number of protective or defensive responses. Young felids and viverrids hiss or spit if disturbed and young *Tupaia* give an explosive snort and also thrust out all four limbs in a way that makes the leaves of the nest rustle. Martin (1966) says that the net effect of this is "quite startling" and certainly the spitting of a young meerkat causes a quite involuntary withdrawal on the part of a disturbing human. We have, however, no information about the effects of these responses on natural enemies.

Most young rodents have a distress squeak, audible, ultrasonic or both, emitted if they are in difficulties—the commonest circumstances evoking it are being accidentally sat on by the mother or being out of the nest. The mother normally responds by approaching and, if the young is out of the nest, carrying it in again. Both rodents and carnivores carry their young in their mouths, but the grips used are not identical. The Carnivora usually carry the young by the nape of the neck but in rodents the orientation is usually less precise and the pup may be taken hold of by the skin almost anywhere between shoulder and flank. The shrew also retrieves young in a similar way, picking them up by whatever part happens to be convenient. In most rodents, however, the grip is normally made towards the dorsal surface. This is recorded in rats, mice, and hamsters (Eibl-Eibesfeldt, 1958) in the dormouse (Koenig, 1960), the field mouse, *Microtus*, the snow mouse, *Chionomys*, (Curio, 1955), and the kangaroo rat, *Dipodomys nitratoides*, (Culbertson, 1946). In *Sciurus*, however, the grip is more ventral and the youngster does not just remain passive but curls its body round and clings to the mother's head (Lang, 1925). A ventral grip is also used by *Meriones* (Eibl-Eibesfeldt, 1951a) and in the muskrat (Curio, 1955); *Apodemus* is unusual amongst the Murinae in using the belly grip. Although the grip is imprecise in its orientation, Carlier and Noirot (1965) have found that as rats become experienced in transporting their young, they tend more and more to take hold near the centre of the body, so that the load is evenly balanced. According to Purohit et al. (1966) the palm-squirrel, *Funambulus pennanti*, uses the neck grip.

Although a cat normally picks up a kitten by the neck, Leyhausen (1956) notes that in a very young mother this orientation may be lacking and the kittens may be picked up more or less anyhow. Here, in contrast to the rats, it appears to be a question of the maturation of a response, rather than of learning. The difference shown by the two female *Suricata* I have studied may reflect the same thing. One had her first litter very early, the other was considerably older when she first reproduced. The latter picked up her kittens by the neck, the former more or less anywhere. Unfortunately I was not able to find out whether the oriented neck hold made its appearance with the younger animal's second litter.

If the mother stands up and moves off while the young are still suckling, it is common for them to be dragged a little way, still clinging to the teats. In some small rodents, if there is a sudden alarm, the female may tow the whole litter with her in this way and in a few cases this is the normal method of carrying the young (Gander, 1929). In the wood rat, *Neotoma fuscipes*, the incisors of the young are specially modified as gripping organs, in adaptation to this means of transport.

The young of a number of species have the ability to retreat from a sudden drop or declivity. Young rats and also young *Cricetomys* will avoid a "cliff" but young *Xerus* show no such response, nor do young *Suricata*. If placed on a table or bench the latter two will promptly walk over the edge and fall to the floor, whereas young *Cricetomys* will retreat from the edge. Since the nests in which they live are all very similar, located in a burrow where there cannot be any danger of falling, the significance of this response is not at once obvious nor is it apparent why it should be present in some species and absent in others. It is, however, common to find that responses mature in advance of the time they are required. We do not, for instance, question the significance of toilet behaviour because it starts at a stage when the youngster is still adequately cleaned by its mother. Young *Rattus* and *Cricetomys* will grow up to be climbers and their world will be a three dimensional one, whereas *Xerus* and *Suricata* are essentially ground dwellers and rarely climb. Caution about descent from heights is therefore of importance to the former but has little relevance to the latter two species. The responses of the young may therefore be related to the difference in their subsequent development, rather than to their immediate needs. It would be of interest to know whether, if information about a wider range of species were available, the correlation between adult climbing ability and juvenile edge response would be maintained.

In many species, either while they are still in the nest or when they reach the stage of starting to explore the world outside, a flight response makes its appearance. In the common shrew this happens at the age of 16–17 days, shortly before the eyes open (Crowcroft, 1957). Previously the young have remained quiet if disturbed in the nest but now they suddenly erupt from it and scatter in all directions. Rongstad (1966) finds that from the age of 10 days, young cottontail rabbits will respond in the same way to the distress squeal of a littermate. Young kittens, when they have reached the stage of leaving the nest, will scatter and take cover each in a different direction, vanishing as though by magic, in response to a growl from the mother, and a similar response is recorded in kittens of the bobcat, *Lynx rufus* (Gashwiler et al., 1961). Piglets, from the age of a week, show a similar response to the sow's alarm bark (McBride, 1963). Young *Xerus* avoid open spaces and will instantly run under the nearest cover if placed in the open. Although *Suricata* are animals of the open and the young show no such avoidance of exposed places, at a sudden alarm they too will flatten themselves under the nearest available cover.

During the period within the nest, there is gradual maturation of sensory and motor abilities. Patterns connected with toilet behaviour make their appearance and play with littermates may begin. The young also become much more adept at finding the teats and, once the eyes are open, soon orient themselves visually to the mother and cease to use the side-to-side searching pattern. In kittens, Schnierla and Rosenblatt (1963) have shown that, although it is superseded by the new visual orientation, the neural basis of the old pattern remains and kittens will revert to it if their eyes are covered.

In a few cases it has been found that a definite teat order is developed: each youngster establishes ownership of one particular teat and soon feeds virtually exclusively from this. This has been found to be true of kittens (Ewer, 1959) and of piglets (Donald, 1937; McBride, 1963). In both cases olfactory stimuli seem to be of importance in teat recognition. In kittens this is suggested by the fact that washing of the mother's ventral surface causes a temporary loss in the kitten's ability to locate its own teat (Ewer, 1961) and, more recently, Kovach and Kling (1967) have shown that kittens rendered anosmic by destruction of the olfactory bulbs are unable to locate or respond to a teat. In pigs, McBride has shown that vision and recognition of neighbours also play a part. The kittens' claws and the piglets' teeth are weapons which could do serious damage if there were prolonged fighting over teats and the main function of ownership appears to be to prevent this. In the social pigs it may also have secondary effects on the later development of rank order and behaviour involved in dominance relations. Although the prevention of injuries may have been the selective pressure responsible for perfecting teat ownership in these two cases, there are species in which this force does not appear to operate and which nevertheless show some degree of teat ownership. Ewbank (1964), for instance, has found that in twin lambs there is a marked tendency for each to use the same teat at successive feeds. Ewen (1956) noted that when hand rearing grysbok* twins, it was necessary to have a separate feeding bottle for each as neither would suck from the teat that "belonged" to the other.

* Species not given, but almost certainly the Cape grysbok, *Raphicerus melanotis*.

Birth and the transition to an extra-uterine environment is the first great crisis in the life of a mammal. For altricial species there is a second crisis, but it is a less dramatic and less sudden one. The youngster must emerge from the safety of the nest into the outside world and it must make the transition from a pure milk diet to solid food. Both these changes can occur gradually and they may take place successively or approximately simultaneously. The young of larder hoarding rodents can start to eat before leaving the nest, but in young carnivores there may be extensive exploration of the outside world before the first solid food is taken.

Leaving the nest is usually a gradual process and the young at first move only a short distance away and then return to their familiar refuge. Their behaviour is, in fact, very much like that of an adult exploring a strange environment. A little later, as they move further afield, they commonly show a strong tendency to follow the parent, or parents. The latter frequently show special behaviour towards them, and will lead them back to the nest if they stray. In the house mouse, Crowcroft (1966) has described both the first cautious exploratory voyages and the mother's leading of the young. Indeed, in one case, Crowcroft saw the male take part in leading the young to safety.

The "caravanning" shown by shrews of the genus *Crocidura* is a more complex form of leading the young to safety and is well adapted to cope with the shrews' rather poor eyesight. Meester (1960) has described this behaviour in *Crocidura hirta*. Typically, the young form a line behind the mother, the first gripping the fur of her flanks in its teeth and the rest similarly holding to the one in front. Meester expresses it as forming "a minute furry railway train with the mother as the engine and its destination the safety of the nest". Frequently, however, the arrangement is less regular and several may cling to the mother or to a fellow, so that a single file is not formed. Frequently, too, the mother "caravans" a single youngster back to the nest. If the latter does not at once take hold, the mother induces it to do so by gripping it by the snout, pulling it in the intended direction of travel and then turning about and presenting her hindquarters. In the litters studied by Meester, caravanning occurred over a period of 9 or 10 days, from about the 8th to 18th days of life. Before this, the mother retrieved scattered young by picking them up in her mouth and by the end of the caravanning period, the young were capable of finding their own way back to the nest.

In the genus *Sorex* caravanning does not occur but if the young stray from the nest, the female leads them back, pausing for them to catch up if they fall behind (Crowcroft, 1957). Crowcroft notes that on their first excursions from the nest, the young are usually pursuing the mother in an attempt to suckle from her. Possibly the caravanning grip has developed from what was originally an attempt to grasp a teat or to hold the mother back so as to be able to suckle.

In *Suricata* I found that up to the age of about four weeks the young could not be enticed out into the garden to forage with their parents: they refused to leave the safety of the familiar house. A little later, they readily followed the parents into strange territory. At this stage in their development two responses are important in preventing their getting lost or injured. The first is a response to the distress call made by any kitten that finds itself alone. I expected this call to summon one or both of the parents to the youngster's aid and was very surprised when neither showed any response to it. Indeed, I was so obsessed with this notion that it took some time to realise what the function of the call actually was. In fact although the parents take no notice, the rest of the litter at once come running to the one who calls and the function of the call is simply to ensure that the young do not become scattered but keep together as a group. Once I had realised this, I turned it to advantage and when necessary I used to summon the young to return to the house by imitating this call. As soon as I thus announced to them that I was left alone, the whole litter would be at my feet in an instant. Young *Xerus* show a similar but less intense response to each others' distress calls but whether in this species the mother also responds is not known.

The young meerkats' second protective response is a simple one to a parental alarm call. If this call is made, they instantly run to whichever of the parents is giving the alarm. If both parents were calling, the young always remained as a single group: I never saw them separate and go some to one parent, some to the other. They appeared to run preferentially to the mother and, having reached her side, would remain pressed close against her flanks, following in a tight group as she led them to safety. Direct defence of the young, however, is primarily the responsibility of the male. In the domestic situation this showed itself in a number of ways; indeed, the male's protective responses to young mammals were first roused by the young of one of the house cats before he had himself fathered a litter. He discovered the litter alone in their nest box and climbed in beside them. When the mother returned, he defended his adopted young against this dangerous predator with great determination, driving her off repeatedly, so that finally it was necessary to remove him. When his own young were born but had not yet become active outside the nest, he would return periodically to them and sniff them over briefly before returning to his own pursuits. The female did not do this, but came back to the nest only when ready to feed her offspring. The male also became much more prone to attack strangers entering the house than he had been before the birth of the young. His role as their defender, however, showed itself most dramatically one day when the whole family were out in the garden. A threat to their safety appeared in the form of a child suddenly irrupting in their midst, in pursuit of a lost ball. Without hesitation, the male hurled himself at this intruder, many times his own size. Had the "enemy" been truly attacking, he would, of course, have been killed but the ferocity of his onslaught was such that it would have delayed even that most dangerous of foes, a jackal, long enough to give the mother and young a good chance of reaching the safety of the burrow.

The transition from an all milk diet to one including solid food, which accompanies leaving the nest, is a complex affair. It involves not only the development of new responses to the external environment which are directly concerned with finding food and eating but also internal physiological changes with which special types of behaviour may be associated. In both these aspects of the transition, the parent may play an important role. It is convenient to deal first with the behaviour related to digestive changes involved in the transition to solid food. The adult digestive system is adapted to function in the presence of a bacterial flora which is particularly important in vegetarian species. In many cases it is known that the establishment of the normal flora is not left to chance: direct inoculation from the gut of the mother is achieved by the simple expedient of the young eating the maternal faeces. This habit has been recorded in many rodents; rats and mice (Harder, 1949), *Cricetomys* (Ewer, 1967), hamsters and guinea pigs (Dieterlen, 1959) and *Phenacomys* (Hamilton, 1962); in lagomorphs (Harder, 1949; Haga, 1960) and in shrews (Crowcroft, 1957). In the koala the process of eating maternal faeces is more highly developed and may be important in the direct nutrition of the young, in addition to inoculating it with the normal gut flora. According to Minchin (1937), for a period of about a month before her youngster begins to feed directly on the eucalyptus leaves that constitute its exclusive food, the mother produces a special type of faeces quite unlike her normal droppings. This consists of a soft brei of half digested leaves which the youngster, putting its head out of the pouch, licks directly from her anus. A feed of this type is given every other day at approximately the same time, between 15.00 and 16.00 hours, i.e. some time before the evening's activity period begins. Routine coprophagy in adult animals escaped notice for a long time and although Minchin was of the opinion that the koala's product was a special "baby food", it seems not impossible that in fact it is simply the caecal contents which may be regularly produced and eaten by the animal itself, as in many rodents and lagomorphs.

A direct inoculation of this type from the mother's intestinal tract is doubtless the most efficient method of establishing the normal flora in the offspring and the caecal contents may also serve as a source of vitamins: but it is a method necessarily restricted to species possessing a caecum. In some other species the young regularly eat earth before they make the transition to solid food. This habit has already been mentioned in artiodactyls, but it also occurs in domestic cats and I have seen the young of *Xerus* do the same thing (Ewer, 1966). I do not consider that this behaviour can be explained as a captivity artefact resulting from dietary deficiency of some sort, as the behaviour is very consistently shown, even when the nutritional status of the animals is excellent.

Another curious habit, whose significance is more difficult to understand, is that of licking maternal saliva. Such behaviour is known in a number of rodents—*Neotoma* (Richardson, 1943), *Micromys*, *Acomys*, *Cricetus* (Dieterlen, 1960), *Glis* (Koenig, 1960), *Cricetomys* (Ewer, 1967) and also *Thryonomys*. Dücker (1957) has recorded the same behaviour in genets. Although juvenile *Suricata*, *Crossarchus* and *Atilax* have not been seen to consume maternal saliva, the readiness of the adults to lick human saliva suggests that they too may have this habit. The significance of this behaviour remains to be elucidated. Since immunoglobin occurs in human saliva, it might be worth finding out whether saliva may not be a supplementary route for the transfer of antibodies to the young. Furthermore, saliva licking may assist in introducing the young to solid food, since the habit would tend to direct their attention to the mother's mouth and thus, incidentally, to the food she eats. Possibly, too, a secondary bond-forming value might be involved, which could even become more important than the primary role.

In vegetable eaters, the young have no particular difficulties in obtaining their first solid food: they need do little more than accompany their mother and feed where she does. As already mentioned, the habit of larder hoarding permits the young to start eating solids before leaving the shelter of the nest. The normal hoarding behaviour would appear to be perfectly suited to coping with the needs of the young and it therefore came as a surprise to me to find that in *Cricetomys* the presence in the nest of young of the appropriate age causes a change in the hoarding behaviour of the female. Her choice of food alters and she becomes more attracted to soft foods and less ready to carry home hard grain, so that the stores accumulated consist of food which is easily chewed and thus suited for introducing the litter to the process of eating solids (Ewer, 1967).

In carnivores, prey catching is a complex business and the young may require solid food before they are capable of procuring it for themselves. Parental behaviour therefore commonly includes some method of supplying this need.

If the prey killed is of suitable size, the female may simply carry it back to her young. This method is adopted by foxes and the racoon dog as well as by the domestic cat. In my observations of two domestic cats I found that food was carried to the young when they were 35 or 36 days old. The female's behaviour is not directly related to how long suckling has been going on but is a response to the characteristics of the young. I found that one of the cats fed foster young older than her own offspring in this way at the correct age. Indeed, the food carrying behaviour may be shown by a female who is not lactating. Leyhausen (1965) kept a group of five female domestic cats and one European wild cat. Two of the former had kittens and, at the appropriate age, all the females brought food to them.

When the prey killed is too large to be carried back whole, as is commonly the case in the Canidae, the parent or parents swallow lumps of flesh and subsequently regurgitate for the young. Martins (1949) found that in domestic dogs the bitch first disgorges food for her young when they are about 3 weeks old. Snow (1967) records the same behaviour in a coyote, starting when her pups were aged 5 weeks. In wolves both parents feed the young in this way (Crisler, 1959; Schönberner 1965). The habit of regurgitating food, however, reaches its highest development in *Lycaon*. Here, as already mentioned, not only do adults of both sexes regurgitate the results of a successful hunt for the young of any member of the group but they also do so when solicited by the adults who have remained behind to guard the young (Kühme, 1964, 1965b, c). Some discrimination is, however, shown; the hunters are less ready to disgorge food to an adult male than to a pup or a female. Also a male who has not participated in the hunt may finally go off and search out the remains of the kill for his meal (Kühme, 1964). According to the size of the prey and the distance the food has to be carried, the lioness may either simply drag the carcase to her cubs or swallow large pieces of meat and regurgitate for them (Carr, 1962).

A captive polecat kept by Goethe (1940) showed a curious form of behaviour. When her kittens were about a month old and their eyes had just opened, she would carry a piece of meat into the nest, then pull it out again and leave it lying.

She thus provided a sort of drag hunt for her young, who would then follow the scent trail and find the food. Whether in natural conditions the polecat does this with the prey she has caught does not appear to be known.

Suricata shows a slightly different method of introducing her young to solid food. Insects constitute the principal food but these are so small and take so long to collect that neither carrying home whole prey individually nor regurgitation is practicable. The young do not receive solid food until they have emerged from the nest and are ready to accompany the mother as she forages. When she succeeds in finding something to eat she now refrains from consuming it herself and, holding it in her jaws, runs to the youngsters and allows them to snatch it from her. If they fail to do so, she will lay it down in front of them. In this way they not only receive food before they are adept at finding it for themselves but can be introduced to a wide variety of foods, for *Suricata* in extremely omnivorous. In captivity conditions, when food is provided in abundance, the female still behaves in this manner; she will pick up the food and try to induce her young to take it from her, regardless of the fact that they are all busily eating already (Ewer, 1963).

The insectivorous grasshopper mouse shows an interesting parallel with meerkat parental behaviour. Although the young are not actually given or helped to catch their first insect prey, there is a change in parental behaviour which results in providing the young with this type of food. The adults normally kill an insect, eat the head and thorax and then, a little later, the abdomen. In the animals studied by Ruffer (1966) the parents developed a tendency to refrain from finishing their meal and leave the abdomens of the prey they killed untouched. These were later found and eaten by the young. Both parents showed this behaviour for a period of 6 days, starting from the time the young were 10 days old. This is when the eyes open and eating of solid food begins.

In addition to their methods of providing the young with meat before they are old enough to catch their own prey, many predators assist their young in their first efforts to make a kill. The domestic cat will bring in a live mouse or rat, often injured so that it is more easily dealt with, and release it in the presence of her kittens. If they fail to kill and the prey seems likely to escape she will catch it again and bring it back to them. Schaller (1967) has described very similar behaviour on the part of the tigress. In this species the mother first brings the young to the kill she has made at the age of 6 weeks and it is not until they are about a year old that they start accompanying her on her hunts. She then helps them to make their first kills. Schaller saw one tigress pull down a buffalo and then leave it for her cubs to kill. It shook them off and she again pulled it down and left it for them and this time they succeeded in despatching it. The cheetah also assists the young to perfect their killing techniques. Kruuk and Turner (1967) describe how a female carried a live Thomson's gazelle fawn to her two cubs and then released it for them to chase. According to Carr (1962) lion cubs remain with their mother for the first 2 years of their life, perfecting their killing techniques with her assistance. In the animals studied by Schenkel (1966a), after the cubs had been introduced to the pride, the mother used to bring them to share in the communal kill. From the age of about 5 months the mothers of two litters encouraged their young to join them in stalking prey. Schenkel was of the opinion that the young were definitely being given a sort of training. When they set forth on a serious hunting expedition, far from encouraging the cubs to join them, the females left them in a very decided manner and the cubs made no attempt to follow. He concludes that "when the lionesses allowed (the cubs) to take part or even stimulated them to do so, they did not really intend to hunt".*

Even the killers of much smaller game may require maternal help in their first efforts. The Canadian otter, *Lutra canadensis*, will catch such things as fishes, frogs or crayfish and summon her young with a special call. When they arrive, she releases the prey and lets the kits catch it again for themselves (Liers, 1951). In this case the difficulties of the first kill are increased by the fact that the prey has to be caught in the water, still

* It should, however, be emphasised that the "training" is of a strictly limited nature. It consists only of making prey accessible to the cubs and thus giving them the opportunity to develop their skill. This they do on the basis of their own experience, not by copying the behaviour of their elders.

a foreign element to the kits. The mother has to coax them into the water as well as help them to catch their prey.

The transition from terrestrial to aquatic locomotion is one which also has to be made by the young pinnipede. How early and how easily this is done varies from species to species. In those where the young have a soft "puppy coat" they do not enter the water until after the first moult. The fleecy coat holds the water and if it were once wetted, the pup would quickly become chilled. In the common seal *Phoca vitulina*, and the hooded seal, *Cystophora cristata*, the puppy coat is shed *in utero* and the pup is born with its adult type pelage. It enters the water and can swim within a few hours of birth. In contrast with these species, the pups of the sea-lion, *Otaria byronia*, are very poor swimmers when they first enter the water and the mother accompanies them and encourages their efforts. In *Arctocephalus gazella*, the pup first enters the water at the age of 10 or 12 days (Paulian, 1964). Before doing so it spends some time crouched on a rock, alternately putting its head into the water and raising it again, as though practising its breathing. It then launches forth and swims perfectly, without any learning or period of tuition being required. It does not, however, venture far from the shore at first and not until it is three weeks old does the pup risk entering the breakers. Coping with these is a learnt skill, which takes a little time to master.

The higher primates are somewhat anomalous and cannot be categorised as either altricial or precocial. Although born furred and with their eyes open, as in precocial species, their locomotory powers are at first quite insufficient for them to keep up with an adult by their own efforts. The grip of the grasping hands and feet, however, is strong enough for the baby to be able to cling to the parent's fur and be carried about, so that the latter's body constitutes a sort of mobile nest. The marmoset, *Hapale jaccus*, is exceptional in that it is the male that looks after the young and carries it about. He hands the baby over to the mother to feed and takes it again once suckling is over (Lucas et al., 1927; Fitzgerald, 1935). In monkeys and apes the newborn youngster is at first permanently in contact with its mother or with some other female whom she has permitted to handle and fondle her baby. The mother grooms the baby and caresses it, as well as suckling it and when she moves off she helps it to take up its position clinging to her fur. As its locomotory powers develop, the youngster makes its first excursions away from the mother's body, only a very short distance at first, then gradually increasing its range, very much as an altricial youngster makes its first excursions from the nest.

The olive colobus monkey, *Procolobus verus*, is exceptional in that for the first few weeks of its life, the mother carries her baby in her mouth. Booth (1957) who described this behaviour regarded it as an adaptation to a combination of anatomical and ecological factors. The pollex in this species is extremely reduced and the fur of the adult rather short—factors which together make it difficult for the youngster to keep a very firm grip. The habit of frequenting dense cover, where the baby is in danger of being brushed off and the mother cannot easily spare a hand to give it support, accentuates the problem. The solution found has been to adopt the method of transport characteristically used by the Carnivora—but with one important difference: the baby is gripped not by the neck, but across the lumbar region.

This obviously secondary evolution of mouth transport in one member of a group otherwise characterised by clinging transport of the young, suggests that it may be worth while at this point to make a slight digression to consider the interrelations of the two methods of carrying the young. The first point to be borne in mind is that the two do not have identical roles. Mouth transport is typically a means of dealing with altricial young before they have attained locomotor competence. Its function is to retrieve accidentally strayed or scattered young or to move the litter to a new nest, should this become necessary. Since at this stage of their development, the young of marsupials can cling firmly to the teats and can easily be carried thus, it is not surprising to find that mouth transport is not known to occur in any marsupial. Clinging transport, on the other hand, is a means of keeping the young with the mother when they are mobile to some extent, but not yet capable of keeping up with an adult or of leading an independent existence without maternal care. The conditions in which this form of transport is both possible and desirable are therefore firstly, that the limbs of the young should not be specialised in a manner which interferes with their grasping ability and secondly, that it should be difficult for the young to keep up with the mother unaided. If there is only one young, then whether mouth transport would impede the mother's progress is also a relevant factor.

It is in arboreal species that the two primary conditions are most adequately fulfilled and it is therefore not surprising that it is in arboreal species that clinging transport is most highly developed. Amongst the placentals it is typical of the arboreal primates and scaly anteaters and also occurs in bats, where the problem of keeping up with mother is even more acute. Amongst the marsupials also, clinging transport is highly developed in the climbing opossums; in terrestrial dasyurids, like *Sminthopsis* and *Dasycercus*, where it plays a minor role, it may represent an inheritance from an arboreal ancestral stage.

It thus seems likely that clinging transport developed at a very early stage of mammalian evolution and that mouth transport is a separate adaptation, which became necessary only with the development in placentals of altricial young, kept in a nest until attaining locomotor competence. Although clinging transport may be primitive, in so far as it was probably evolved earlier, it is not in any sense "ancestral" to mouth transport. The two are separate adaptations to different situations: neither has evolved from the other and, in different species, either may be emphasised or lost, depending on the special circumstances, as may be seen amongst the Lemuridae (Petter, 1965). Indeed, they are, to some extent, complementary and there is no reason why both methods should not be found in the behavioural repertoire of a single species: for example, it would be no surprise to find both patterns in an arboreal placental species whose young are first kept in a nest and later accompany the mother. This may actually be the case in Bushbabies. At first the young are retrieved and returned to the nest using the mouth grip but, according to Shortridge (1934), the young of *Galago senegalensis* at a later stage accompany the mother, clinging to her fur and do so until they are at least half grown. Sauer (1967), who describes mouth transport in detail, however, saw such clinging once only. This was in *Galago crassicaudatus*, on an occasion when the mother was suddenly frightened out of the nest.

This digression has carried us some distance from our consideration of juvenile behaviour in higher primates, to which we must now return.

Since much of the interest in studies of higher primates is related to their possible relevance to human social behaviour, a great deal of attention has been devoted to recording the details of how the infant becomes gradually less attached to its mother, how play with age-mates gradually becomes its dominant interest, and how and when behavioural differences between the sexes make their appearance. Analysis of the stages by which the juvenile reaches its final position in the social group has also been carried out (see, for instance, Jay 1963; DeVore 1963; Hinde and Spencer-Booth, 1967). Relatively little is, however, known about specific types of maternal behaviour related to such things as the protection of the young and their introduction to solid foods. Since they are predominantly vegetable feeders, finding food is relatively simple and even in the apes, the young do not appear to receive very much maternal assistance. According to Schaller (1965c) young gorillas learn what to eat by feeding where they see the parents eat, combined with trial and error. He only once saw an infant take a piece of food from its mother's hand and once a female removed and threw away a leaf which its offspring was attempting to eat. On five occasions he saw infants trying a food not eaten by the adults and in three cases the trial was followed by rejection and the food was spat out again. In chimpanzees maternal assistance appears to be a little more pronounced. Goodall (1965) saw infants take half chewed food from the mother's mouth and eat it and she also saw them take food from her hand. A female chimpanzee that gave birth in the Berlin zoo regularly passed food from her own mouth to her offspring's as soon as he had cut his incisor teeth (Heinroth-Berger, 1965). Goodall also mentions that she never saw a chimpanzee mother steal food from her child, although this occurs commonly with monkeys. As yet, little is known about how the young learn the specialised feeding techniques of fishing out termites and killing of prey which Goodall saw used by the group of chimpanzees she studied, but her observations (1964) suggest that, at least in tool-using, watching the parents plays an important role.

In studying the relations between mother and young, we are dealing with a highly complex evolutionary product: not merely with the adaptedness of an individual to its environment but also with the mutual adaptations of two sides of a highly asymmetrical partnership. Furthermore, the relations between the partners are not static but dynamic, continually adjusted to the changes taking place in the developing young. Even in much simpler relationships, since the total environment, biotic and inanimate, is itself not static, no species can be expected to evolve until its adaptations reach perfection' and then remain unchanging. It is therefore of some interest to ask whether there may not be cases in which adaptation is just a little "sub-standard" or in which there is some degree of mismatch between parental behaviour and the characteristics of the young. The places where incomplete adaptation is most likely to be found are where there has recently been some change which impinges on the mother-young relationship. The most obvious are alterations in the environment into which the young are born or in the stage of development at which they enter the world.

The Mount Kenya hyrax, *Procavia johnstoni*, has its young in an extremely well protected refuge in a cleft or crevice in the rocks. In such a situation one would expect the young to be altricial, but instead they are highly precocial; born fully furred and with their eyes open, they start nibbling solid food when they are only 2 days old. Sale (1965) interprets this as indicating that the use of rock refuges is a habit of recent origin, which has not yet been fully exploited in terms of shortening the gestation period, which is approximately 7 months.

In the golden hamster, Richards (1966a) has found that new born pups are less effective releasers of maternal behaviour than are slightly older ones, that is, pups aged 6–10 days. In this species, the gestation period is extremely short, only

17 days, as compared with 20–25 days in most related small rodents. Richards suggests that the shortened gestation period is of relatively recent evolution and that the maternal response has, so to speak, lagged behind, being still greatest to youngsters of the "normal" birth age.

Acomys has been mentioned as anomalous amongst the Muridae in having precocial young, which suggests that the lengthening of the gestation period may be relatively recent. *Acomys* is unusual in that eating of the placenta is not restricted to the period immediately following parturition and is also remarkable for the fact that the female does not go apart from her fellows to give birth, as well as for the strength of the drive to care for a pup immediately before parturition. Possibly these peculiarities are related to the recent evolution of the long gestation period (38 days) and clearly a detailed study of the relations between hormones and behaviour in this species would be extremely interesting.

The great apes' habit of carrying the placenta about until it rots is another example of behaviour that seems distinctly maladaptive. This is a puzzling case and cannot be disposed of as a captivity artefact, in view of Goodall's observations in the wild. It seems as though the usual habit of eating the placenta is being abandoned but no satisfactory alternative method of disposing of it has been evolved. If so, what selective pressure operates against the normal placentophagy? Two possibilities suggest themselves. The mother may have been inclined to eat too enthusiastically and injure the baby in the process. Possibly the disadvantage of maternal over-eagerness lies not in the danger of direct damage to the baby but in depriving it of the blood in the placenta. As Barcroft (1946) has shown for other species, if given a little time, this blood will drain into the body of the foetus and give it an increased reserve of iron with which to start its independent existence. Only further observations can elucidate this problem.

Another recently described case which may represent lagging adaptation is that of the grey seal (see E. A. Smith (1968), *Nature Lond.*, **217**, 762–763). This species apparently originally bred on the ice floes, in pairs or very small groups. More recently it has taken to breeding further South, in large crowded rookeries. In this situation one would expect the female's maternal attentions to be strictly limited to her own pup, but this is not the case. There is considerable promiscuity in feeding arrangements, resulting in heavy infant mortality, since many of the weaker pups fail to get enough milk and die of starvation. The simplest explanation of this state of affairs would seem to be that the attachment of the female to her own pup, which is characteristic of normal rookery breeders, is missing in the grey seal because in its recent past there is no selective pressure favouring its evolution. The maternal behaviour suited to its solitary or near solitary rearing of the young has not yet been brought into harmony with the new habit of rookery breeding.

Such occasional imperfections as have been described, however, serve only to emphasise how strikingly in general the relations between the mammalian mother and child illustrate the adaptiveness of behaviour and the essential role it plays in ensuring the survival of the species. The self-evident fact that the adaptation must needs be close, that in the course of evolution every increase in juvenile helplessness must have been accompanied or preceded by a compensatory increase in maternal solicitude, in no way lessens the interest attached to understanding the details of the patterns of mutual interaction that have been evolved.

CHAPTER 11

Play

No mention was made in the last chapter of one very characteristic type of juvenile behaviour—play. This was partly because, although typical of young animals, play does not necessarily cease with maturity but also because play is a sufficiently complex subject to require a separate chapter to itself.

Although play in many species of mammals is briefly described in numerous writings, detailed studies of the play of individual species are relatively few and are mainly concerned with carnivores. One may cite, for example, Eibl-Eibesfeldt's (1950) description of play in the badger, Schiller's (1957) in the chimpanzee, Tembrock's (1958) in the fox, Rensch and Dücker's (1959) study of two species of *Herpestes* and Ludwig's (1965) of the Boxer dog. General theoretical treatments of the subject of play may be found in Bally (1945), Beach (1945), and Meyer-Holzapfel (1956). Play is also discussed at some length by Leyhausen (1965). There is a very considerable degree of agreement amongst these authors as to what are the main characteristics of play. At first sight this might seem a very simple matter. Any cat or dog owner will use such expressions as "he is playing" or "look, he wants me to play with him"; clearly taking it that play is a perfectly definite thing which is easily recognised. But if we ask exactly what is play and what criteria do we use to identify it, the matter becomes more difficult.

Play is not something which, like most other activities, can be identified on the basis of the form of the movements used. One can, for instance, at once tell that a cat is ambushing prey or threatening a rival simply from the attitude adopted, without having to see the mouse or the other cat. If the animal starts to scratch a hole in the ground, one knows that micturition or defaecation will follow, or if it inclines the head slightly obliquely and raises one fore paw, one knows that face washing is starting. In play, however, we do not find special actions specific to play but a whole series of movements with which we are already familiar in some other context, where they obviously fulfil some biological end: prey catching, fighting and escape behaviour are the commonest sources from which the movements used in play are drawn. It is, however, quite obvious that although the movements used may be the same, the animal's motivation is different and the normal objectives of the actions are not now relevant. This difference may be crudely expressed by saying that the essence of play is that it is not "in earnest". What this implies may be more objectively stated by making a comparison of exactly what does happen in the play situation and in the genuine one. The successful end of a genuine flight consists either in outdistancing the pursuer or finding safety in some refuge: a successful genuine pursuit leads to attack. In play, however, the roles of pursuer and pursued alternate without reference to the attainment of these ends, just as they do in a child's game of tag. In fighting play, biting is inhibited, claws are sheathed and no serious hurt is inflicted. "Mistakes" do occur now and then, but the sudden yelp of pain of a puppy whose ear has been bitten a little too hard merely serves to emphasise that this is an exceptional occurrence. A kitten that has taken up the defensive threat posture in earnest remains almost as though frozen long after the danger has passed and only slowly regains its equanimity. In play it will switch instantly to some other action.

Almost all the characteristics of play are related to this basic fact, that the motivation is not that of the "in earnest" situation. When used in earnest, a series of actions are linked together as a functional unit and we can thus group them as belonging, one set in the context of prey capture, another set as constituting intra-specific fighting behaviour, etc. In the earnest situation the actions normally follow in a regular sequence and the whole series has a definite end point, whose function is usually easily discerned. Once the end action is achieved, the series comes to an end—the prey is killed and eaten, the enemy defeated and driven away. Since the animal is normally quite capable of renewing its attempts, should the first fail, it appears that it is the achieving of the end point that causes the actions to come to an end. Consummatory stimuli are received which act as negative feedback and inhibit further action. In play, none of these constraints operates. The actions do not follow each other in a fixed order and even those from different functional contexts are interwoven. A kitten will ambush a littermate as though he were prey one moment, attack him as a conspecific the next and then promptly flee from him as a predator.* Since the genuine motivation is absent, there is no defined end point and the actions may be repeated over and over again; in practice it is this repetition that constitutes one of the most obvious characteristics of play. The absence of an end action also makes it difficult to decide what is the "point" of the whole performance and one is driven to conclude that play has no biological objective other than its own performance—it is an end in itself, not a means of acquiring a full stomach, safety, cleanliness or the like. This is, of course, not to say that play has no function—merely that it has no simple clearly defined and easily discerned goal in relation to the immediate external situation. The question of its function will be considered later.

Three other characteristics of play must also be mentioned. The first is a very simple one—namely that although performed in a not very clear order and without relation to an objective, the actions are none the less oriented. The kitten ambushes a littermate or a dead leaf, the ground squirrel runs to and from the mouth of its burrow, young ungulates chase round a bush or hillock, the puppy runs towards and away from his master or worries an old shoe. Play is thus clearly not simply a matter of a deficiency in inhibitory control permitting immature patterns to discharge in an uncontrolled manner. The second

* It is, however, an exaggeration to say that there is absolutely no ordering in play. From watching species as diverse as carnivores, rodents and marsupials I have come to the conclusion that different types of play may have different thresholds. While some normally appear at the beginning of a play period, others seem to require a preliminary "warming up" period and do not usually appear until the general level of activity is high. Moreover, although switches from one functional context to another frequently occur, short sequences of acts performed in the normal order are by no means uncommon.

is that, although the same instinctive movements as those of the earnest situation are used in play, they are often a little exaggerated, carried out with just a little more energy than is strictly necessary. The result is a sort of exuberance so that the whole demeanour of the playing animal expresses eagerness, excitement, in fact, enjoyment. The absence of this exuberance and their monotonous lack of variation at once distinguishes from play the neurotic stereotyped movements that are often performed by captive animals kept in too small an enclosure. Thirdly, play in young animals generally occurs as a regular part of the daily routine. For each species there are times of day when play is most likely to occur: these are commonly after the young have fed, following their arousal from a period of sleep.

Another characteristic of play given by Meyer-Holzapfel (1956) is that it occurs only when no instinctive pattern is activated in earnest. This may be true, but it is not very helpful. We have no way of judging when an instinctive pattern is "activated", apart from seeing the resulting behaviour. In practice therefore, Meyer-Holzapfel's criterion is equivalent to saying that if an animal is behaving in earnest, then it is not playing. It is, however, true to say that play—like toilet behaviour—has a low priority rating in the general organisation of behaviour and can easily be switched off, to give place to an earnest activity. I have even seen a young *Thryonomys* cease play-fighting for a moment to scratch itself. The reverse is not true. One does not see an animal break off an earnest activity *in medias res* in order to play. The formulation used by Meyer-Holzapfel is that play occurs only when no genuine instinctive pattern is activated. I take this to mean no more than that play is suppressed if the motivation for a functional end act sets the animal into action in earnest. Leyhausen (1965), however, takes it that Meyer-Holzapfel really means what is implied by the formulation used—that play itself is not an instinctive action—and he is at some pains to controvert this view. In this he is certainly correct. The actions used in play are the same endogenous movements as those of the earnest situation; play and the need to play mature in the same way as any other innate type of behaviour and play is just as species specific in its form. That learning can be interwoven with play is no argument against its basically innate nature for, as we shall see presently, the same is true of many innate patterns.

If all the characteristics that have been mentioned are shown, then we may be sure that we are dealing with play. Where only some are present, the problem becomes more complex and it may be difficult to decide where to draw the line and how the behaviour in question should be classified. For the meantime, however, let us consider only the typical cases, where all the criteria of play are fulfilled. Such behaviour is sufficiently common, especially in young animals, for it to be clear that play is a definite form of behaviour, recognisable by a series of definite characteristics. This being so, two further questions arise. Firstly, is there a special motivational state, corresponding to readiness for play? Meyer-Holzapfel (1956) is inclined to attribute readiness for play not to a specific play mood, but to a generalised drive towards activity. This is seen as finding its outlet in different actions, determined on the one hand by the external situation and also, to some extent, by weak activation of normal drives which, although insufficient to call forth the earnest behaviour is sufficient to channel the general activity drive in a particular direction. However, since play is not a series of specific actions but can take a wide variety of forms, there is no very clear distinction between a generalised activity drive, for which a variety of outlets is available and a specific motivation for play. Both imply that there is a particular condition of the animal's central nervous system, a particular mood, which is conducive to play. Leyhausen (1965) puts the emphasis in the opposite way from Meyer-Holzapfel. He considers that each action of a functional unit (and he is particularly considering prey catching) has its own independent motivation which builds up and is discharged in play. The discharge is facilitated by a state of general hyperexcitability which is particularly common in young animals. This hyperexcitability may well reflect an immaturity of inhibitory controls, whose later development contributes to the decline in playfulness of adult animals. Thus, while for Meyer-Holzapfel play will be maximal when no other motivation is present and the general activity drive has free rein, for Leyhausen the motivation for play arises from the accumulated motivations for individual instinctive acts. In the mature animal these motivations for the individual acts do not normally build up in the same way, since their use in the earnest context provides them with an outlet. Furthermore, according to Leyhausen, it is only when motivation for the normal end action is in control that the subordinate acts are ordered in the normal sequence. In play this genuine motivation is lacking and the individual acts are therefore free to be discharged without relation to the normal sequence. Thus not only does motivation for the individual acts build up and require outlet, but they are also free to be discharged in any order. According to this scheme, the general hyperexcitability of the young does no more than facilitate the discharge of the accumulated motivations.

Meyer-Holzapfel is assuming a single motivation for each functional activity complex. In her scheme, if motivation for prey capture is high, the animal will start hunting in earnest; it will not indulge in prey catching play. To account for the latter, it is therefore *necessary* to postulate some separate drive which is responsible for activating the animal to play and which can do so only when the earnest motivation is absent or low. Leyhausen's postulation of independent motivations for the separate actions of the activity complex of prey capture permits the drives for the preliminary actions to be high while the motivation for the end acts of killing and eating is low or absent. In his scheme, while there is no *necessity* to postulate it there is, at the same time, nothing incompatible with the existence of a separate independent motivation for general activity or for play. There are two lines of evidence which suggest that such independent motivation does exist. Firstly, the fact that appetitive behaviour towards play is shown and secondly, the ease with which a learnt form of play becomes a favourite game, even if it does not utilise the movements that are commonest in natural play. Appetitive behaviour towards play is familiar enough in cases where an animal invites or incites another to play with it. Fighting play is often invited by leaping at the partner, delivering a harmless inhibited bite and springing back; or chasing play by running up to the partner and away again or dashing past close in in front of him. A human companion may also be invited to play; Eibl-Eibesfeldt's tame badger, for instance, would invite him to play by galloping up with a very stiff-legged gait and barking, stopping abruptly and waving his head from side to side, circling about and then dashing off, only to stop again abruptly and once more wave his head.

Learnt forms of play are familiar to everyone who has kept a dog and there are few carnivores that, if kept as domestic pets, do not invent games for themselves. A learnt series of actions may also be used in a play invitation as, for example, when a dog fetches a ball, lays it near his master and crouches beside it, wagging his tail and barking. This invitation includes an attempt to create the situation in which a learnt mode of playing with a substitute object normally occurs. The behaviour of a pair of tame rats* that I once kept included an invitation of this type and it also provides an example of a learnt game using unorthodox movements. The animals were frequently allowed out of their cage and then had the free run of a laboratory bench. A form of play gradually developed, starting by the rats attacking the duster with which I used to clean the bench. This became a tug-o-war game, in which I let the rats win and carry the duster to "their" corner of the bench. I would then remove it and the game was repeated. The next stage consisted of their stealing a handkerchief from my pocket as I sat on the bench and the usual game followed. The final complexity was reached when, if I tired of the game first and refused to recapture the handkerchief from them, one of the rats would pick it up in his mouth, carry it to me and thrust it against my hand. The thing about this performance which is difficult to explain without recourse to some type of special motivation is that although fighting games are the commonest form of play in rats, the movements used in this invented game most closely resemble those used in collecting and carrying home nesting material—not those used in fighting. Moreover, despite the presence of a conspecific potential sparring partner, the animals chose the complex performance of attempting to make me "play the game", rather than the direct one of play fighting together. This seems most simply described by postulating some general play drive

* Laboratory *Rattus norvegicus* of the hooded pattern, no details of strain known.

which the animals had learnt to channel into this particular form of invented game.

It is, of course, quite possible that the relative importance of general play motivation and accumulated motivations for specific actions varies quantitatively from species to species and from time to time. Certainly it is not intended to controvert Leyhausen's analysis of play in Felidae by quoting examples from the behaviour of rats. If his view is correct, then inducing a kitten to a prolonged bout of chasing play with a substitute object should make little difference to its readiness to fight, ambush etc. On the other hand, if a general play drive is an important factor, such a procedure should drastically reduce the animal's readiness for other types of play. I know of no analysis of this type of the play of any species. Until such time as further information of this sort is available for a number of different species, it is necessary to be cautious in drawing general conclusions. The evidence in favour of a special mood conducive to play is strong, but not conclusive. Whether one should call this mood a specific play mood or a general activity drive is not of great importance, although against the latter may be set the fact that the types of play in which animals indulge do not necessarily involve maximal activity. The terms, however, are no more than two attempts to find a suitable name for a drive that has the peculiarity of not leading to any clear end action: they do not reflect any fundamental theoretical difference. The accumulation of motivations for individual actions, while almost certainly a factor of great importance in the causation of play, appears to be, in isolation, inadequate, since it does not account for the enthusiasm with which learnt games may be performed, even if they do not use the movements that are commonest in normal play.

It is now time to turn to consider those cases where some but not all of the characteristics of typical play are shown. Of these, one of moderately common occurrence is the immature discharge of separate components of complex patterns in an incomplete or unordered manner. Indeed Kruijt (1964) on the basis of a study of the ontogeny of fighting behaviour in the jungle fowl has been led to question whether the concept of play is necessary—is play perhaps merely the performance in an incomplete manner, of innate patterns whose neural basis has not yet fully differentiated? In mammals at least, one can answer these questions quite definitively. Play is not a dispensible concept but refers to a perfectly genuine category of behaviour and it is not synonymous with the partial, precocious performance of innate patterns. Two things make this clear. Firstly, play in mammals can go on long after the relevant earnest patterns are fully matured and in use. Secondly, the precocious discharge of incomplete patterns does occur, but is quite distinct from play. In such a case the animal usually acts hesitantly, often with insufficient decision or muscular power to make the act effective; the performance is transitory and not repeated. In all this it is the direct antithesis of play. If one sees the same action first discharged precociously and then occurring in play, the difference is unmistakable. This is very obvious in the behaviour of young *Dasycercus*. In this species the principal form of play is related to escape behaviour. It consists of rapid dashes hither and thither, with lightning quick changes of direction, very frequently in and out of some refuge. It is remarkably similar to the "jinking play" described in the ground squirrel, *Xerus erythropus* (Ewer, 1966). Just before the animals are old enough to play, one may see this pattern being discharged prematurely—and there is no doubt that although the same action is involved, what is happening is different from the play that subsequently appears. In the first case, the animal may be moving about slowly; then, quite suddenly and for no visible reason, it accelerates and makes a little dash, exactly like the quick darting movements that will very soon appear in play. But there is a difference. One's subjective impression is that in play the animal knows perfectly well what it is about and is in full control of what it does, whereas in the first immature discharge it simply has no idea what is going on and is in fact surprised at what happens. Analysing what it is that creates this impression, I realised that the main point is that the immature discharge lacks the orientation which has previously been mentioned as characteristic of play. When playing, the animal runs to and from some definite points, whereas the immature discharge not only bears no such relation to particular places but may be so divorced from it that the animal may inadvertently run through a food or water dish. I saw something exactly comparable in the case of fighting play in this species. Before this type of play had come into being a youngster emerging from under a piece of bark suddenly came face to face with another and in an instant they were reared up, not playing at fighting, but fighting, quite violently and with accompanying threat vocalisations. This lasted only a moment and again I had the impression that they literally "did not know what had come over them". True play fighting is preceded by an apparently deliberate approach to the partner and is not accompanied by vocal threat. In both cases, although the patterns of movements used may have their initiating neural substrates in the mid-brain, their use in play appeared to be under cortical direction, whereas the immature discharge was not. It is also interesting that in young *Sminthopsis*, an immature discharge of escape darting may occur, much resembling what happens in *Dasycercus*, but this never develops into play.

There is a second category of behaviour, shown by adult animals, whose relationship to typical play is more difficult to decide. This is the discharge to a substitute object of a pattern which has been denied its usual outlet. My male meerkat, for example, was still sufficiently playful to join in the games of the youngsters of the first litter he fathered. Thereafter he became much more staid and ceased playing in the typical manner. However, he would frequently seize a duster, towel or the like in his jaws, bite at it and worry it. I interpreted this behaviour as working off his accumulated killing (or possibly fighting) motivation, which in the domestic situation was denied its usual outlet. The behaviour differed from typical play in that it was carried out with considerable ferocity and was not varied with other movements. If Leyhausen is right in his analysis of the causation of play, wherein does this differ from ordinary play? The same considerations apply to a form of behaviour known as dowsing, commonly shown in captivity by the raccoon, *Procyon lotor*. This will be described more fully in the next chapter; it will suffice here to say that this behaviour appears to be a sort of substitute for prey catching. The marsh mongoose also dowses food and a tame female would sometimes dowse inanimate objects, carrying them to her water dish and putting them in, scraping them about with her paws and then lifting them out again in her mouth. Should this type of behaviour be regarded as playing at food catching? To carry it one stage further—a vacuum activity arises from abnormally accumulated motivation; should it too be classified as play? Although they may have something in common with play, these performances do not usually "look playful". This is true, at least, of the meerkat's towel worrying. The marsh mongoose's dowsing of inedible objects, on the other hand, often did have the appearance of play. Unsatisfactory and unscientific as this formulation is, if an experienced observer can detect a difference between two phenomena, then the chances are that they are not the same, even if he cannot explain clearly just wherein the difference lies. In the present case, it is most likely a matter of rather subtle differences in muscle tensions, particularly the presence or absence of gratuitous exuberance of movement, which it is tempting to interpret as reflecting whether the specific play mood is or is not in operation.

The behaviour of adult Felidae, however, is rather different. Even an old female will sometimes play with her kittens, which suggests that here, although play motivation in the adult may be diminished, it is by no means totally atrophied or suppressed. "Play" with prey is also common, in wild species as well as in domestic cats, even in aged animals. Either live or dead prey may be played with. In play with live prey the cat refrains from killing, when it could perfectly well do so, and instead repeats over and over again the movements of chasing the prey, clutching it with the paws, hooking it out from a refuge etc. Leyhausen (1965), for instance, describes how an adult male golden cat, *Profelis temmincki*, would delay in killing a rat until he had chased it about and hooked it out with a paw from various refuges several times. To a casual observer this performance gave the impression that the rat was too clever for the cat and the latter simply did not know how to kill it. This might go on for as much as half an hour; then quite suddenly, after a short pause, the cat's whole demeanour would change. He would go slowly and purposefully towards the rat; a couple of quick movements with his paw and the rat was seized with the neck bite, was dead and being carried over to the favourite eating place—all in a matter of seconds. The switch over from play to real killing behaviour was instantaneous and unmistakable. Another of Leyhausen's animals—a female serval cat—even discovered how to put a mouse back into a crevice, so that she might hook it out again with her paw. She would pick up the live mouse gently, carry it to the right place and push it in with her paws and then repeat the process of hooking it out. In terms of Leyhausen's analysis, these forms of behaviour are easily explicable. In the captivity situation, the actions used in prey capture, particularly those concerned with the preliminary stages of the hunt, are utilised much less frequently than they would be in nature. If each action has its own "reservoir" of motivational energy, then a need to perform these actions will build up and when prey is available, the end action of killing is deferred until some of this accumulated excess motivation is dissipated. Play with dead prey is often the sequel to a rather difficult battle, in which the prey was not easily overcome. Apparently the cat's general excitement and its determination to make a kill have been raised by the struggle to a very high level: then suddenly the prey is dead and all is at an end. The cat is not able to "switch off" instantaneously and continues its prey catching behaviour with the corpse, very commonly rising high on its hind legs, tossing up the prey and clutching it again with its paws.

Behaviour of these types, with live or with dead prey, is popularly referred to as play. It is usually carried out with the sort of exuberance characteristic of play and may involve the performance of a number of different prey catching actions. Where we decide to draw the line between the extremes of typical juvenile play at one end of the spectrum and vacuum activities at the other is very much a matter of taste. If we decide to accept the popular verdict that "play-with-prey" *is* play, a good case may still be made for excluding from the category of play vacuum activities and such things as the meerkat's towel worrying, where only a single action is involved. These latter types of behaviour seem to be captivity artefacts and the only thing they share with play is that the discharge of excess motivation to a substitute object, or in a substitute situation, is involved. It is very doubtful whether in the wild deprivation of natural releasers could ever be so acute as to produce these manifestations. The criterion of whether, to an observer experienced in the ways of the species, a performance does or does not look playful may not be a very objective or satisfactory one to use in deciding where the line should be drawn. It may, for all that, be the best available at the present stage of our knowledge. Regardless of these difficulties it remains true, as Leyhausen (1965) points out, that the existence of the doubtful cases is no argument for concluding that the concept of play is not a useful one.

A final point that remains to be considered is the function or functions of play. Since play involves physical activity it will obviously have the same incidental effects as any other form of exercise; strengthening muscles, increasing the efficiency of pulmonary ventilation and circulation etc. But play is something more than a means of keeping fit. One has only to watch carefully to see that during play, particularly play with objects, something further is involved. Play results in discoveries (Plate VIa). This can be seen very clearly in a kitten's play with a ping-pong ball. A cautious approach to the unfamiliar object is followed by the usual tentative pat with the paw. The ball moves and provides the kitten with the ideal releaser of chasing behaviour: "small object moving away". Very soon the kitten is involved in a new game, in which it dribbles the ball with increasing speed and skill, using alternate movements of the fore paws to keep it in motion. The kitten is here using exactly the same movements as it does when pursuing a mouse in what Leyhausen (1956) calls "inhibited play". To the inexperienced young cat the mouse is still a cause of some anxiety and although the kitten chases, the death bite is not at once delivered. The mouse is not, however, permitted to run freely, but is restrained by alternate movements of the paws. What the playing kitten has done is learn how to use these movements out of its innate repertoire to control the ping-pong ball. In short, in play, the animal is experimenting with the relationships between its own actions and the external world and it is learning all the time as a result. What it learns may be very simple; merely how much exertion is required to jump accurately from A to B, how fast it is possible to run and still successfully make a right angled turn into a refuge or the like. However, it may involve quite complex manipulations; the animal may even, in the course of its play, develop new types of movement. My rats discovered how to turn their nest building patterns into the tug-o-war game: Eibl-Eibesfeldt's badger (1950) discovered how to turn somersaults; porpoises learn very quickly how to retrieve a rubber ring and throw it back to a human play fellow; an Indian rhinoceros discovered how to manipulate a wooden ball so as to make it act as a substitute sparring partner (Inhelder, 1955); a young honey badger, one of whose toys was an old motor car brake drum, discovered how to use a piece of wood as a lever to raise it up enough for her to get her paws under it and she would then heave it up and turn it over (Sikes, 1964). The young of higher primates in their play make innumerable manipulative discoveries. *Herpestes* (Rensch and Dücker, 1959) and *Suricata* (Ewer, 1963) develop a form of play in which they crawl under some object, a mat, a newspaper or the like, and then run about with it on their backs. Rensch and Dücker regard this "tortoise play" as the result of the animal's having found out that these objects can be made to produce a substitute for pleasurable sensations normally experienced when moving about in a burrow. The propensity of both *Suricata* and *Dasycercus* to run up trouser legs apparently reflects the same thing.

These performances are certainly captivity artefacts, in so far as wooden balls, brake drums and so forth are not parts of a natural environment; but their very unnaturalness serves to emphasise the experimental and inventive aspect of play. It also makes clear how play teaches the animal all sorts of things about its environment that it could not otherwise have known. As Leyhausen (1965) points out, it thus enriches its experiential world far beyond what would be possible if its activities were strictly limited to the direct and immediate needs of existence. In Eibl-Eibesfeldt's metaphor, the animal's repertoire of innate movements is like a workshop rack full of tools available for its use. In its play it learns what effect is produced by using them in a variety of situations upon a variety of objects.

From watching the play of young mammals I am convinced

that at first they literally do not know what the results of their actions will be. This, however, is gradually learnt so that, after a time, although the kitten may have no notion as to why he should be about to aim a blow at his littermate, he does know that the blow is aimed and that it is likely to knock him over. Without this sort of knowledge, it is difficult to see how patterns of prey catching and fighting, involving quick responses to a moving object, could function effectively in the earnest situation. The kitten will presently launch a genuine attack against real prey—but it knows in advance how fast it must move to intercept the mouse before it reaches its hole and how hard it must launch itself to reach it. The ground squirrel fleeing from a hawk, or the *Dasycercus* from a wedge-tailed eagle, knows in advance how fast it can dodge and not overshoot the entrance to its burrow; how to dash round *this* clump of grass to reach *that* particular entrance hole. The fighting behaviour of adult *Thryonomys* has been described in Chapter 6. Youngsters show all the adult fighting patterns in play fights: nose pressed against nose they push, strain, twist their heads in the "boring" movement, leap away and thwack each other with their rumps. Watching the performance it is difficult to believe that this experience will not be useful to them in the genuine fights that lie ahead. They know in advance what it feels like to be pushed back, how to give ground without losing balance, how to utilise their weight effectively in a well aimed rump thwack or brace themselves to receive one.

This is the general accepted view of the function of play and it seems an eminently reasonable one. It must, however, be noted that it is no more than a plausible assumption. No one has, to date, shown that (non-specific general effects on health and well being excluded) an adult not having had the opportunity to play fight as a youngster is later at a disadvantage compared with one who has had this experience.

In social species, the experience gained in play may have other far reaching effects of a more general character. Scott (1962) working with dogs and Harlow (1962; Harlow et al., 1963) with rhesus monkeys have both found that an animal deprived of contact with its own species during youth grows up into an asocial neurotic: a sort of behavioural cripple, with its social responses to its fellows distorted or missing altogether. In the natural situation social contacts for the developing youngster come from two sources; those provided by maternal care and those made with age mates in play. The two types of social experience seem to act synergically and play with its fellows can largely compensate for deprivation of maternal contact.

In polytocous species, play may have yet another function—that of ensuring that littermates stay together. In most Carnivora the young must remain under parental care long after the time when they have become physically active and mobile and in solitary species they may have to be left on their own while the mother goes hunting. The fact that they are each other's playmates gives them a reason to seek each other's company and so creates a bond between them. The extreme playfulness of young Felidae and the fact that their play is so largely *with* each other may be related, in part at least, to this bonding effect of play.

One may well ask, if play is then so advantageous, why do not all mammals play? The answer to this question involves three factors. Firstly, the value of the sort of experience gained in play depends on the mode of life of the species. It will be maximal where there is requirement for quick movement, accurately oriented in relation to the variable external world, with failure of a first attempt carrying a heavy penalty. These criteria apply most cogently to predators and to fast moving arboreal species and, to a lesser extent, to species that depend for their safety on swift flight in open terrain. We need not therefore expect to find play particularly well developed in species whose modes of life do not have these requirements.

Secondly, not all species have the opportunity for play. Play requires that there be a period during which the neural basis for a number of patterns has matured, but the need for their use in the earnest situation does not yet arise, or does so rarely: in other words, a childhood period under adult protection. In small species that mature very rapidly and are launched early into the serious business of life, there is literally no playtime. This is shown very clearly by comparing the two dasyurids, *Sminthopsis* and *Dasycercus*. Young *Sminthopsis* become fully independent almost as soon as they emerge from the nest and I have never seen any indication of playfulness in this species. In *Dasycercus* the young remain together and in association with the mother for much longer and during this period they are playful, while the larger *Sarcophilus* is still more frolicsome (Plate VIb and c). Amongst the rodents, the species listed by Eibl-Eibesfeldt (1958) as not playing are *Mus musculus*, *Micromys minutus*, *Clethrionomys glareolus*, *Microtus arvalis* and *M. agrestis*: these too are all small, rapidly maturing species.

Lastly, if the main advantage of play lies in the opportunities for learning it provides, we can expect play to be characteristic of species possessing the ability to learn relatively quickly. A certain degree of learning capacity is a necessary precondition for play to have selective value. This is another factor which, like their quick maturing, will tend to restrict the playfulness of small species for, other things being equal, smaller species have less learning ability than their larger relatives (Rensch, 1956). It is no mere coincidence that Primates, Cetacea and Carnivora are the most playful of mammals as well as the most intelligent. We cannot, however simply say that they play a great deal because they are intelligent, nor yet that they are intelligent because they educate themselves in play. Selection for the two must go hand in hand. The more intelligent it is the more the animal is able to learn by playing and the greater the selective value of playfulness. The more playful it is, the more advantageous becomes its capacity for profiting by experience. This does not, however, result in an ever-accelerating positive feedback loop. Play has its disadvantages too: by making themselves conspicuous, the playing young are in danger of attracting the attention of predators. Selection for playfulness will therefore cease at the point where this negative factor balances the positive advantages that accrue from play.

We may, of course, also expect that selection will operate upon play qualitatively as well as quantitatively and, within the limits set by the innate behavioural repertoire, emphasise in play those actions whose performance brings most gain. Such qualitative differences are obvious if we compare the play of a prey species and a predator. The ground squirrel,

Xerus erythropus, does show fighting play, but this is of minor importance and play derived from escape behaviour predominates (Ewer, 1966). In cats, the relative importance of the two is reversed and fighting play is more important than flight play. In this connection the play of *Dasycercus* is of some interest because it so closely resembles that of *Xerus*. Despite the fact that *Dasycercus* is carnivorous, the principal form of play is escape play, and fighting play is much rarer. This I interpret as being related to two factors: firstly, being small and inhabiting open terrain, *Dasycercus* must be vulnerable to aerial predators; secondly, its prey catching behaviour is extremely simple and its repertoire does not include the variety of complex movements characteristic of placental Carnivora. It literally has not got the patterns available to utilise a littermate as substitute prey in its play, as the kitten does, and its play is correspondingly impoverished in comparison. There is also another factor, which restricts the use of intraspecific fighting patterns in play. Meyer-Holzapfel (1956) notes that in lower species "misunderstandings", in which fighting play goes over into a genuine quarrel with real hurt inflicted, are much commoner than in higher forms. In *Dasycercus* this very often happens: fully playful fights do occur and they are silent, but very often the fight becomes real and the threat vocalisation is then given. In these circumstances play fighting must often be an unpleasant experience and it is not surprising that it does not become a favourite game. Von Ketelhodt (1966) mentions escape play as being typical of the aardwolf and of the bat-eared fox, *Otocyon megalotis*. Although these species belong to the Carnivora, both have reduced dentitions and escape behaviour probably plays a more important part in their lives than does fighting.

To summarise, from the data available it appears that the necessary condition for the appearance of play is a superfluity of motivation for endogenous movements. This situation arises most frequently in young animals, since the neural basis of patterns matures in advance of the need or opportunity to use them in the earnest situation. The conditions making for elaboration of play by natural selection are twofold: firstly, that the animal's mode of life be such that what it can learn in play is later of use to it; secondly, that its learning capacity be sufficient for it to be able to take advantage of the experience it gathers in play. And here we may repeat the *caveat* that in fact the usefulness of this experience remains to be demonstrated. What remains much less well understood is the central nervous organisation responsible for play, particularly whether a special motivation for play does exist and if so, whether its importance varies from species to species. More detailed quantitative studies will be necessary before further light can be shed on this problem. We are also very ignorant of the relationship between the motivation for a particular type of behaviour in the earnest situation and the use of the pattern in play. The different views of Meyer-Holzapfel and Leyhausen on this subject have already been explained. There are a number of questions relating to this problem which await investigation. For instance, if, in a species that does play a new endogenous movement is evolved, does it automatically become part of the play vocabulary? It would be extremely interesting to make quantitative studies of the play of different breeds of dogs that have been selected for different attributes; for example, a terrier, a greyhound and a retriever or spaniel. Does the puppy's play foreshadow the special abilities of the adult right from the start, does it gradually begin to do so, or do they all play alike? Learnt forms of play also require a much more detailed analysis than they have so far received. How far is it true to say that the movements used differ from those of innate play and how far is the learnt part of the activity restricted to an appetitive phase—a phase in which the animal uses its learning ability to find out ways of manipulating the situation so that, in the end, a substitute object can be made to provide the opportunity for the discharge of the innate patterns characteristic of normal play? The detailed analysis of play still offers a wide field for further investigation.

CHAPTER 12

Ethological theories and mammalian behaviour

The foregoing chapters have given some idea of the richness and diversity of mammalian behaviour concerned with the major activities which ensure individual and genetic survival. It is from observations of this sort—not, of course, restricted to mammals—that the ethological approach outlined in the first two chapters is derived. It is now time to turn to more analytical studies and ask how far this approach is valid, with reference to mammals.

The central thesis of the ethologists is that what they have called the animal's fixed action patterns form the basis upon which its behaviour as a whole is constructed and upon which learning can operate. These patterns are innate, in the sense of being a function of the characteristics and organisation of the animal's sensory, nervous and muscular systems and are as much subject to genetical control as are structural features. It follows that during the course of evolution they have been adaptively moulded by selection in exactly the same way as has structure.

The second thesis is that the performance of such patterns is determined not only by the external stimuli impinging directly on the organism, but by internal motivational factors of a particular type, having the characteristic of varying in a manner related to the time elapsed since last the particular pattern was performed.

Before proceeding to examine these two theses, one point should be made clear. If the first proves valid, then the evolutionary approach which it has stimulated is legitimate, regardless of the details of the motivational control of the genetically determined neural mechanisms underlying the patterns. For this reason the former will first be considered.

It might appear to be a very simple matter to decide whether or not fixed action patterns have a genetically determined structural basis, but in practice, this turns out to be very complex, as witnessed by the fact that it has been possible seriously to advance the view that there is no such thing as innate behaviour (Lehrman, 1953). The direct genetical approach includes two main types of investigation. Different strains of a single species may be studied and behavioural differences found. That the differences are indeed genetically controlled may be confirmed by orthodox breeding experiments. Similarly, starting from a single stock, selection in opposite directions for behavioural characters may be applied, to find out whether this will produce strains showing clear behavioural differences.

As examples of the latter approach may be cited experiments in which rats have been selected for maze learning ability (Tryon, 1942), for "emotionality" as manifested by the tendency to defaecate when placed in strange surroundings (Hall, 1951) and for wheel-running propensity (Rundquist, 1933). Mice have similarly been selected for readiness to explore a strange open area (McClearn, 1959) and for wheel-running performance (Bruell, 1962). In each case lines showing significant quantitative differences in performance resulted. It must, however, be noted that Searle (1949) came to the conclusion that the difference between Tryon's maze-bright and maze-dull rats was largely one of timidity. The "dull" animals were slow to solve the problem of the maze mainly because they were more upset by the experimental conditions than their "bright" relatives.

The two most extensive investigations in the first category are those of Jakway (1959) and Goy and Jakway (1959) on mating behaviour in guinea pigs and Scott and Fuller's (1965) extensive researches on the inheritance of behavioural characters in different breeds of dogs. The former workers investigated male and female responses separately, crossing strains which differed quantitatively in various aspects of their mating behaviour. In the female, latency and duration of heat and responsiveness to injected oestradiol appeared to be under the control of a single genetic mechanism, while duration of lordosis and frequency of male-like mounting behaviour each had an independent separate control mechanism. In the male, the preliminary courtship actions of circling round the female, sniffing and nuzzling her and mounting were found to be under a separate genetical control mechanism from intromission rate and frequency of ejaculation.

Scott and Fuller (1965), whose experiments followed on a study of factors influencing social behaviour in dogs, chose for their main subjects two breeds, cocker spaniel and basenji, which differ in a great many characters that can be quantitatively assessed. Their results showed clearly that genetical factors did underlie the behavioural differences; in most cases several genes were implicated in the control of each behavioural characteristic. The basenji, for instance, shows greater wariness than the cocker and the breeding results suggested that differences in two genes are involved. As in Tryon's rat experiments, one of the facts that emerged was the importance of timidity in determining the animals' performance in tests designed to assess other characteristics. Basenjis are much more agile than cockers. A test devised to measure this ability required the animals to climb a step-ladder like structure of boxes in order to get food. The cockers, in the event, made the higher scores because, although poor at climbing, they were undisturbed by the strange structure and, at the same time, eager to get the food. They therefore simply ran at the ladder and somehow, anyhow, got up and got the food. The basenjis on the other hand, with their greater wariness and caution, were suspicious of the unfamiliar object and hesitant about approaching it; they therefore took longer to get the reward, despite their superior climbing ability.

Experiments of this sort, although they clearly demonstrate genetical effects on behaviour, are not very relevant in the present context. Since only closely related animals can be crossed, the behavioural differences involved are usually quantitative rather than qualitative and the results do not tell us much about the more complex patterns that concern us here. Mutations causing loss or major alteration in some complex pattern have received little study. Their detection is much more difficult than in the case of mutations affecting structural characters and since the control of a complex pattern probably involves a number of genes, single mutations would be expected, as a rule, to have only small effects. An interesting case has, however, recently come to light. Eibl-Eibesfeldt (Tinbergen, 1965b) has found a female red squirrel which completely lacks the normal ability to open nuts. It has proved impossible to teach her to do so, even by cracking nuts for her and allowing her to eat the kernel. If breeding from this animal is successful, the results should be extremely interesting.

There is another quite different approach to the problem of the existence or otherwise of innate patterns which has not yet been developed very far. If we take an animal which has not yet performed a certain fixed pattern and find a location in its brain, such that stimulation by implanted electrodes causes the performance of the pattern, then clearly the latter is innate. The stimulus is simple and unpatterned and the patterned motor output can therefore be a function only of a corresponding nervous organisation. The following example illustrates exactly such a case. If cats do not make a kill during a sensitive period in their youth, they grow up as non-killers, who fail to attack a mouse. Roberts and Kiess (1964), working with such non-killer animals, were able to find an area in the hypothalamus whose stimulation caused them to attack and kill with the normal feline oriented neck bite. Here, there can be no question of the orientation having been learnt by trial and error, since no previous kill had been made.

The information provided by this type of study is, however, still very meagre and it is therefore necessary to turn to a different technique—the deprivation experiment. In theory this should be simple. If an animal performs a fixed action pattern in the manner normal for its species without having had any opportunities for learning how to do so, then the pattern must be endogenous (or innate) and is the result of the way the animal's nervous, sensory and muscular systems have been constructed under the direction of its genotype. In practice, however, there are complications and it is necessary to begin by formulating clearly exactly what question is being asked in the deprivation experiment. Lorenz (1961) has made his viewpoint abundantly clear and Eibl-Eibesfeldt (1963) adopts the same position. More recently Skinner (1966) has elaborated Lorenz's approach at great length. Briefly expressed, this is as follows. Fixed action patterns are normally adapted in a very close and detailed manner to a certain situation. Short of special creation, this adaptation can have come about only through a process of selection. This may have happened in either of two ways. The animal, as a result of its previous experience of the situation, by the neural processes of habituation, conditioning and trial and error, or on the basis of insightful use of experience gained in comparable situations, may have selected out the responses leading to the adaptive end results. The alternative possibility is that, in the course of its phylogenetic history, natural selection has operated in an entirely analogous way to preserve responses leading to success and so build up the neural organisation capable of producing the adaptive end result. The technique of operant conditioning, developed by Skinner and his associates, can very quickly lead to the production of remarkably complex learnt responses. We need therefore be in no way surprised or incredulous if the analogous process of natural selection has also produced complex built-in patterns. In order to decide between the two possibilities that have been outlined, it is necessary only to rear the animal in such a way as to deprive it of experience relevant to the particular situation and, when it has reached the age at which we know the response can be shown by normal animals, present it with the relevant situation and see whether it can or cannot perform the adapted response.

Lorenz (1961) has been at some pains to elaborate the procedure to be followed in this type of work. The two most important considerations are as follows. Firstly, it is necessary to ensure that the animal is deprived only of experience which is relevant to the situation in which it will subsequently be tested. Apart from this, it must be, as far as possible, a healthy and normal animal. Secondly, in the test situation, all the relevant stimuli must be provided and, equally important, no disturbing or inhibitory elements must be present. If the pattern is shown in a normal manner, the meaning of the result is clear. It is, however, so difficult to be certain that the conditions outlined above have been adequately fulfilled that the interpretation of a negative result is not always simple. Bad rearing conditions may lead to behavioural deficiency or we may, through ignorance, have omitted some essential element in the test situation.

Some authors have adopted a slightly different approach and have assumed that it is desirable not merely to deprive the experimental animal of experience relevant to the test situation but to rear it in an environment as impoverished as is compatible with survival. This is a procedure which can lead only to confusion since it confounds all sorts of non-specific effects with the one which it is desired to study. It stems from accepting a definition of what constitutes learning that is so wide as to be in practice useless. Thorpe (1956) has defined learning as "adaptive changes in individual behaviour as a result of experience". Taken at its face value this would include such things as the improvement in motor performance resulting from muscle hypertrophy as well as all the multifarious ways in which one part of a developing organism can affect another. That Thorpe had no intention of including these effects under the heading of learning is apparent from the phenomena which he proceeds to discuss.

A number of workers have claimed that early experience can affect the development of the nervous or endocrine systems and these, in turn, may have very generalised effects on later behaviour. Bennet et al. (1964), for instance, compared the brains of rats reared in enriched and in impoverished environments. The former were found to have heavier brains, with a higher cholinesterase level and a thicker cerebral cortex than the latter. Levine and his co-workers (see Levine and Mullins, 1966) found that in the rat the first few days of life constitute a critical period, during which lasting effects may be produced on nervous or endocrine organisation. The systems which they have found to be affected include the brain mechanisms determining the cyclic or acyclic patterns of sexual responsiveness in the two sexes and the feedback systems operating between the pituitary gland and the thyroid and adrenals. They found the adrenal system to be influenced by the amount of stimulation received during the critical period. There are, however, differences of opinion between the various workers in this field and effects found by one have not always proved repeatable by others. Barnett and Burn (1967) summarise the position and themselves produce evidence strongly suggesting that cutaneous stimulation, whether provided by the mother licking the young or by some externally applied treatment, may have some influence on the general development of the nervous system. Similarly Harlow, dealing with monkeys, and Scott with dogs, have shown that absence of social contact during a critical period may result in major disruption of social behaviour—a point to which we will return later. To include all these phenomena in a single category together with such things as conditioning, trial and error and insight is hardly likely to assist our comprehension. As Marler and Hamilton (1966) point out, "there is some danger in allowing the concept of learning to become too broad and vague, or it may become as useless and diversionary as the term instinct".

Nevertheless, having made a test in the manner outlined by Lorenz and having received a positive answer (i.e. the animal does perform the normal pattern), it is then a perfectly legitimate procedure to go both forwards and backwards in time and enquire, on the one hand, whether subsequent experience brings about any modification and, on the other, whether further less specific deprivations affect the ontogeny of the pattern. It is, for instance, not legitimate to suggest, as does Lanyon (1960), that a bird's song pattern cannot be regarded as innate if it is performed by a bird reared in auditory isolation; the bird must also be deafened. It is, however, perfectly legitimate to follow up the finding that the species specific song is performed by birds reared in isolation by asking a new question—what is the effect of deafening? The answer will tell us something about the form in which the endogenous pattern is represented. If the bird will still sing, but sings incorrectly, we can conclude that some form of matching of the auditory feedback received against a genetically determined "blue print" is involved. If the song is still correct, then the innate mechanism consists of a motor programme in which any matching up that goes on must be in terms of the feedback received from the vocal muscles. Cutting afferent nerves from the vocal organs would then be required to find out whether any such matching process is involved.

This long digression has been necessary only in order to make clear that Lorenz's criteria are being accepted here, and what this implies. A fixed action pattern is regarded as endogenous (or innate) if it is shown at the appropriate time in the appropriate circumstances by an animal so reared that only the experience relevant to the performance of the particular pattern has been withheld from it. We may now proceed to examine the results of experiments of this type that have been performed on mammals. Before doing so, however, it should be noted that the patterns of behaviour shown by newborn mammals fall into the category of innate, as defined above, since the animals have had no opportunity to gather information relevant to the outside world. The same is true of the parturitional behaviour of the female with her first litter and of her first responses to her young. These too are patterns which, in most species, the female has had no opportunity to learn. Although experience may add to the confidence and dexterity shown, the basic patterns all appear in the appropriate situation without the need for previous experience. These types of behaviour, however, have been dealt with in Chapter 10 and there is no need here to dwell on them further.

The most extensive investigations which have been made on mammals, using the deprivation technique, are those of Eibl-Eibesfeldt (1963) who studied a variety of patterns in a variety of species. The most intensive study is that of Leyhausen (1956, 1965) on the ontogeny of prey catching in Carnivora, particularly cats. It is not necessary to consider in detail all the available information: a representative sample will suffice. Eibl-Eibesfeldt's study of nut burying in the red squirrel, *Sciurus vulgaris*, will be dealt with first. Adult squirrels are most likely to bury nuts if they are full fed and further nuts are still available. The technique used comprises five distinct actions: it is very regular and shows little or no individual variability. The squirrel picks up the nut in its mouth, runs off and chooses a burying place, usually close to some vertical object. It then digs a hole with the fore paws (1), lays the nut down in the hole (2) and rams it down with a few blows of the incisor teeth (3). The nut is then covered over by scraping earth over it from the sides with the fore paws (4) and tamping it down firmly (5). Eibl-Eibesfeldt's animals were reared on liquid foods in cages where there was no moveable object which could be picked up and carried about. The animals were completely tame and accustomed to their owner. At the age of 2–2½ months, the first test was made. Each squirrel was given shelled hazelnuts, one after another, as many as it would eat. In each case, after some had been eaten, the squirrel picked up the next nut and ran off with it. Many of them came out of their cages and searched about in the laboratory; finally they came to a stop in a corner and tried to bury the nut. Despite the fact that no hole could be dug in the floor, some of the animals went through the entire sequence of actions involved in normal burying, while others omitted some steps in the sequence. Eighteen animals were tested, with the following results.

Actions 1–5 or 1–4	8 animals
Actions 1–3	7 animals
Actions 2 and 3	3 animals

The squirrels were then allowed access to earth and all showed the complete pattern.

Several points of interest emerge from this study. Firstly, the normal movements are not only present from the start but they are linked together in the right order. This linkage is so strong that even in the absence of the natural feedback, some animals went through the whole sequence. Subsequent experience, however, is not without effect. The inexperienced animal may chance to turn slightly away from the nut before covering it. It will then scrape the earth which should have covered the nut in the wrong direction. If this happens it will look at the nut, sniff it and repeat the covering process, often first repeating the ramming down. Occasionally an animal picked up the nut and started the burying process again from the beginning. With experience, mistakes of this sort vanish. Clearly there is present in the naive squirrel's brain not only a motor programme, but a responsiveness to the consummatory situation of the nut vanishing, which permits it to learn how to achieve this end more economically. It is also of some interest that the two actions carried out by the worst original performers on the bare floor were putting the nut down and ramming it with the incisors. If a squirrel is given as many nuts as it will bury it does not simply stop after a certain number have been buried perfectly. The digging and covering processes first begin to deteriorate; next the covering is omitted altogether, then the digging vanishes and the last nut or two will merely be set down and rammed. These two actions thus are the most firmly entrenched in the nervous system and their motivation the most resistant to exhaustion. They are also the only two that alone can result in some degree of hiding of the nut and it is precisely these two movements that constitute the normal burying in *Atherurus*. The conclusion lies very close that these two movements are phylogenetically the oldest and that the squirrel's elaboration of making a preliminary hole and ending by scraping the earth on top are later evolutionary improvements.

The ground squirrel, *Xerus erythropus*, buries food with an almost identical technique (Ewer, 1965), the only difference being a minor one in the covering movement and the still further refinement of often camouflaging the cache by scraping a dead leaf or a stone over it. A point of some interest is that with my tame *Xerus* the favourite hoarding material was not a nut but a mouthful of maize grains. These were rammed down in the usual way with a couple of blows of the incisors—a method entirely suited to a single hard object like a nut but less efficient with a little pile of grain. To tamp the latter down with the paws would be more effective and this is a movement which the animal has at its disposal and is, in fact, about to use in a moment's time to pat down the covering earth. There was not the slightest suggestion of any modification of the pattern in this direction, which again speaks for a rather rigid central nervous linking of the component actions.

It must not be assumed from what has just been said that once a complex fixed action pattern has been released, it runs its full course without further reference to the external situation. Although the squirrels were capable of "burying" a nut on the bare floor in the absence of many of the normal stimuli, their behaviour shows that when the latter were provided they were not without influence on the animals' performance. In the same way, a cat will make the appropriate scraping movements to "bury" the smell of faeces on a bare floor, but is capable of burying her droppings neatly either in light sawdust or in heavy damp garden earth. She could not make the correct movements against such variable resistance, were she not responding all the time to afferent feedback.

In general, the patterning of a fixed action pattern is something that arises within the central nervous system, not something imposed from outside by a pattern of external stimuli: nevertheless if it is to function effectively in any real environment, it must be governed by stimuli which orient it in the first place and continue to modulate and steer it as it proceeds. This is particularly obvious in social behaviour involving a partner whose own behaviour may be variable, as may be seen very easily in any interactions involving courtship or threat and fighting. For example, the male cane rat stands in front of the female he is courting and wags his tail towards his left or right side, depending on his orientation towards the female at the time. Here optical stimuli provide the basis for adjusting the setting of the tail as the wagging movement is carried out. The fighting of adults of this species has already been described and the young use the same technique in fighting play. Here not only is there an initial orientation to the partner's nose but as the action proceeds each continually adjusts to his partner's actions. In the tame youngster who used to play fight against my knuckles this was particularly obvious, as one could deliberately "steer" his actions. Such cases make it clear that the fixity of a fixed action pattern refers more to the basic central nervous mechanism than to the ultimate motor output, which may show considerable flexibility in relation to the stimuli which orient it and monitor and regulate its performance. This distinction between the basic pattern and its modulation in relation to the environment makes it easy to appreciate that although the orientation and control of the movements may improve with practice, this is not in conflict with the idea that their origin and patterning is the product of a hereditarily controlled neural mechanism.

With these considerations in mind, we may now return to Eibl-Eibesfeldt's experiments and take as the next example his study of nest building in rats, which demonstrates this type of improvement with experience. As in the case of the squirrels' nut burying, distinctive movements are used. An experienced rat first picks up pieces of nesting material and carries them to the home—normally the burrow. They are then heaped up, using pushing movements of paws and snout. The heap is next hollowed out by turning round in the centre, pushing the bedding aside the while and tamping it down with the paws. Scattered pieces are either picked up in the mouth and gathered in or raked in with the paws. The final result is a ring-like nest, with a wall round the outside and a hollow in the centre. If necessary, in addition to the actions described, thick pieces of straw are split lengthwise in the teeth and bits of paper are torn up. The nest-building movements, unlike those of the squirrel burying its nut, do not follow in a regular sequence but are adjusted to the momentary needs of the situation.

The test animals were reared in conditions where they had no opportunity to carry anything about, much less attempt to build a nest. The cages were floored with wire mesh so that droppings could not accumulate and the food was given in powdered form. Since rats will often pick up their own tails and carry them, if nothing else is available, the animals' tails were amputated. In a bare cage some rats will adopt one corner as a sleeping place but others fail to establish such a home base and sleep sometimes in one place, sometimes in another. Eibl-Eibesfeldt therefore reared some of his animals in bare cages but others he provided with a partitioned off corner, so as to encourage the adoption of a home base. Since the nest is normally built in the home, these rats might be more inclined to build than "homeless" individuals. In all, 94 virgin female rats were used in the tests: 39 were reared in the bare cages and 55 in the cages with partitions. At the age of 2–3 months nesting material was provided and the responses of the animals to it were observed. The results are summarised below.

	Bare cage	*+ partition*
Built nest within first hour	13	45
Built nest within 5 hours	20	9
Failed to build by following morning	6	1

The results show that the majority of rats will build a nest without the necessity of previous experience of carrying things about: there were in all only 7 failures out of 94. They also demonstrate clearly the importance of the home base: the difference in performance between the two sets of animals is highly significant ($\chi^2_{(2)} = 23$). It is his failure to appreciate that a rat does not nest randomly, just where it happens to find suitable material, that renders Riess' (1954) experiments meaningless. Tested in an unfamiliar cage, as his rats were, even an experienced animal will not build, a fact which Eibl-Eibesfeldt also demonstrated.

Eibl-Eibesfeldt, however, did more than merely record whether a nest was or was not built: he watched (and filmed) how it was done. He found that the inexperienced animals used all the normal movements, but they did so in a rather ill-organised manner and with much hesitation; frequently they stopped to scratch themselves or eat. The whole performance had a "senseless" air about it, in contrast to the adept behaviour of an experienced animal. For instance, a paper strip might be picked up and dropped on the way to the nest, but the rat completed the journey home although nothing was thereby achieved, or it might make the patting down movements without actually touching the nesting material. It seems therefore that while the individual movements are innate, the rat must learn how to relate them to the situation, so as to produce the proper end result. That it can and does do so implies that here too, as in the case of the nut-burying, there is a responsiveness to the correct end situation, a recognition of the "rightness" of the proper type of nest.

In the eating behaviour of a young Tasmanian devil I have seen an exactly comparable case of the necessity for the animal to discover for itself the function of a definitive movement. When first given a rat, the devil began to eat it from the head down. After a few chewing movements and before the skin at the neck had been cut through, his paws came up and, with extended claws, executed a few quick downward stroking movements. These *should* have the effect of peeling back the skin, however, not only were they made too soon, but the claws did not even contact the skin of the prey. In subsequent trials the skinning movement came to be made effectively.

In cases of this type I believe that a very important element in the conversion of the first "senseless" performance of an innate movement into its fully adaptive use, is the development of confidence. In the unfamiliar first time situation, the movement is tentative and partly inhibited. With experience, this inhibition vanishes and the movement is performed fully and decisively. This was certainly the case in the behaviour of a female *Cricetomys* when first faced with the situation of a youngster out of the nest (Ewer, 1967). She moved towards the babe, made an intention movement of picking it up in her jaws but drew back. In the end she raked it back into the nest with her paws. Her every movement and attitude were eloquent of anxiety and nervousness. Later on she retrieved her young with no hesitation, picking them up in her mouth. It can hardly be argued that by successfully raking the youngster back with her paws, the rat learnt how to pick them up in her mouth. What she did learn was not to be disturbed by the situation of young-out-of-the-nest. The experienced animal will use the paw raking method for pulling in a babe that is merely lying at the far side of the nest, but the stronger stimulus provided by the babe outside the nest is answered by picking up in the mouth. The inhibited inexperienced female reacts to the strong stimulus with the same response as the experienced one makes to the weaker stimulus. I believe that a considerable part of the hesitancy and incompleteness of innate movements at their first appearance stems from this factor of partial inhibition resulting from the unfamiliarity of the situation.

Be this as it may, Eibl-Eibesfeldt's (1963) observations on the nut opening techniques of squirrels provide an example where there is no doubt that it is learning by trial and error how to utilize innate movements to the best advantage that plays the major role in the elaboration of the final pattern. Most experienced squirrels open a nut by gnawing a furrow along one side from base to tip and often also a counter furrow on the opposite side; a hole is made near the tip, the incisors are inserted and with a deft twist of the paws and jaws in opposite directions, the nut is neatly split open. A few animals make a series of furrows approximately at right angles to each other and end by jerking out the centre piece with the incisors in what Eibl-Eibesfeldt calls the hole cracking technique. This variability at once suggests that individual learning plays an important role. Eibl-Eibesfeldt tested 16 animals reared with hard objects available, so that they were accustomed to gnaw and the jaw muscles developed their normal strength. They had, however, no access to anything which could be opened. At 4–6 months nuts were given, one every day until each had had twenty or thirty and the development of their techniques was noted. All the squirrels took their first nut at once, turned it about in their paws and started to gnaw. At first they made furrows more or less randomly but all attempted to insert the incisors and twisted the nut in the normal splitting movement. Gradually they discovered how to make a furrow in the usual place, working parallel to the fibres of the nut and unnecessary gnawing was eliminated. Two animals first got the reward of eating the kernel by making a hole in the bottom of the nut and for some time they retained this method. From this they went over to hole cracking and one finally switched to the orthodox splitting method which all the rest had mastered much earlier. Here the special splitting movement was clearly innate but the animals had to learn how to produce, in the most economical way, the situation in which it is effective.

Another instance in which the innate nature of specific movements is particularly clear appeared in the behaviour of my *Xerus*. When they retire for the night into their burrow, these animals block up the entrance behind them, using the normal digging actions of scraping the earth loose with the forepaws and then shoving it out behind with the feet. In making the block, the loose soil is pushed up towards the entrance in this way but not all the way out, so that it accumulates to form a barrier. Youngsters hand reared from a stage when the eyes and ears were shut and locomotory powers very feeble, showed this pattern at a very early age. They were kept in a small cage, with a nest box provided with a piece of knitted woollen material to act as bedding. When they reached the stage of moving about actively they "closed the door" on retiring for the night by getting part of the blanket across the entrance of the nest box. The simplest way to achieve this would have been to pick it up in the mouth and pull it into position—an action they were physically easily able to perform. This, however, was not how it was done: the squirrels used the correct movements, first scraping with the paws and then pushing with the hind feet until finally the end situation of "entrance closed" was attained. This occurred a considerable time before any genuine digging was done, and before they had had access to earth.

Leyhausen's (1956, 1965) study on prey catching in carnivores, particularly felids, is of special interest because the behaviour concerned is complex and the observations he has made are very extensive. In cats the movements used in stalking and ambushing prey appear in play before the kitten is ready to make its first kill and there is also extensive use of the paws in play fighting with littermates and in play with inanimate objects. Another response which makes its appearance is that of chasing after any small moving object, particularly if it is moving away. The oriented neck bite, however, is not shown, even in a partially inhibited form. When the kitten is 2½–3 months old, it normally encounters its first live prey, brought in for it by its mother. As soon as the prey runs, the kitten will chase after it and may then make a kill with a perfectly oriented neck bite. Often, however, there is hesitancy and it may be some time before the kill is made. At first the factors responsible for the variation in performance remained mysterious. By filming and analysing exactly what happened, Leyhausen was finally able to show that before the oriented response to the constriction at the back of the prey's head can be released, a certain level of excitement must be reached. This is required to overcome the inhibitory influences exerted by the unfamiliar and possibly dangerous prey. This excitation may be built up by prolonged "play" with the prey but in the natural situation, it is usually provided by competition with littermates. Leyhausen's assistant, familiar to his animals as their own littermates would be, could sometimes induce a hesitant animal to kill by pretending to be about to capture the mouse herself. I once inadvertently did the same with a young meerkat that had previously chased a mouse, but never made any attempt to bite it. The result was the same; a sudden lightning quick release of the oriented neck bite.

As the cat grows older, it becomes increasingly difficult to raise her excitation to the level necessary for this first releasing of the neck bite response, so that there is a sort of critical period, when the first kill can most easily be made. The experiments of Roberts and Kiess (1964), which have already been mentioned, show that even in cats where normal release of this response has never occurred, the neural basis of the oriented neck bite is nevertheless present and requires only some extra excitation to be activated. The difficulties Mrs. Adamson (1960) experienced in inducing the tame lioness, Elsa, to make a kill appear to reflect the same thing. Elsa was beyond the normal age for a first kill and it was therefore very difficult to generate in her sufficient excitement for the activation of the normal patterns.

Once it has been released for the first time, killing becomes progressively more easily elicited and after a few kills have been made, the stage is set for learning to start taking a hand in the process of shaping the responses.

Although hunger potentiates killing in an experienced cat, Leyhausen found it to be virtually without effect on an inexperienced one; apparently the connection between making the kill and having a meal requires to be learnt. The young cat's confidence increases after it has made a number of kills "correctly", almost always refusing to attack except in the optimal situation, with the prey running away so that the strike can be made obliquely downwards from behind. It now starts to attack in more difficult situations. Because of this, its performance may seem to have deteriorated and killing bites may now be imperfectly oriented. The cat is, in fact, learning how to utilise the movements in its repertoire to deal with prey not in the orthodox position, prey that fights back or succeeds in getting into some refuge. Its early released innate response to the neck constriction is no longer essential. The cat can aim its attack at the correct spot from any direction in an apparently deliberately controlled manner. At the end of this learning process, the cat has perfected a killing technique adaptable to almost any situation. One of Leyhausen's animals (the hybrid offspring of a cross between a male Bengal cat, *Prionailurus bengalensis*, and a female domestic cat) learnt how to "field" a mouse in her paws when one was thrown to her as she sat on a high shelf. As she did so, she twisted it and brought her head down to make the killing bite in the correct position just behind the skull. Her speed and dexterity were quite astounding and it was only after seeing her repeat the performance several times that one was convinced that she really *could* do it and her success was not just an incredibly lucky fluke. The learning of this type of skill does not affect the innate responses; their neural basis remains unchanged and they may still be released by the appropriate situation.

With other carnivores the situation is less clear. Working with polecats Eibl-Eibesfeldt (1963) came to the conclusion that the oriented neck bite had to be learnt, whereas Wüstehube (1960) reached the opposite conclusion. Eibl-Eibesfeldt related the difference in performance of his and Wüstehube's animals to the fact that his were reared in isolation while hers had the company of littermates. Since young polecats do grip each other by the neck in play fighting, Eibl-Eibesfeldt concludes that they were in this way able to learn the correct orientation. While not contending that the experience gained in play is without effect, Leyhausen (1965) suggests that much of its effect may be non-specific. According to him, analysis of Eibl-Eibesfeldt's film showed that the animals that bit at the hindquarters of their first prey prefaced this by a couple of tentative intention movements of biting at the neck—exactly as my *Cricetomys* made to pick up her first displaced young in her teeth, but lacked the confidence to complete the action.

Numerous other observations have been made, showing that specific patterns of movement in mammals are endogenous or innate, in the sense already outlined. Dieterlen's (1959) observations on golden hamsters reared in isolation from conspecific odours may here be mentioned again. These animals developed the normal patterns of smell marking and showed the usual threat behaviour when presented with the mark of another individual. Marler and Hamilton (1966) quote unpublished work by L. Wilsson with beavers, in which motor patterns, including those used in tree-felling and dam building, were found to develop normally without going through a prolonged learning period. Similarly there is no record of any case in which the vocalisations characteristic of the species have been found not to develop normally in a mammal reared in isolation from its fellows. To multiply examples of this sort would, however, serve no useful purpose.

Major deprivation experiments which produce widespread non-specific effects on subsequently differentiating behaviour, although of considerable importance in their own right, are not strictly relevant here and will therefore be mentioned only very briefly. Such effects are particularly important in social animals and have been found in species as widely separated as dogs and primates. In both cases, deprivation of contact with conspecifics during infancy may result in major disturbance of adult social behaviour. In dogs, Scott (1962) found that a relatively short sensitive period, roughly from 3–10 weeks, was involved. This covers the time from when the puppy starts active play with littermates to the time when it begins to react with avoidance to unfamiliar objects. If deprived during this period of contact with its own species, it will respond to them with such exaggerated fear and caution that its normal behaviour is completely disrupted.

Harlow and his co-workers (Harlow 1962; Harlow and Harlow, 1965; Mason, 1960), studying rhesus monkeys, believe that deprivation of contact with conspecifics during infancy can lead to gross disturbances of adult sexual and parental behaviour. The effect appears to be relatively unspecific, for contact with age mates in play and contact with the mother can very largely substitute for each other. Moreover, it appears that visual and vocal contact alone are sufficient to produce adult behaviour which is very much less disturbed. Meier (1965a, b), using monkeys which were reared in isolation but could see and hear the animals in neighbouring cages, found that as adults, his animals could mate and rear young successfully. The interpretation of Harlow's results is not in fact so simple as he assumes. His animals were not deprived merely of social contact with conspecifics and kept in an otherwise normal environment: they were kept in very restricted space and bereft of virtually all the variations in other types of stimulation that characterise the natural situation. Even the final testing situation was so restricted that the failure of his animals to mate can hardly be taken as proof that they were actually incapable of doing so. The experiments therefore do not distinguish between the specific effects of contact with fellows and the results of general sensory deprivation.

In rodents, of both social and solitary species, deprivation of contact with conspecifics has less effect on the development of social responses. Male mating behaviour in animals reared in isolation has been most extensively studied. In general, sexual responses have been found to be normal, although initially there may be defective orientation. Eibl-Eibesfeldt (1963), found that in rats 5 out of 28 males first showed incorrect orientation but quickly learned to correct this. Beach (1942) and Kagan and Beach (1953) found exactly the same thing. In hamsters, Eibl-Eibesfeldt also found some initial disturbance of orientation, and in guinea pigs, although the orientation was normal, the inexperienced males did not at first achieve intromission and ejaculation. They also at first attempted to mount females that were not on heat but learnt not to do so on being repulsed. Dieterlen's (1959) golden hamsters reared in olfactory isolation, showed normal sexual behaviour. It must, however, be pointed out that the results of different workers have not always been in agreement. Gerall (1963), for instance, found considerable disturbance of sexual behaviour in male guinea pigs reared in isolation, as did Gerall, Ward and Gerall (1967) in rats. It is not at present possible to decide how far the discrepancies are attributable to differences in techniques and how much to differences in the strains of experimental animals used.

While dealing with the subject of male orientation to the female in mounting, it is of some interest that Leyhausen's studies have provided a case in which an innate orientation proved to be modifiable but within highly specific limits. A young male cat, when making his first mounting attempts, grips the female by the skin at the back of the neck. This, apart from its effect on the female, results in orienting him in such a way as to facilitate intromission. Leyhausen had a male blackfooted cat, but, having no female of the same species, decided to attempt to mate him with a domestic cat. Although the male responded normally to the female, his first attempts at copulation failed because, being much smaller, it was impossible for him to achieve intromission while maintaining the normal neck grip. After a number of attempts, however, the male was able to work his way back until, gripping the female's fur behind her shoulders instead of on her neck, he mated successfully and fathered a litter (see Plate VII). It is important to note that the initial orientation was never modified. The male always first gripped the female "correctly" by the neck. What he learnt was not to change this innate movement but to follow it up by working his way back until he reached the position where the next act in the sequence, intromission, could be successfully performed (Leyhausen, 1962).

This, however, is a digression. The point at issue is that complexities in the interpretation of deprivation experiments are created by the fact that alterations in the animal's early experiental world may have very generalised effects. This emphasises the importance of Lorenz's insistence on the need to keep deprivation strictly limited to the relevant factors, if the results are to be meaningful in terms of the question of innateness, as originally posed. The findings which have been discussed are sufficient to show that for mammals, as for other species, the ethologists' first thesis is valid: to wit, that complex form-constant movements, adapted to some particular set of circumstances, develop endogenously without the necessity for concurrent experience in the relevant situation. They are prepared in advance and ready for use on the first encounter with that situation.

There is, however, another point which emerges from the studies that have been described. Just how much becomes programmed into the central nervous organisation is highly variable from case to case, even within a single species; this in a manner that is itself highly adaptive. A nut is a relatively standard object and the sequence of movements used by the squirrel is adequate to bury any nut—indeed, an almost identical pattern in *Xerus* copes well enough with a mouthful of rather different food. The entire sequence can therefore be built in and the requirement for learning kept minimal. In fact it is sufficient only to correct the mistakes that occasionally occur when some distraction prevents the programme from running through uninterruptedly from start to finish. The other patterns discussed are rather different, in that they must function effectively in a wide variety of situations. A squirrel must be able to get the kernel out of any fruit-stone or nut it finds; a rat is better off if it can build its nest from a variety of materials in a variety of situations and a carnivore if it can kill a wide variety of species of prey. This is achieved by the operation of two principles in the programming. The first is one of parsimony. What is built in is the minimum required to ensure that the correct end result is attained, together with an innate responsiveness to the latter, so that once attained it is "recognised". This permits individual learning to perfect the pattern in relation to the detailed circumstances of the situation. The second principle is that this result can be achieved on the basis of innately determined individual movements. The necessary learning is thus vastly simplified. The animal does not have to find out how to attain the satisfactory end state without guidance. The correct movements are already to hand and it must learn only how to use them. The end result is a performance which is stereotyped, in that the movements show little variation, but at the same time has the required flexibility in relation to the external situation.

In cases where learning involves the sensory mechanism rather than the motor output, the same adaptability may be achieved by having an innate response to a stimulus which is originally very generalised but can later be tailored to fit the individual's experience. In *Suricata*, for instance, the young respond with escape behaviour to any bird overhead and learn presently to cease doing so with harmless species.

The importance of learning in shaping the final adaptive use of innate movements has been repeatedly emphasised. It must also be borne in mind that the learning of which any species is capable is itself a function of the genetically determined characteristics of its nervous system; and this in a very precise and detailed manner. It is not merely that "how much" an animal can learn depends on what sort of brain it has. The hereditary organisation determines what sort of things it can learn and the exact form the learning can take. A particular species is not simply a good or a poor learner in general. Each species is adapted to learn rapidly certain things that are relevant to it but may in other things appear quite "stupid". The cat that learns so fast how to get its prey into position to make the killing bite does not learn that its offspring's droppings will not get covered by scraping on the floor beside them. Even under the stress of extreme calcium need, the little marsupial *Sminthopsis* cannot modify its feeding behaviour and chew bits off a piece of cuttle-bone too large to be held in the paws, but the young do learn the location of places of refuge on their first night away from the home nest. The dependence of learning on hereditary make-up, however, goes much deeper than this. Much of learning depends on the ability of the animal to respond appropriately to rewards and punishments. What should constitute a reward or a punishment, however, is not something which the animal has to decide for itself and, indeed, on what basis could it do so? This "knowledge" is part of its innate equipment and is itself a product of long selection in remote phylogenetic history. The idea of an animal that, for example, treats food as punishment but jumps with enthusiasm into a bramble bush and scratches out both its eyes, is so manifestly absurd that we tend to take the appropriate type of response for granted, failing to see that to do so is equivalent to assuming a sort of special creation which Lorenz (1961) calls a "prestabilised harmony between the organism and its environment". The individual animal cannot even learn that eating is desirable by finding out that it dies of starvation if it fails to do so, but this is precisely what the operation of natural selection "taught" the ancestral populations as long ago as the first evolution of holozoic nutrition. The results have been preserved ever since and are embedded in the reward systems governing the learning processes of their extant descendants.

In this connection the work of Roberts and his co-workers is of considerable interest. His experiments with Kiess (1964) on brain stimulation which produced killing in non-killer cats have already been mentioned. In similar experiments with rats, Roberts and Carey (1965) elicited gnawing by hypothalamic stimulation. In both cases performance of the appropriate action was found to be "rewarding", in that the animals learnt to traverse a simple maze in order to find a suitable object for the discharge of the pattern activated by the stimulus—a rat in the one case, a block of wood in the other. Similarly Delgado and Anand (1953) have shown that in the rat stimulation of the lateral hypothalamic feeding centre causes the animal not merely to eat, but also to search for food, while Andersson and Wyrwicka (1957) found that on stimulation of the hypothalamic drinking centre, a goat would walk up some steps in order to get a drink.

A number of important consequences follow from these experiments. Firstly, what the hypothalamic stimulation has activated is not a simple patterned motor output but a readiness or need to perform a certain pattern. The animal does not go through the motions of killing, gnawing or drinking; it searches for a situation in which the response can be performed in relation to the normal objective. Delgado (1966) draws the same conclusion from his studies of aggressive responses in rhesus monkeys elicited by hypothalamic stimulation. There was no constancy of motor pattern; what the animal did was related to the general situation. As Delgado puts it "the offensive purpose was constant, while strategies and motor activities varied".

In Roberts and Carey's experiments the rat will not accept a piece of wood covered with sheet metal and will learn to run its maze only if by so doing it finds a suitable block of wood. The stimulus for initiating the gnawing appears to be a projecting edge and for its continuation the fact that a fragment can be torn off against a moderate resistance. The reward might be considered to be constituted purely by internally generated feedback from the muscles involved in performing the relevant action. Roberts and Carey judge that in view of "the powerful control exerted by sensory cues over the initiation and continuation of the gnawing" it is much more likely that it is constituted by "sensory feedback from the interaction with the board". This, of course, corresponds with Lorenz's concept of the "innate teaching mechanism". The latter is simply a statement of the fact that reward consists not merely of doing the "right" thing but of doing it so as to produce the right effect. This was apparent even in the case of the squirrels' nut burying, where the effect of external stimuli, although relatively small, was not absent. The animals spent considerable time searching for the right situation before they would bury on the bare floor and when burying in earth, if the covering was not successful, failure of the nut to vanish was noticed and corrected. The relevance of all this to the natural situation is obvious. If the animal were in fact rewarded merely by performing the correct movements, it would be continually accepting substitute objects instead of searching out the appropriate ones or even simply performing functionless vacuum activities. In fact, the latter are very rare and occur only in situations of extreme deprivation.

Some further points of interest are made by Roberts and Carey. As we have seen, their interpretation of their results involves two concepts: that the motivation generated by the stimulation consists of readiness to act in a certain way and that doing so in such a way as to produce the consummatory stimuli created by the correct interaction with the appropriate object is what constitutes reward. They point out that such a system is one which will result in the types of learning that we have seen are normally at work to increase the efficiency with which innate patterns are performed—learning affecting the appetitive phase so that the correct situation is rapidly found in the familiar environment; learning how to organise individual innate movements so as to lead most rapidly to the rewarding end situation and, finally, even improvement in the fine details of the innate movements themselves.

Roberts and Carey also found that in the rat, the loci from which gnawing could be elicited lay close to the hypothalamic eating centre. In the same immediate neighbourhood they also found loci from which paper gathering and paper shredding could be elicited. They suggest that this may indicate that all these other responses involving the use of the mouth have been derived originally from feeding behaviour, that they "involve cells that have differentiated out of the original primitive feeding mechanism and become adapted for functions having greater species specificity".

It is now necessary to turn to the much more complex problem of the second thesis—that form-constant movements, or patterns composed of them, while they may be released by specific external stimuli, are also dependent on endogenous motivation. This motivation is regarded as arising in the relevant control centre and is considered to be characterised by the properties of build up or accumulation during rest and exhaustion during the performance of the activity. If a new pattern is evolved, then this will inevitably have, as one of its essential and integral components, a motivation of this sort (Lorenz, 1966b).

Observation of an animal's behaviour can distinguish between motivations produced by external stimuli and by changes occurring within the organism, but cannot tell where inside it the latter take place. It is true that a motivational change must affect the nervous system before it finds expression, but in certain cases it is clearly established that the causal chain originates elsewhere. Selection must operate to produce an animal that behaves appropriately and does the right thing at the right time, but there is no *a priori* reason why all motivational systems should function identically. Indeed, one might well expect the reverse; considerable diversity in the detailed manner in which internal changes affect the performance of different activities. We ought therefore to ask two questions—does the sort of endogenous motivation postulated by the ethologists really exist at all; if so, is it a universal characteristic of all fixed action patterns?

Before proceeding any further, it is desirable to distinguish between the two senses in which the word "motivation" is in practice used. Motivation for a particular action consists of the sum total of all the factors which are conducive to its performance. However, an ethologist, studying overt behaviour, when he speaks of *motivation analysis* is concerned with the relationships between the motivations for different activities. If he finds that a certain type of threat is most frequently followed by attack, whereas another threat normally precedes flight, he will say that the former is an *aggressive threat*, including as an important constituent of its motivation the tendency to attack; whereas in the motivation of the latter, the tendency to flee is more important. This, of course, tells us no more than that the situation producing aggressive threat is also likely to produce attack. We cannot, without further information, tell whether we are dealing with two distinct neural mechanisms, both independently affected by a number of the same stimuli or whether there are two mechanisms which act upon each other, or some combination of these possibilities. All we can say is that there is something in common.

The physiologist, on the other hand, working from within the organism, when he speaks of motivation is thinking in terms of the changes brought about in some definite part of the brain by alterations in nervous input along definite pathways, or in the characteristics of the blood which supplies it. Expressed in these terms, the ethologist's thesis is that the motivation for any innate pattern includes excitation generated by some locally occurring process, possibly some type of neurosecretion, in which an excitatory substance is used up during activity and regenerated during a rest period.

Unfortunately, the information is not to hand to answer adequately either section of the two-part question which has just been posed about endogenous motivation. Ethologists have amassed information which provides evidence for endogenously generated changes in responsiveness and have shown in a number of cases, that readiness to perform a particular pattern is not directly related to the need for attaining its biological end. However, apart from one rather unorthodox example, to be described later, I do not know of any case relating to mammals where a detailed quantitative relationship between readiness to perform an action and time since its last performance has been established, except in cases such as eating and drinking, where the effects can be related to the feedback resulting from changes produced in parts of the body other than the central nervous system itself. On the physiological side, our knowledge of the factors involved in regulating activities is still very meagre. Even those governing the processes of eating and drinking, on which much effort has been concentrated, are not yet fully understood. In this situation, the best that can be done is to summarise briefly; on the one hand the physiological information available as to sources of motivation and, on the other, the ethological evidence for independent motivation of individual innate actions, and see what conclusions emerge.

The activities which have been most intensively studied by physiological techniques are eating and drinking. In both, the factors responsible for terminating the activity are as important as those which initiate it. In feeding, two centres in the hypothalamus are involved: a ventromedial one, the satiety centre, in which there is summation of influences tending to inhibit feeding and which acts antagonistically to the second, the more laterally situated feeding centre. Activity of the latter initiates both food seeking behaviour and eating. No comparable separation of excitatory and inhibitory centres has been found in the case of drinking behaviour. Possibly this reflects a greater complexity of control of eating which has made a 2-stage process necessary, with excitatory and inhibitory influences first summed separately before their final integration.

Larsson (1954) and Anand (1961) review the factors influencing food intake and Marler and Hamilton (1966) give summaries of the present state of our knowledge of the physiological control of both eating and drinking. For the present purpose, it is more useful to tabulate the types of motivational factor which have been found to be involved. These fall into a number of categories.

(i) *Negative feedback from receptors activated as a result of activity*

In this category come the impulses from stretch receptors in the walls of the stomach which are relayed via the vagus nerve to the satiety centre (Paintal, 1953; Sharma et al., 1961). Paintal has found that these receptors may also be activated chemically, which may account for the fact that the effect of filling the stomach is not simply related to the volume of food consumed but is also related to its quality. Receptors in the pharynx may also exert a similar effect, since an animal with an oesophageal fistula does not eat indefinitely but feeds intermittently, although no food enters the stomach (Grossman, 1955). In the absence of any direct evidence for the existence of these receptors this could also be interpreted as resulting from exhaustion of excitation in the feeding centre which requires some time for recovery. Introspectively, however, one is aware that a not inconsiderable part of the pleasure of eating arises from the sensation of swallowing the food and it seems very likely that whatever receptors mediate this sensation also relay to the satiety centre.

(ii) *Positive feedback from tonically active receptors*

Russek and Morgane (1963) have produced evidence that peripheral receptors, probably located in the liver, are involved in the control of feeding. From experiments in which the satiety centre was destroyed, they conclude that these receptors are tonically active and cause excitation of the feeding centre. Their discharge is inhibited by a rise in glucose level so that after a meal their activity is reduced, resulting in a lowered level of excitation in the centre.

(iii) *Feedback resulting from changes in blood composition produced by metabolic activity*

The characteristics of the circulating blood have been considered to act directly on both feeding and drinking centres in the hypothalamus. Andersson et al. (1960) have shown that in goats hydration raises the threshold of the hypothalamic drinking centre to direct electrical stimulation, but the effects of blood characteristics on the centres concerned with feeding are less clearly understood. Alterations in glucose level, in temperature and in some factor related to depot fat have all been considered important by various authors.

It should be noted here that although physiologists have shown that stimulation of the feeding centre can activate appetitive behaviour for eating as well as eating itself, they have not discovered a reverse effect, which certainly does exist, to wit, that food catching behaviour can lower the threshold for eating. I have seen this most clearly in viverrids. My meerkats would not normally eat millipedes but would do so if the prey had been dug out from a difficult crevice after a considerable struggle. The extra excitement generated in the battle to secure the prey apparently overrode the inhibition produced by its somewhat distasteful characteristics. Similarly my *Atilax* would not normally eat a convolvulus hawk moth. She might chase it and possibly even eat off the head, but the rest was rejected. However, if she chanced to catch one conveniently near her water bowl, she would pick it up, carry it to the water, put it in and go through the process of catching it: she would then eat it. Large *Periplaneta* were also rather distasteful and were eaten only after an aquatic chase. This was not a question of something distasteful being washed off the outside, since the cockroach was not eaten if removed from the water and given to her on the floor. She showed the same sort of behaviour with a piece of cooked mutton, of which she was not very fond.

Physiological investigation of the hypothalamic feeding and

satiety centres has also produced evidence that, important though they may be, these are not the only structures controlling motivation for feeding behaviour. A rat in which the satiety centre is destroyed over-eats until it becomes obese. A limit is, however, reached and the animal again begins to limit its food intake, regulating its weight at a new and higher level. It has also now become more finicky (Wiepkema, 1963) and is less ready to accept slightly distasteful foods than a normal animal. Similarly Baillie and Morrison (1963) have found that if the feeding centre is destroyed, rats that will no longer eat will still press a lever to feed themselves by a stomach fistula.

(iv) *Specific effects of hormones*

Changes in hormone concentration have been found to affect differentially the thresholds for different activities, the most striking effects being those of sex hormones on sexual and reproductive activities. Fisher (1964), for example, found that micro-injection of testosterone into the brains of rats evokes either male sexual behaviour or maternal responses, according to the site of injection, regardless of the sex of the animal. The paradoxical effect of a male hormone evoking female behaviour he interprets as due to the weak progesterone-like action which testosterone is known to possess. Terkel and Rosenblatt (in press) have recently shown that in virgin female rats injection of blood plasma from a female that has just given birth accelerates the development of her maternal responses to pups, whereas similar injections from a dioestrous female cause a delay. More conventionally, the role of hormones in reproductive and sexual behaviour has been demonstrated by studying the effects of castration and subsequent replacement therapy (Lehrman, 1961; Young, 1961). That sex hormones may also affect aggressive behaviour is also well known. Fighting between males is commonly either restricted to, or greatly enhanced in the breeding season, and the increased aggressiveness of a female with young has already been mentioned.

None of these sources of motivation corresponds with the ethologist's concept of endogenous motivation. It should, however, be noted that the experiments which have been carried out are not such as to have revealed the latter type of motivation. Andersson and his co-workers have shown that hydration raises the threshold of the drinking centre and that altering the body temperature affects that of the hypothalamic heat regulating centres (Andersson, Persson and Ström, 1960). No attempt, however, has been made to study variations in threshold in relation to the time that has elapsed since the relevant behaviour was last naturally evoked.

There is, nevertheless, one case in which motivation of just the type postulated by ethologists does appear to occur: sleeping. It may be objected that sleep is not an activity but an absence of activity. There is, however, a growing tendency among physiologists to regard sleep as something much more positive than a mere dying out of activity, as a phenomenon in its own right with its own causal mechanisms. The animal itself certainly behaves as though this were the case, seeking out a special sleeping place and, in many cases, adopting a special sleeping posture. Cohen and Dement (1966) have shown that if cats are deprived of the deep sleep that has come to be known as *REM sleep* (rapid eye movement) by waking them as soon as the characteristic eye movements begin, then they compensate for this by a recovery period in which more time than usual is spent in REM sleep. If, however, the animals are given electroconvulsive shock, REM sleep is reduced and there is no compensatory recovery period. Their interpretation is that "the intense activity accompanying the seizure in some manner altered the brain levels of substances, or their precursors, that induce REM sleep". Jouvet (1967) suggests that two substances are involved: 5, hydroxy-tryptamine (serotonin), secreted by cells of the medullary raphe system and adrenaline, secreted in the dorsally situated *locus coeruleus* at the level of the pons. These two substances are supposed to accumulate during waking periods and are then liberated by neurosecretory pathways to cause switching off of the midbrain reticular activating system. Considerably more work is required before such interpretations can be regarded as established. They do, however, correspond exactly with the Lorenzian concept of motivation accumulating during a period when the particular activity is not shown and being used up in its performance.

We must now turn to the question of the behavioural evidence for endogenous motivation for fixed action patterns. The clearest cases relate to complex appetitive behaviour patterns concerned with obtaining food. Leyhausen's (1965) analysis of play with prey in adult felids, which was discussed in Chapter 11, depends entirely on the assumption that separate motivation for the individual acts used in prey capture does show independent accumulation. It is very difficult to account for the observed behaviour without making this assumption. One can invent a formal scheme to cover much of the behaviour shown by postulating firstly, that subliminal excitation of the feeding centre, although insufficient to lead to eating, can yet activate the earlier links in the appetitive chain sufficiently to allow external stimuli to trigger them off, and secondly, that the resulting activity then generates more excitation in the feeding centre. A scheme of this sort, however, will not account for all the findings. Quite apart from the paradox of an animal that is hungry enough to search for food but not hungry enough to eat, it will not account for the persistence with which one particular action is shown. For instance Leyhausen's cat that repeatedly put the mouse back in the crevice was not being set into action by the stimuli provided by the mouse: it was making the mouse provide the specific stimulus of "mouse needing to be angled out of crevice". Moreover, Leyhausen has shown that the connection between prey capture and eating is not established until after the first kills have been made, but prey catching play matures much earlier. Similar considerations apply to the racoon's dowsing behaviour, briefly mentioned in the last chapter. It has been known for many years that in captivity raccoons often take food from their feeding dish and dowse it in their water supply, but no corresponding behaviour has ever been seen in the wild. We owe to Lyall-Watson's (1963) careful analysis an understanding of this curious habit, which is not only of interest in itself but is a beautiful demonstration of the reality of endogenous motivation. The racoon has two types of food finding behaviour. It searches for insects and the like in dead leaves and litter and it also fishes in streams for crayfish and similar delicacies, feeling for them in the water with its extremely sensitive paws. In captivity, scraping the food out of its feeding dish provides a substitute for the first pattern, but the fishing pattern finds no outlet. The dowsing of its food is not an endogenous pattern for, although they will dabble their paws in the water, young reared in isolation will not dowse their food until somehow—by chance, or by the experimenter's design, they find food in the water and can at last perform an approximation to their normal fishing pattern. Once this has happened, they will promptly start to dowse; they have learnt that in this way they can give themselves the chance to discharge their pent up fishing pattern. The dowsing has no other function than this: to permit the animal to perform its normal pattern. The fact that it is done so readily and so regularly admits of no other interpretation than that the fishing behaviour is endogenously motivated. The pattern *requires* to be performed, for internal physiological reasons, regardless of the absence of any external need and the animal quickly learns a method of meeting this requirement. As already mentioned, captive *Atilax* will also dowse their food, and in the mountain beaver, *Aplodontia rufa*, Wandeler and Pilleri (1965) have seen a somewhat similar type of behaviour. They found that soft vegetable food such as lettuce was often put in the water dish and there "kneaded" with the paws. Since *Aplodontia* normally lives in the vicinity of streams, it is quite possible that this behaviour is related to a pattern of collecting some form of aquatic vegetation for food, in exactly the same way as the raccoon's dowsing is to its natural fishing habits. This type of appetitive food catching behaviour may be shown with no relation at all to eating. *Atilax*, for instance, will dowse inedible objects and will even go and fetch them from some distance in order to do so: in fact the animal shows appetitive behaviour for an appetitive behaviour pattern. The various mongooses that used stones and other inedible things as substitute objects for their egg breaking patterns also provide examples of patterns showing a need to be performed without relation to their normal biological role.

My male meerkat was obtained from a zoo, where for some months he had been kept in a concrete enclosure so that scratching for food was impossible. As already mentioned, this produced in him an almost insane need to scratch, to cope with which I finally provided him with a piece of soft-board fixed against a wall and renewed it when he had scratched a hole in it. I regard his behaviour as the result of disturbance of normal motivational control so that hyper-normal accumulation occurred, followed by hyper-normal discharge. Recovery from this condition took a considerable time. Similarly *Sminthopsis* when first moved from small cages to a large enclosure where they had the opportunity to forage for food spent so much time in this activity that I was at first afraid they might starve, although food in dishes was available in abundance.

Eibl-Eibesfeldt (1963) has produced evidence of a different type for the existence of motivation for nut opening in squirrels independent of the reward of eating the kernel. He presented naïve animals with nuts that had been split open, the kernel removed and halves glued together again. The squirrels continued to open the nuts and perfected their splitting technique in the same way as animals presented with normal nuts do. The performance of the splitting must therefore have been rewarding in its own right and did not require to be reinforced by eating. This does not, of course, mean that the eating is not also rewarding. With normal nuts at its disposal, a squirrel learns to distinguish a worm-eaten or empty nut from a sound one and, after some experience, rejects the former after a brief sniff.

Hoarding patterns are also performed regardless of need or with substitute objects. My *Cricetomys* would hoard inedible objects, empty tobacco tins being a favourite collectors' item*, and Eibl-Eibesfeldt (1963) records one of his squirrels removing all the decorations from a Christmas tree and burying them in the flower pot, despite their inedible nature.

In species showing extensive toilet behaviour, this activity also appears to occur regularly as a result of internally generated motivation and not simply in relation to afferent stimuli from the skin. I have found this to be true of *Cricetomys* and Barnett (1963) finds the same thing in *Rattus*. Furthermore, in a quantitative study of grooming in the former

* It is, of course, possible that this behaviour reflects the need to collect bedding material, rather than to hoard food.

species, I was unable to produce any scheme which would account for the findings without postulating some degree of independent motivational control of the different actions involved in the toilet (Ewer, 1967).

E. E. Shillito (1963) found that in her voles, exploratory behaviour did not require the stimulus of any alteration in the environment. It appeared to be performed spontaneously and she judged its motivation to be endogenous. It is, however, only fair to add that Sheppe (1966) regards similar behaviour in the deer mouse, *Peromyscus leucopus*, as a response to stimuli arising from the novelty which once familiar environmental features acquire, as memory of their details fades. I am inclined to give more weight to Shillito's opinion, since it is the result of watching what the voles did whereas Sheppe's conclusions are based on the interpretation of records showing the places visited by his mice.

In social animals amicable behaviour patterns are transferred to a human conspecific substitute with such ease and regularity as to suggest a definite motivation for performing the pattern as a more important factor than any basic similarity in the releasers provided. King's (1955) tame prairie dog, for instance, would always "kiss" his hand on first approaching him and the invitation to groom is made to a human owner by a wide variety of species. The need to show social responses may be very strong in some cases and this, coupled with lack of timidity, appears to account for the extraordinary ease with which some species become tame. My male meerkat, when first received from the zoo, had not been handled and would not permit me to touch him. He did, however, investigate me carefully by smelling and even crawled across my lap if I remained still. On his first night in my company and away from his fellows, he had crept into my bed to sleep in contact with me within five minutes of my going to bed. This can hardly be regarded as a response to warmth, since his movement to the bed was far too fast for him to have discovered that it was warm, but he did already know from his previous investigations that some sort of fellow mammal was now available as a sleeping partner. I interpret his behaviour as showing that the need for the normal social contact was sufficient to over-ride all caution and suspicion. Without some such assumption his behaviour is quite incomprehensible.

Behaviour connected with fighting is a difficult case. Marler and Hamilton (1966) are of the opinion that aggressive behaviour is normally a response to external stimuli and is not the result of endogenously arising appetitive behaviour, although searching for fights may develop as a learnt pattern, if some reward is thereby attained; Scott (1966) expresses very similar views. This may be so, but there is abundant evidence that the performance of an aggressive action can be its own reward and can generate motivation for further repetition. This is the case in the killing of mice by rats. Rats that do kill a mouse will continue to do so, even if denied the opportunity to eat their victims (Myer, 1964) and will learn a discrimination for the reward of being allowed to kill (Myer and White 1965). They also develop a sort of addiction and an experienced killer is less easily deterred by punishment than a beginner (Myer, 1967).

Viverrids are extremely pugnacious animals and provide examples of seeking opportunities to show aggressive behaviour. An adult female *Atilax* regularly invited me to a bout of playful fighting every evening when taken out for exercise and a male *Crossarchus* used to indulge in very rough fighting play before bed time. Both male and female meerkats used to seek for the opportunity to indulge in prolonged threatening against a pair of Alsatian dogs and when taken out for a walk would deliberately cross the road and head for the house where they had learnt that this experience was likely to be forthcoming. The male's towel-worrying has already been mentioned. This I regard as the use of a substitute object to discharge a pattern which may relate either to fighting or to attacking large prey.

Eibl-Eibesfeldt (1951b) found that male squirrels showed fluctuations in agression in relation to the time of the rut. Outside the breeding season, males would play-fight with him, showing the usual inhibition and not biting in earnest. As the breeding season started, they became gradually more aggressive, first attacking strangers and finally even their familiar owner. Such attacks were not playful, the bites were delivered with full force and were accompanied by threat. The animals would also sometimes threaten with no visible external cause: the squirrel would be sitting quietly, then suddenly lay its ears back and gnash its teeth. Outside the breeding season, such behaviour was not shown.

Another example of aggressive behaviour arising without any very clear external cause is provided by *Cricetomys* (Ewer, 1967). In this species the young, at the age of 11–12 days, start to show violent aggressive behaviour towards each other. This happens only when the mother has left the nest and ceases instantly on her return. The young have previously remained together with no signs of hostility but now, as they move about, one will suddenly bite viciously at a littermate who in turn attacks whatever is within his reach and so a chain reaction of infuriated nipping is set off and the whole litter appears to be overcome by an outburst of insensate rage. The eyes do not open until about 10 days later and the first bite made appears to be a response to movement by a littermate. Movement previously evoked no such reaction and some quite sudden maturation of attacking behaviour appears to be going on. It would be interesting to find out if this behaviour coincides with histologically visible changes in the adrenals. Enraged fighting occurs for a few days only and is gradually replaced by grooming instead of biting at littermates. Even a solitary youngster showed corresponding changes in behaviour at the same age: both previously and subsequently tractable, for a period of a few days he bit viciously when handled. It is not known whether a solitary animal that was never handled would show redirected aggression and worry the bedding, or even bite the mother, nor yet whether absence of opportunity to show this precocious aggression would have any effect on adult agonistic responses.

All these observations suggest that Marler and Hamilton's concept of aggressive behaviour as purely a matter of response to external stimuli may be too sweeping and too abstract. An animal in whose life intra-specific aggression is important must react to the remoter stimuli signifying the possible presence of a rival not with avoidance but by searching for the source of the stimulus in readiness to fight. Whether this should be described as a response to external stimuli or as appetitive behaviour towards fighting involves a rather tenuous distinction. Similarly, if an animal returns to the place it has previously found food, this would usually be interpreted as appetitive behaviour in which learning plays an orienting role. Is there any justification for describing differently the return to a place where a fight has occurred?

Scott (1966) regards the fact that fighting may be rewarding as of little importance and says that it does not demonstrate the existence of internal motivation "any more than the fact that most people find the odor of roses pleasant indicates that there is spontaneous internal stimulation to go out and smell flowers"—but of course, the latter is exactly what the animal does do; it goes looking for the opportunity to behave aggressively. This, of course, is precisely what we mean by describing something as "rewarding", for the animal has no other means of telling us that it "finds something pleasant". What may well prove to be true is that an aggressive response may require external stimuli to evoke it for the first time. Failing such release, there may be no appetitive behaviour and the threshold for spontaneous discharge may never be reached. This is exactly what happens with prey killing in the cat, but the fact that release by external stimuli forms a normal part of the ontogeny of the neural mechanism tells us nothing at all about whether or not it subsequently develops an ability to respond to endogenous motivational influences.

The degree to which aggressive tendencies are self motivated may vary very much from species to species, and one would expect that, while some species might require periodic outlet for aggressive tendencies, others might not. The conditions in which the animals live may be a relevant factor, for Barnett (1955) notes that strange male rats are tolerant of each other if no female is present and if they are introduced simultaneously to a strange environment, so that none can establish ownership in advance of the others. He never states explicitly that the tolerant rats showed no aggressive behaviour whatever and his formulations imply that this was not the case. He speaks, for instance, of males living together without killing each other, sometimes even without injury (1955) and of a clash between the animals being "unusual" (1963). Minor forms of threat are not mentioned. Similarly Scott (1966) speaks of rats and also mice as being able to live in communities for long periods "in a relatively peaceful manner". At the time of writing, a male *Thryonomys*, after driving away a rival, has lived with two females and their young for over a month without fighting. He does, however, now and then threaten and chase the youngsters, although he does not attack them and minor squabbles with the females over food also occur. These activities may provide the animal with all that is required in the way of outlet for its aggressive tendencies. Certainly it is perfectly clear from this animal's behaviour that one cannot equate absence of visible injuries, or even of overt fighting, with absence of aggressive behaviour.

Although further examples could, no doubt, be produced, the ones cited are sufficient to demonstrate that many instinctive actions do have endogenous motivation which is not directly related to their normal biological function and do show a need to be performed regardless of this function. But they also demonstrate something else: that the intensity of this motivation and the consequent urgency of the need to be performed varies from species to species and from one activity to another in relation to the role played by the particular type of behaviour in the life of the animal concerned. This has already been noted in the case of aggressive behaviour but it is also true of patterns concerned with getting food. It is when complex patterns are involved that the need for them to be performed is most obvious; and this applies as much to the squirrel's nut opening as to the complex prey capturing techniques of carnivores. This may be partly a matter of the complex pattern being more obvious to the human observer and therefore more likely to be recorded, but it is not the whole explanation, for there is at least one case of an adaptive absence (or extreme reduction) of such build up of endogenous motivation for patterns concerned with procuring food. This concerns *Cricetomys*. In the very similar pouched tree mouse, *Beamys major*, B. Morris (1962) reports that if the larder is well stocked, an animal may remain in its burrow for up to four days, living on the proceeds of its previous industry and avoiding exposing itself to the perils of the outside world. Although there is no information on the subject, it is extremely likely that a male *Cricetomys* will do the same—to judge from captivity observations. This species is reputed to be solitary but a male and female can be kept together without trouble. The female is much the more enthusiastic in hoarding. My pair were allowed the freedom of the laboratory in the evening and the female, who became very adept at raiding my stores, always kept the nest piled up with food. In this situation the male became obese and lethargic. He ate enormously but very rarely carried in any food and his excursions from the cage were brief. Dr. C. K. Brain (personal communication) observed the same thing in even more acute form with a pair which he kept. The female maintained a constant supply of food in the nest. The male ate to excess and scarcely ever left the nest so that finally he resembled in his obesity an animal with its hypothalamic satiety centre destroyed. He showed no signs of being driven forth to forage for himself as a result of accumulation of motivation for food collecting.

It thus seems that while endogenous motivation may be a characteristic of innate patterns, the degree of accumulation shown is highly variable and this in a manner adapted to the mode of life of the species concerned. This is hardly surprising; there is no reason why this particular aspect of behaviour should be exempt from the effects of natural selection. Complex food procuring patterns, as we have noted, are remarkable for a high degree of endogenous motivation. Leyhausen (1965) and I (Ewer, 1963) have independently pointed out the adaptive significance of this arrangement. Such patterns must show a high degree of persistance, for the animal may have to hunt repeatedly before it succeeds in capturing prey. They must also be self-rewarding, so that they are not extinguished by negative conditioning when no food is caught. Lorenz (1966a,b,) as we have already noted, has argued that bond forming patterns show this characteristic very strongly and by virtue of it they drive the animal to seek and find a social partner; they are not merely responses evoked by the presence of a partner. The adaptive significance of the motivational build up is here equally apparent.

If we consider what must go on in the central nervous system when a new innate pattern is evolved, it at once becomes obvious that it must inevitably have some form of independent motivation. The pattern may have begun as an intention movement, a displacement activity or a trial and error solution to a problem. In the course of time, the genetic basis for a distinctive motor pattern is evolved. This may result in extensive alteration in the original movements, with reduction or elimination of some components and exaggeration or repetition of others. The evolution of such a new motor sequence implies the elaboration of something new in the central nervous system, forming the structural basis of the new pattern. That such changes in the brain do occur is shown by the work of Brown and Hunsperger (1963). They were able to show by brain stimulation experiments that in cats threat patterns have their own central nervous representation and are not merely evoked simultaneous activation of loci for attack and for escape behaviour. The conclusions drawn by Roberts and Carey (1965) from the positions of the hypothalamic loci from which various patterns involving the use of the mouth could be elicited in rats are also relevant here.

The new pattern cannot consist of output alone; if it is to be activated, it must also have its controlling and directing input system; in other words, it must have motivation. We must expect selection to act not only on the output, making the final motor pattern adaptive, but also on the input side. This will involve selection for appropriate responsiveness to initiating stimuli, so that the pattern becomes linked with the most suitable releasers. Similarly there will be selection for responsiveness to whatever stimuli may be necessary for governing the course of the action and bringing it to an end when completed. Selection will also affect the persistence with which the action is shown. If high persistence is a desideratum, then some type of motivational storage capacity is necessary; and if the action is particularly important to the animal, an appropriate appetitive component may be included —which again implies the ability to store and maintain motivation. It is only to be expected that the degree to which these latter characteristics are shown should be adaptively related to the part the particular pattern plays in the life of the animal.

The simplest way to achieve a large storage capacity would appear to be some form of neurosecretory process. There is, however, no need to assume a complete uniformity of mechanism in all cases. The accumulation of neurosecretory substances in the relevant centre may be one answer. In the case of such a short period activity as breathing, rhythmically discharging cells whose pattern can be modulated by other influences may suffice. For the feeding centre a dependable source of motivation may possibly lie in tonically discharging receptors in the liver (themselves dependent on the liver's ability to store glycogen). In the patterns of defaecation and micturition, reflex excitation initiated by normal metabolic processes may be so dependable as to require little or nothing further as a source of endogenous motivation.

Before concluding this section, it is necessary to refer more explicitly to the motivational relationships that exist between different activities. Clearly no behavioural control system could be expected to operate in an integrated manner without there being some form of interaction between its component parts, and in fact there is good evidence for the existence of both inhibitory and excitatory linkages. As might be expected, allied activities may show facilitatory effects, frequently in a manner related to their normal temporal sequence: those of prey catching on eating and of eating on hoarding may be cited as examples. Another familiar one is the proneness of play, particularly playful fighting, to lead to sexual behaviour. Inhibitory interconnections, on the other hand, are essential if the animal is to be able to switch rapidly from one activity to another; the ease with which behaviour relating to escape or defence can inhibit any other activity is a case in point.

From his extensive studies on aphids, Kennedy (see 1966a, which gives references to earlier papers) has concluded that the facilitatory and inhibitory effects which one centre may exert upon another are extremely important in determining alterations in motivational level. Elsewhere (1954, 1966b) he has argued that there is no basic difference between the alterations in readiness to perform a fixed action pattern under discussion here and the changes in threshold of a spinal motoneurone which, as Sherrington (1906) showed long ago, characterise the reflex activities of the vertebrate spinal cord. These latter changes have been shown to be a function of the relative rates at which excitatory and inhibitory impulses are impinging on the motoneurone and the way in which these affect the polarisation of the membrane (Eccles, 1957). It seems to me, however, that it is a far cry from the way in which a spinal motoneurone can be held, by a suitable rate of input stimulation, at an excitatory level where a very small excitatory increase will cause it to fire, to the behaviour of Leyhausen's cat, repeatedly putting the mouse back in the crevice so as to be able to hook it out again. The time scales are very different and my own investigations on toilet behaviour and food hoarding in *Cricetomys* (1967) appear to require the postulation of a long term process in addition to short term effects, the latter probably of the type envisaged by Kennedy. My own objection to his attempt to relate all motivational changes to the single process of variation in rates of excitatory and inhibitory input impinging on summating membranes is based partly on a mental recoil from the colossal and, I believe, unnecessary impulse traffic that this demands. To perform the summation by means of a neurohumor of some type, which is steadily accumulated and not liberated until required, seems so much more economical than to use transmitter substances which are frittered away continually as impotent impulses, that it is difficult to believe natural selection would have been so inefficient as not to have "discovered" this. Furthermore, an accumulated chemical also provides a material basis for the differences in persistence which characterise the mechanisms governing the discharge of fixed action patterns just as clearly as do alterations in threshold for their release. This is a point which is often neglected and is concealed if accumulation of motivation is referred to merely as *threshold lowering*. To explain the differences in persistence in Sherringtonian terms, we must devise yet another mechanism which controls the input lines to our summating membrane, and which is itself less amenable to being turned off the longer it has been kept in action.

It thus appears that at present the most rewarding approach might be to devote attention to discovering the physiological basis for the observed changes in motivation level. For this it would be desirable to select for investigation a type of behaviour which shows very clearly the properties of accumulation and is characterised by major changes in persistence as well as in threshold. Carnivore prey catching patterns suggest themselves as possibly the most promising, but rodent nest building might also repay investigation, if it should prove possible to find the relevant brain locations. Roberts and Carey's (1965) brief mention of eliciting such actions as paper gathering and paper shredding indicate that this may not be very difficult. I would, however, suggest that the laboratory rat is not the animal of choice for this type of work. Selection for cage life may well have caused reduction in the capacity to store motivation as an adaptation to conditions where this ability could lead only to increasing frustration. If the form in which the accumulated motivation is stored can once be established—whether as increased rate of firing in certain cells or circuits or in some more direct chemical form—then it will be possible to turn to the problem of the immediate causation of the changes in level that occur. In the meantime, while it is almost inevitable that some workers will prefer to believe in the local origin of endogenous motivation in the relevant centre and others to refer its origins elsewhere, it is desirable to avoid being over dogmatic. The answer may ultimately prove to be neither one nor the other but both.

The ethologist's work, however, is done once he has shown that endogenous motivation is a reality, whatever its sources and forms of storage, and once he has carried out detailed quantitative observations on particular cases. The latter are necessary to map out what it is that requires to be explained and to detail the properties shown by the system as a whole. To elucidate the underlying mechanism then becomes the task of physiology.

If the ethologist in the past has oversimplified the matter and has often spoken as though there could be only one sort of physiological basis and one source and origin for endogenous motivation, of equal importance in all cases, he must be forgiven. He is, after all, only exemplifying one of his own principles: if a signal is to be effective and to succeed in transmitting its message, it must be noticed and interpreted. It must therefore be clear and unambiguous and, in attaining this end, simplification and exaggeration are valuable aids.

CHAPTER 13

The behaviour of mammals

The most striking behavioural characteristic of mammals is their learning ability. Compared with other animals, mammals can learn faster, learn more, remember more and show more insight. Correlated with these abilities is the possession of well developed sense organs, particularly the distance receptors concerned with olfaction, hearing and vision. These make it possible for behaviour to take into account and to be adapted to, an enlarged world, not limited to the objects in contact with or very close to the animal. Learning and memory enlarge this world still further, adding a time dimension and so taking in the past as well as the present.

All this, however, does not mean that the innate basis of behaviour has a diminished importance. The dependence of learning on the adapted innate basis has already been discussed in the last chapter. It does, however, imply certain characteristics in that innate basis. As we have already seen, complex behaviour which is yet sufficiently flexible to cope with a variety of circumstances has in many cases been achieved on the basis of built-in units which are small, often individual movements. These can be adaptively tailored by learning to fit a multiplicity of situations. Before such units can be functional, however, two further things are required. Their organisation within the central nervous system must be such that they can be joined together as the appetitive links in a chain of events leading to an end action that serves the biological needs of the species. There must also be a responsiveness to the end situation once attained, so that the stimuli then received terminate the activity. The functional organisation of the systems investigated by Roberts and his co-workers is of exactly the sort outlined. In the last chapter we have seen that this type of organisation leads the animal to discover how best to attain the correct end result and so to perfect its skill. A system of this sort is, of course, suitable only if the young can be protected and cared for by the parents during the learning period. Amongst carnivorous forms this sort of perfecting of technique is characteristic only of species in which long parental care is provided. In *Sminthopsis*, by contrast, where the young are virtually on their own from the moment they leave the nest, prey capture is very simple, based on a few innate responses and there is little development of individual skill (Ewer, 1968).

The sort of learning just discussed is, however, characteristic only of a certain type of behaviour; behaviour which is concerned with manipulating the environment—including the other species that may serve as food. There is also an equally important category to which these considerations do not apply. This comprises the movements used in intraspecific communication. The complex and conspicuous displays that occur in courtship and in threat and the smaller expressive movements that continually give information to conspecifics about current mood, all develop without going through a learning period of the sort described. Moreover, they develop in the species-specific form, even in animals reared from infancy in the company of species other than their own. I have seen this very clearly in the case of a meerkat reared with domestic cats and Fentress (1966) reports the same thing of a wolf reared with domestic dogs.

Even here, however, learning may have a part to play. It is not uncommon, in other vertebrate Classes, to find that signalling movements show what Morris (1957) originally called "typical intensity" and Tinbergen (1964) later "typical form". The movement is highly stabilised and shows a sort of all-or-nothing character: if it is performed at all, it is done fully and there is virtually no intensity scale. The gestures of a trained traffic policeman are an excellent example of deliberate use of the principle of typical form, in the interests of unambiguity. Normal mammalian expressive movements (see Plate VIII), in contrast, have the opposite characteristic and show every gradation from the barely perceptible to full intensity. Moreover, as noted in Chapter 2, they also show superposability; both constituents of a mixed motivation may be simultaneously expressed, with consequent enrichment of the vocabulary. The same is true of vocalisations; both gradation and superposition are common. Workers on primates have often stressed the gradations that exist both in facial expression and in vocalisations (Bolwig, 1959; Hinde and Rowell, 1962; Andrew, 1963; Schaller, 1963, 1965c; Hinde, 1966) and the same thing is also shown very clearly in the Carnivora. The scale of vocal gradation could not be more strikingly exemplified than it is in the threat call of *Atilax*, ranging as it does from the low "waauk", signifying no more than slight dissatisfaction, in a continuous spectrum to the truly appalling screech of high intensity threat. Tembrock (1963b) has demonstrated that canid vocalisations show superposition as well as gradation and two sounds may be combined or delivered in rapid alternation. The same applies to viverrids for the threat screech of *Atilax* may be mixed with growling and in *Suricata* I interpret the difference between the alarm notes given to an aerial and to a terrestrial predator as resulting from the fact that the former expresses pure fear while in the latter the fear call has a harshness derived from growling and reflects fear mixed with aggression (Ewer, 1963).

With such an extraordinarily sensitive expressive repertoire, it seems very difficult to believe that there could be a fully innate recognition of the significance of all the subtle variations that are possible. Workers on primates have suggested that an appreciation of the finer shades of meaning is the result of experience. This is gained during the juvenile period when the young are still treated with tolerance and a mistake or misunderstanding need not have very serious consequences. It seems very likely that the same is true of social carnivores. There may be in dogs an innate recognition of tail wagging as a friendly gesture but it seems too much to expect an appropriate built in response to suit all the intermediates that link the slow, large amplitude wag with the tail held low, that signifies a mood of relaxed amiability, with the rapid, tense, small amplitude vibration with the tail held high that carries a complex message of readiness to be friendly coupled with some anxiety or uncertainty as to the intentions of the opposite number, and at the same time indicates a readiness for defence if necessary. The fact that our dogs can learn to interpret our unconscious expressive movements suggests that the learning required would be easily within their powers. The history of Kluger Hans shows that the same considerations apply to horses.

The mammalian mode of reproduction, leading to the evolution of parental care, has provided the young with a learning period during which they can perfect the techniques required for dealing with the external world, but it has done more than that. The contact which originates in the relationship of parent and child has also led to the evolution of the intricate and sensitive signalling systems on which are based the complex and mutable forms of social relationship that characterise the mammals. Viviparity may have begun purely as a reproductive adaptation, but it provided also the preconditions for the evolution of a type of behavioural organisation richer and more flexible than that of any other vertebrate class.

The contrast just discussed between the expressive communication movements and the movements used in manipulating the environment shows itself in our own behaviour as well as in that of other species. Apart from a few neonatal patterns and possibly some connected with copulation, our facial expressions and our non-articulate vocalisations are the most complex built in movement patterns we possess and our responsiveness to them the most universal and least obviously dependent on learning. Our manipulative behaviour, on the other hand, seems to be almost totally synthesised by learning out of very simple elements. Leyhausen (1965) has pointed this out and regards the human condition as simply one stage beyond that of the cat. According to him, one should not consider man as particularly impoverished in innate patterns; it is rather than the quanta upon which his motor learning can operate are very small. Just as the cat can learn to perform a skilled action by directing the movement patterns at its disposal, so too can we—with the difference that our control is finer and our range of possible movements much wider. The difference, of course, reflects itself in the relative sizes of motor cortices and pyramidal tracts in cats and men. Moreover, just as the young monkey or puppy may learn the finer details of the innate expressive language of its species, so our own young face and master in addition the vastly more complex task of learning an articulate language. Here, however, there is a major discontinuity. The vocalisations of all other species carry information about one thing only—the current mood of the sender. In higher primates the messages carried may be very finely graded but there is nothing at all that foreshadows even the beginnings of the major characteristic of human speech—the freeing of the message from the mood. In fact Goodall (verbal communication, 9th international ethological conference) has commented on the chimpanzee's inability in this respect. The excited calls given when members of two groups meet summon any others within hearing; but a lone individual seeking for its fellows is quite incapable of making this call, apparently because its mood has nothing in common with the mood that follows the meeting.

Maybe we should not be surprised at this deficiency in our nearest relatives, for it seems not unreasonable to ask, "What would an ape have to say, if he could talk?" In other words, if there were great selective advantage in sending messages about things other than one's own current mood, would not something have turned up, upon which selection could begin to operate and so start at an ever accelerating tempo the process of the evolution of true language? There is, of course, one context in which actions normally occur divorced from the earnest mood—to wit, in play. It is tempting to suggest that discoveries made in vocal play may have had a hand in the process. It might, for instance, afford infinite amusement to a teenage proto-human if he discovered, by a happy chance, that by "playing" an alarm call he could send his revered grandparent running in an undignified manner to the nearest tree. The sad fate of the little boy who cried "wolf, wolf" once too often may be the oldest of all cautionary tales.

This flight of fancy is not intended to be taken over-seriously. There is, however, one further point to be made about the way in which mammalian behaviour is synthesised on the basis of the mutual adaptation of the innate and the learnt components. So far we have treated learning as something built out of, or on top of, the basic innate characteristics;

but I believe the relationship includes something further. In the first chapter it was suggested that, by a process akin to genetic assimilation, patterns that began as learnt ones or as direct responses to stimuli might come to be genetically determined. Some observations on marsupials give unexpected support to this hypothesis. In a recent study I have found amongst the Dasyuridae a number·of cases in which a piece of behaviour occurs as a direct response to stimulation, while amongst the placentals one can find parallel cases in which the behaviour has been incorporated as a component part of a larger sequence and requires no direct external trigger of its own.

In both *Sminthopsis* and *Dasycerus* shaking of small prey does not form a standard part of the catching and eating patterns but occurs only if the prey struggles. In contrast, viverrids shake small prey as they pick it up, whether it resists or not and even dead food is treated in this way. In *Sminthopsis* there is normally no stalking and ambushing, but if large or unfamiliar prey is pursued and there is some degree of intimidation, then the cautious flattened approach, punctuated by a pause before the final rush, is surprisingly reminiscent of the standard felid technique. *Dasycercus*, although a skilled burrower, does not block up the entrance when it retires inside, as is the routine practice in a variety of placental burrowers. One of my animals was tame enough to come and take food from my hand, as I sat in her enclosure, provided she was hungry. Once she had fed, however, she would usually return to the burrow and she then often blocked it behind her. The response was thus conditional upon the presence of something potentially dangerous in the immediate neighbourhood. In rats, Baillie and Morrison (1963) have shown that post-prandial face-washing does not require afferent stimuli from food particles on face or paws; it can be initiated via the hypothalamic feeding centre by impulses from the stomach. In *Sminthopsis* I found the occurrence of face-washing after eating a mealworm to be so sporadic and random that I could only conclude its appearance must depend on whether the animal did or did not happen to dirty its face slightly as it ate.

In these two marsupials we therefore have four cases where

an action which occurs as an individual response to a direct external stimulus and may or may not be shown in a certain context can be matched in placentals by corresponding behaviour which is built in as an essential component of the relevant total pattern. Few though they may be, it should be made clear that these examples came to light in the course of a very brief study on a single species, *Sminthopsis*, together with some incidental observations on a second, *Dasycercus*. Furthermore, the work was started in the expectation of finding exactly the opposite situation—that the marsupials would prove to have highly complex and rigid built-in sequences of behaviour. The future will show whether the behaviour of these two species is typical of marsupials in general or not, but the observations made do suggest the possibility that the corresponding placental patterns began in the same way, as direct responses to external triggers and that the first stage in their incorporation into more complex built-in sequences was to learn to perform them in advance of need. If, for instance, you find that a scorpion is more easily overpowered if you shake it—why not shake the next one before it gets the chance to struggle? If there is continued selection for quicker learning of the shaking response then, just as Waddington's (1953) selection for maximal expression of the crossveinless phenotype in response to temperature shock, produced lines of *Drosophila* which were crossveinless even in the absence of the morphogenetic trigger, so too one would expect that ultimately the behavioural response would occur as part of feeding behaviour without requiring a separate stimulus of its own.

There seems to be general agreement that marsupials are poorer learners than placentals (Sälzle, 1936; Neumann, 1961; Munn, 1964; Pollard and Lysons, 1967). The difference in brain organisation that makes them slower to learn than advanced placentals may also have made them slower to incorporate adaptive learnt behaviour into their innate repertoires.

In this connection, Eibl-Eibesfeldt's (1963) observations on the red squirrel's method of collecting nesting material are of interest. An experienced squirrel gathers a mouthful of dead grass or the like; then with a stroking movement of the paws downward from above the eyes, it pushes the loose ends into its mouth and grips them with its teeth. A neat bundle is thus produced and there is no danger of tripping, on the climb home to the nest. A naïve squirrel, however, gathers its mouthful of bedding and at once attempts to return to the nest with it. As soon as it stumbles, it promptly goes through the correct sequence of movements of the bundling pattern and presently learns to do so without having to trip on the loose ends first. Here, although the external trigger is still required, the distinctive sequence of movements is already built in; the squirrel does not have to find out by trial and error how to make a neat bundle: but, can one doubt that its ancestors had to do so—or that its descendants will no longer have to trip before they bundle?

What is being suggested here is that the relationship between innate and learnt is not a unidirectional traffic, with learning replacing innate patterns in higher forms, but a circular one in which the innate directs and provides the basis for the learnt. The learnt also provides the basis for new innate patterns and facilitates the accumulation of the necessary genetical changes, under the directing influence of selection. One would not expect this process to produce a larger and larger repertoire of innate patterns. Disruptive mutations are far commoner than advantageous ones and a pattern would not be likely to outlast its utility for very long.

What has just been said must not, of course, be taken to imply that a built in pattern is in general superior to a learnt one. It can be more advantageous only in cases where the problem to be solved recurs again and again in a virtually constant manner and where "right first time" does confer a real advantage over a trial and error solution. The behaviour of the cane rat provides an illustration. The basic feeding patterns* used in dealing with the grass stems and leaves which constitute the main food, are built in and do not have to be learnt. The youngster has therefore "got what it takes" to start him off, eating the food which is certain to be available. The ripe seeds of a number of species of grasses are a much appreciated delicacy but they are to be had only now and then and the structure of the flowering head varies from species to

* Described in Chapter 3.

species. Adult cane rats have a highly skilled way of dealing with the ripe heads of certain grasses. Holding the stem in their paws, they pull it through the mouth, hand-over-hand, stripping off the seeds, exactly as one may pull them off oneself between finger and thumb. This technique, however, is not innate and the young learn it gradually, trying at first to eat the seed bearing tips in the orthodox manner and discovering more or less by chance how easily they can be pulled off.

From what has been said about the relations of the learnt and the innate, it follows that in mammals, and particularly in placentals, behavioural evolution should have been particularly rapid and should have proceeded with an accelerating tempo wherever learning ability was most highly developed. Moreover, it implies that behavioural rigidity is most unlikely to have exerted any restriction on mammalian adaptive possibilities. It is no surprise to find that there are purely vegetarian Carnivora, a rodent that preys upon insects and two genera of bats that have taken to fishing, not to mention the changes in food finding behaviour that have characterised our own evolutionary history.

Whether or not all the types of process suggested have played a part, it remains true that the richness and diversity of mammalian behaviour provides a vast and rewarding field for further study. We have still much to learn at all levels, descriptive as well as analytical. Let us only hope that we succeed in directing our own behaviour in such a way as to call a halt to the process of elimination of our fellow mammals. If we fail to do this, we deprive our descendants of something that is not only a source of scientific information which may prove of value to us in coming to comprehend our own species, but also a source of almost infinite delight and pleasure; for, although we may subdivide and analyse, our studies of physiological and evolutionary mechanisms are pointless if we do not also appreciate (in the words of the famous physiologist, Professor A. V. Hill) "their biological synthesis into the complete and intelligent, the wonderful and beautiful, living creature".

GLOSSARY OF COMMON NAMES

In each case, the Order and Family are given as well as the scientific name. In the case of a few of the larger families of rodents and artiodactyls, the subfamily, instead of the family, is used.

AARDWOLF	*Proteles cristatus*	CARNIVORA: HYAENIDAE
ACOUCHI, green	*Myoprocta pratti*	RODENTIA: DASYPROCTIDAE
AGOUTI	*Dasyprocta aguti*	RODENTIA: DASYPROCTIDAE
ANT-EATER, spiny	*Tachyglossus aculeatus*	MONOTREMATA: TACHYGLOSSIDAE
ASS, wild	*Equus asinus*	PERISSODACTYLA: EQUIDAE
BABOON		
anubis	*Papio anubis*	PRIMATES: CERCOPITHECIDAE
chacma	*P. ursinus*	PRIMATES: CERCOPITHECIDAE
hamadryas	*P. hamadryas*	PRIMATES: CERCOPITHECIDAE
gelada	*Theropithecus gelada*	PRIMATES: CERCOPITHECIDAE
BADGER		
European	*Meles meles*	CARNIVORA: MUSTELIDAE
honey	*Mellivora capensis*	CARNIVORA: MUSTELIDAE
BANDICOOT, long-nosed	*Perameles nasuta*	MARSUPIALIA: PERAMELIDAE
BAT		
fishing	*Noctilio leporinus*	CHIROPTERA: NOCTILIONIDAE
Mexican guano	*Tadarida brasiliensis*	CHIROPTERA: MOLOSSIDAE
BEAR		
grizzly	*Ursus arctos*	CARNIVORA: URSIDAE
polar	*Thalarctos maritimus*	CARNIVORA: URSIDAE

BEAVER	*Castor fiber*	RODENTIA: CASTORIDAE
mountain	*Aplodontia rufa*	RODENTIA: APLODONTIDAE
BOAR		
European wild	*Sus scrofa*	ARTIODACTYLA: SUIDAE
BOBCAT	*Lynx rufus*	CARNIVORA: FELIDAE
BUSHBABIES	*Galago crassicaudatus, G. senegalensis*	PRIMATES: LORISIDAE
BUSHBUCK	*Tragelaphus scriptus*	ARTIODACTYLA: BOVINAE
BUSHPIG	*Potamochoerus koiropotamus*	ARTIODACTYLA: SUIDAE
CARACAL	*Felis caracal*	CARNIVORA: FELIDAE
CAT		
African wild	*Felis lybica*	CARNIVORA: FELIDAE
Bengal	*Prionailurus bengalensis*	CARNIVORA: FELIDAE
blackfooted	*Felis nigripes*	CARNIVORA: FELIDAE
domestic	*F. catus*	CARNIVORA: FELIDAE
fishing	*F. viverrina*	CARNIVORA: FELIDAE
golden	*Profelis temmincki*	CARNIVORA: FELIDAE
serval	*Felis serval*	CARNIVORA: FELIDAE
wild	*F. silvestris*	CARNIVORA: FELIDAE
CHAMOIS	*Rupicapra rupicapra*	ARTIODACTYLA: CAPRINAE
CHEETAH	*Acinonyx jubatus*	CARNIVORA: FELIDAE
CHIMPANZEE	*Pan troglodytes*	PRIMATES: PONGIDAE
CHINCHILLA	*Chinchilla* spp.	RODENTIA: CHINCHILLIDAE
CHIPMUNKS	*Tamias, Eutamias*	RODENTIA: SCIURIDAE

COATIS	*Nasua narica, N. rufa*	CARNIVORA: PROCYONIDAE
COTTONTAIL	*Sylvilagus floridanus*	LAGOMORPHA: LEPORIDAE
COW	*Bos taurus*	ARTIODACTYLA: BOVINAE
COYOTE	*Canis latrans*	CARNIVORA: CANIDAE
COYPU	*Myocastor coypus*	RODENTIA: CAPROMYIDAE
DEER		
axis	*Axis axis*	ARTIODACTYLA: CERVIDAE
black-tailed	*Odocoileus hemionus*	ARTIODACTYLA: CERVIDAE
Chinese water	*Hydropotes inermis*	ARTIODACTYLA: CERVIDAE
fallow	*Dama dama*	ARTIODACTYLA: CERVIDAE
red	*Cervus elaphus*	ARTIODACTYLA: CERVIDAE
roe	*Capreolus capreolus*	ARTIODACTYLA: CERVIDAE
white-tailed	*Odocoileus virginianus*	ARTIODACTYLA: CERVIDAE
DEVIL, Tasmanian	*Sarcophilus harrisi*	MARSUPIALIA: DASYURIDAE
DOG		
African hunting	*Lycaon pictus*	CARNIVORA: CANIDAE
bush	*Speothos venaticus*	CARNIVORA: CANIDAE
domestic	*Canis familiaris*	CARNIVORA: CANIDAE
prairie—see Prairie dog		
raccoon	*Nyctereutes procyonoides*	CARNIVORA: CANIDAE
DORMOUSE	*Glis glis*	RODENTIA: GLIRIDAE
DOUROUCOULI	*Aotus trivirgatus*	PRIMATES: CEBIDAE
DRILL	*Mandrillus leucophaeus*	PRIMATES: CERCOPITHECIDAE

DROMEDARY	*Camelus dromedarius*	ARTIODACTYLA: CAMELIDAE
ELAND	*Taurotragus oryx*	ARTIODACTYLA: BOVINAE
ELEPHANT		
African	*Loxodonta africana*	PROBOSCIDEA: ELEPHANTIDAE
Indian	*Elephas maximus*	PROBOSCIDEA: ELEPHANTIDAE
ELK	*Cervus canadensis*	ARTIODACTYLA: CERVIDAE
FENNEC	*Fennecus zerda*	CARNIVORA: CANIDAE
FERRET	*Mustela furo*	CARNIVORA: MUSTELIDAE
FOSSA	*Cryptoprocta ferox*	CARNIVORA: VIVERRIDAE
FOX		
Bat-eared	*Otocyon megalotis*	CARNIVORA: CANIDAE
European red	*Vulpes vulpes*	CARNIVORA: CANIDAE
GAZELLE		
Grant's	*Gazella granti*	ARTIODACTYLA: ANTILOPINAE
Thomson's	*G. thomsoni*	ARTIODACTYLA: ANTILOPINAE
GEMSBOK	*Oryx gazella*	ARTIODACTYLA: HIPPOTRAGINAE
GERBIL, Persian	*Meriones persicus*	RODENTIA: GERBILLINAE
GIBBONS	*Hylobates*	PRIMATES: PONGIDAE
GIRAFFE	*Giraffa camelopardalis*	ARTIODACTYLA: GIRAFFIDAE
GLIDER, sugar	*Petaurus breviceps*	MARSUPIALIA: PHALANGERIDAE
GOAT		
domestic	*Capra hircus*	ARTIODACTYLA: CAPRINAE
mountain	*Oreamos americanus*	ARTIODACTYLA: CAPRINAE
GORILLA	*Gorilla gorilla*	PRIMATES: PONGIDAE

GROUND-SQUIRREL		
African	*Xerus erythropus*	RODENTIA: SCIURIDAE
Mexican	*Citellus mexicanus*	RODENTIA: SCIURIDAE
Uinta	*C. armatus*	RODENTIA: SCIURIDAE
GUANACO	*Lama guanicoe*	ARTIODACTYLA: CAMELIDAE
GUINEA PIG	*Cavia porcellus*	RODENTIA: CAVIIDAE
GRYSBOK, Cape	*Raphiceros melanotis*	ARTIODACTYLA: ANTILOPINAE
HAMSTER		
common	*Cricetus cricetus*	RODENTIA: CRICETINAE
golden	*Mesocricetus auratus*	RODENTIA: CRICETINAE
HARE		
blue	*Lepus timidus*	LAGOMORPHA: LEPORIDAE
brown	*L. europaeus*	LAGOMORPHA: LEPORIDAE
HEDGEHOG	*Erinaceus europaeus*	INSECTIVORA: ERINACEIDAE
HIPPOPOTAMUS	*Hippopotamus amphibius*	ARTIODACTYLA: HIPPOPOTAMIDAE
HOG, forest	*Hylochoerus meinertzhageni*	ARTIODACTYLA: SUIDAE
HORSE,		
domestic	*Equus caballus*	PERISSODACTYLA: EQUIDAE
Przewalski	*Equus przewalskii*	PERISSODACTYLA: EQUIDAE
HYAENA,		
spotted	*Crocuta crocuta*	CARNIVORA: HYAENIDAE
striped	*Hyaena hyaena*	CARNIVORA: HYAENIDAE
HYRAX, Mt. Kenya	*Procavia johnstoni*	HYRACOIDEA: PROCAVIIDAE
IMPALA	*Aepyceros melampus*	ARTIODACTYLA: ANTILOPINAE
JACKAL		
black-backed	*Canis mesomelas*	CARNIVORA: CANIDAE

golden	*C. aureus*	CARNIVORA: CANIDAE
JAGUAR	*Panthera onca*	CARNIVORA: FELIDAE
KANGAROO		
grey	*Macropus giganteus*	MARSUPIALIA: MACROPODIDAE
red	*Megaleia rufa*	MARSUPIALIA: MACROPODIDAE
tree	*Dendrolagus leucogenys*	MARSUPIALIA: MACROPODIDAE
KOALA	*Phascolarctos cinereus*	MARSUPIALIA: ?PHALANGERIDAE/ ?PHASCOLOMIDAE*
KOB, Uganda	*Adenota kob*	ARTIODACTYLA: HIPPOTRAGINAE
KUDU	*Tragelaphus strepsiceros*	ARTIODACTYLA: BOVINAE
KUSIMANSE	*Crossarchus obscurus*	CARNIVORA: VIVERRIDAE
LANGUR	*Presbytis entellus*	PRIMATES: CERCOPITHECIDAE
LEMMING		
Norwegian	*Lemmus lemmus*	RODENTIA: MICROTINAE
steppe	*Lagurus lagurus*	RODENTIA: MICROTINAE
LEMUR		
gentle	*Hapalemur griseus*	PRIMATES: LEMURIDAE
lesser mouse	*Microcebus murinus*	PRIMATES: LEMURIDAE
ringtailed	*Lemur catta*	PRIMATES: LEMURIDAE
LEOPARD	*Panthera pardus*	CARNIVORA: FELIDAE
LION	*Panthera leo*	CARNIVORA: FELIDAE
LORIS, slow	*Loris tardigradus*	PRIMATES: LORISIDAE

* The koala is normally included in the Phalangeridae: I consider its affinities are probably closer to the wombats and would therefore rather include it in the Phascolomidae.

LYNX	*Lynx lynx*	CARNIVORA: FELIDAE
MANDRILL	*Mandrillus sphinx*	PRIMATES: CERCOPITHECIDAE
MARA	*Dolichotis patagona*	RODENTIA: CAVIIDAE
MARMOSET	*Hapale jaccus* (= *Callithrix jaccus*)	PRIMATES: CALLITHRICIDAE
MARMOT		
European	*Marmota marmota*	RODENTIA: SCIURIDAE
yellow-bellied	*M. flaviventris*	RODENTIA: SCIURIDAE
MARSUPIAL "MOUSE"		
fat-tailed	*Sminthopsis crassicaudata*	MARSUPIALIA: DASYURIDAE
yellow-footed	*Antechinus flavipes*	MARSUPIALIA: DASYURIDAE
MARTEN, pine	*Martes martes*	CARNIVORA: MUSTELIDAE
MEERKAT	*Suricata suricatta*	CARNIVORA: VIVERRIDAE
MOLE	*Talpa europea*	INSECTIVORA: TALPIDAE
MONGOOSE,		
banded	*Mungos mungo*	CARNIVORA: VIVERRIDAE
Cape grey	*Herpestes pulverulentus*	CARNIVORA: VIVERRIDAE
crab-eating	*H. urva*	CARNIVORA: VIVERRIDAE
dwarf	*Helogale undulata* (= *H. parvula*)	CARNIVORA: VIVERRIDAE
Indian	*Herpestes edwardsi*	CARNIVORA: VIVERRIDAE
marsh	*Atilax paludinosus*	CARNIVORA: VIVERRIDAE
yellow	*Cynictis penicillata*	CARNIVORA: VIVERRIDAE
MONKEY		
bonnet	*Macaca radiata*	PRIMATES: CERCOPITHECIDAE
capuchin	*Cebus apella*	PRIMATES: CEBIDAE

howler	*Alouatta palliata*	PRIMATES: CEBIDAE
olive colobus	*Procolobus verus*	PRIMATES: CERCOPITHECIDAE
patas	*Erythrocebus patas*	PRIMATES: CERCOPITHECIDAE
rhesus	*Macaca mulatta*	PRIMATES: CERCOPITHECIDAE
spider	*Ateles sp.*	PRIMATES: CEBIDAE
squirrel	*Saimiri sciureus*	PRIMATES CEBIDAE
titi	*Callicebus moloch*	PRIMATES: CEBIDAE
MOOSE	*Alces alces*	ARTIODACTYLA: CERVIDAE
MOUFLON	*Ovis ammon*	ARTIODACTYLA: CAPRINAE
MOUSE	*Mus musculus*	RODENTIA: MURINAE
deer	*Peromyscus leucopus*	RODENTIA: CRICETINAE
field	*Apodemus spp.*	RODENTIA: MURINAE
	Microtus spp.	RODENTIA: MICROTINAE
grasshopper	*Onychomys torridus* *O. lencogaster*	RODENTIA: CRICETINAE
hare	*Lagidium peruanum*	RODENTIA: CHINCHILLIDAE
Australian hopping	*Notomys spp.*	RODENTIA: MURINAE
pocket	*Perognathus pacificus*	RODENTIA: HETEROMYIDAE
snow	*Chionomys* (= *Microtus*) *nivalis*	RODENTIA: MICROTINAE
spiny	*Acomys cahirensis*	RODENTIA: MURINAE
striped	*Lemniscomys striatus*	RODENTIA: MURINAE
pouched tree	*Beamys major*	RODENTIA: MURINAE

"MOUSE", marsupial,—see Marsupial "mouse".

MULGARA	*Dasycercus cristicauda*	MARSUPIALIA: DASYURIDAE
MUNTJACK	*Muntiacus sp.*	ARTIODACTYLA: CERVIDAE
MUSKRAT	*Ondatra zibetica*	RODENTIA: MICROTINAE
NUMBAT	*Myrmecobius fasciatus*	MARSUPIALIA: DASYURIDAE
OCELOT	*Leopardus pardalis*	CARNIVORA: FELIDAE
tree	*L. wiedi*	CARNIVORA: FELIDAE
OKAPI	*Okapia johnstoni*	ARTIODACTYLA: GIRAFFIDAE
OLINGO	*Bassaricyon sp.*	CARNIVORA: PROCYONIDAE
ONAGER	*Equus hemionus*	PERISSODACTYLA: EQUIDAE
OPOSSUM, Virginia	*Didelphis marsupialis*	MARSUPIALIA: DIDELPHIDAE
ORANG-UTAN	*Pongo pygmaeus*	PRIMATES: PONGIDAE
OTTER		
Canadian	*Lutra canadensis*	CARNIVORA: MUSTELIDAE
Cape clawless	*Aonyx capensis*	CARNIVORA: MUSTELIDAE
sea	*Enhydra lutris*	CARNIVORA: MUSTELIDAE
PACA	*Cuniculus paca*	RODENTIA: DASYPROCTIDAE
PACARANA	*Dinomys branickii*	RODENTIA: DINOMYIDAE
PANGOLIN, tree	*Manis tricuspis*	PHOLIDOTA: MANIDAE
PECCARY	*Tayassu sp.*	ARTIODACTYLA: TAYASSUIDAE
PIG, domestic	*Sus scrofa*	ARTIODACTYLA: SUIDAE
PIKAS	*Ochotona spp.*	RODENTIA: OCHOTONIDAE
PLATYPUS	*Ornithorhynchus anatinus*	MONOTREMATA: ORNITHORHYNCHIDAE

POLECAT		
European	*Mustela putorius*	CARNIVORA: MUSTELIDAE
lesser African	*Poecilogale albinucha*	CARNIVORA: MUSTELIDAE
PORCUPINE,		
American	*Erethizon dorsatum*	RODENTIA: ERETHIZONTIDAE
brush-tailed	*Atherurus africanus*	RODENTIA: HYSTRICIDAE
giant or crested	*Hystrix cristata*	RODENTIA: HYSTRICIDAE
Indian	*H. leucura*	RODENTIA: HYSTRICIDAE
POSSUM, brush-tailed	*Trichosurus vulpecula*	MARSUPIALIA: PHALANGERIDAE
POTOROO	*Potorous tridactylus*	MARSUPIALIA: MACROPODIDAE
PRAIRIE DOG	*Cynomys ludovicianus*	RODENTIA: SCIURIDAE
PUMA	*Felis concolor*	CARNIVORA: FELIDAE
RABBIT		
common	*Oryctolagus cuniculus*	LAGOMORPHA: LEPORIDAE
swamp	*Sylvilagus aquaticus*	LAGOMORPHA: LEPORIDAE
RACCOON	*Procyon lotor*	CARNIVORA: PROCYONIDAE
RAT		
black	*Rattus rattus*	RODENTIA: MURINAE
brown	*Rattus norvegicus*	RODENTIA: MURINAE
cane	*Thryonomys swinderianus*	RODENTIA: THRYONOMYIDAE
giant	*Cricetomys gambianus*	RODENTIA: ?MURINAE/ ?CRICETOMYINAE*
kangaroo, giant	*Dipodomys ingens*	RODENTIA: HETEROMYIDAE

* Although normally classified in the Murinae, *Cricetomys* also has affinities with the cricetids. Petter (1964), largely on the basis of tooth cusp pattern, has proposed to include the genus in the Cricetidae, in a separate sub-family, the Cricetomyinae.

Fresno	*D. nitratoides*	RODENTIA: HETEROMYIDAE
wood	*Neotoma fuscipes*	RODENTIA: CRICETINAE
REINDEER	*Rangifer tarandus*	ARTIODACTYLA: CERVIDAE
RHINOCEROS, black	*Diceros bicornis*	PERISSODACTYLA: RHINOCEROTIDAE
Indian	*Rhinoceros unicornis*	PERISSODACTYLA: RHINOCEROTIDAE
white	*Ceratotherium simum*	PERISSODACTYLA: RHINOCEROTIDAE
SABLE	*Martes zibellina*	CARNIVORA: MUSTELIDAE
SASIN	*Antilope cervicapra*	ARTIODACTYLA: ANTILOPINAE
SEA-BEAR	*Arctocephalus gazella*	PINNIPEDIA: OTARIIDAE
SEA-LEOPARD	*Hydrurga leptonyx*	PINNIPEDIA: PHOCIDAE
SEA-LION		
Galapagos	*Zalopus wollebaecki*	PINNIPEDIA: OTARIIDAE
South American	*Otaria byronia*	PINNIPEDIA: OTARIIDAE
Steller's	*Eumetopias jubata*	PINNIPEDIA: OTARIIDAE
SEAL, common	*Phoca vitulina*	PINNIPEDIA: PHOCIDAE
elephant, northern	*Miroungua angustirostris*	PINNIPEDIA: PHOCIDAE
elephant, southern	*M. leonina*	PINNIPEDIA: PHOCIDAE
grey	*Halichoerus grypus*	PINNIPEDIA: PHOCIDAE
harp	*Phoca groenlandica*	PINNIPEDIA: PHOCIDAE
hooded	*Cystophora cristata*	PINNIPEDIA: PHOCIDAE
northern fur	*Callorhinus ursinus*	PINNIPEDIA: OTARIIDAE

SHEEP		
bighorn	*Ovis canadensis*	ARTIODACTYLA: CAPRINAE
domestic	*Ovis aries*	ARTIODACTYLA: CAPRINAE
Stone's	*O. dalli*	ARTIODACTYLA: CAPRINAE
SHREW		
common	*Sorex araneus*	INSECTIVORA: SORICIDAE
elephant	*Nasilio brachyurus*	INSECTIVORA: MACROSCELIDIDAE
otter	*Mesopotamogale* (= *Micropotamogale*) *ruwenzorii*	INSECTIVORA: POTAMOGALIDAE
red musk	*Crocidura hirta*	INSECTIVORA: SORICIDAE
short-tailed	*Blarina brevicauda*	INSECTIVORA: SORICIDAE
Turkestan desert	*Diplomesodon pulchellum*	INSECTIVORA: SORICIDAE
SIFAKA	*Propithecus verreauxi*	PRIMATES: LEMURIDAE
SITATUNGA	*Tragelaphus spekei*	ARTIODACTYLA: BOVINAE
SKUNK		
striped	*Mephitis mephitis*	CARNIVORA: MUSTELIDAE
spotted	*Spilogale putorius*	CARNIVORA: MUSTELIDAE
SQUIRREL		
European red	*Sciurus vulgaris*	RODENTIA: SCIURIDAE
grey	*S. carolinensis*	RODENTIA: SCIURIDAE
palm	*Funambulus pennanti*	RODENTIA: SCIURIDAE
pine	*Tamiasciurus hudsonicus*	RODENTIA: SCIURIDAE
TARSIER	*Tarsius sp.*	PRIMATES: TARSIIDAE
TAYRA	*Eira barbara*	CARNIVORA: MUSTELIDAE

TENREC		
common	*Tenrec ecaudatus*	INSECTIVORA: TENRECIDAE
hedgehog	*Echinops telfairi*	INSECTIVORA: TENRECIDAE
TIGER	*Panthera tigris*	CARNIVORA: FELIDAE
TOPI	*Damaliscus korrigum*	ARTIODACTYLA: HIPPOTRAGINAE
TREE-SHREW	*Tupaia sp.*	PRIMATES: TUPAIIDAE
VICUÑA	*Vicugna vicugna*	ARTIODACTYLA: CAMELIDAE
VOLE		
bank	*Clethrionomys glareolus*	RODENTIA: MICROTINAE
mountain	*Alticola strelzovi*	RODENTIA: MICROTINAE
short-tailed	*Arvicola terrestris*	RODENTIA: MICROTINAE
WALLABY, swamp	*Wallabia bicolor*	MARSUPIALIA: MACROPODIDAE
WARTHOG	*Phacochoerus africanus**	ARTIODACTYLA: SUIDAE
WATERBUCK	*Kobus ellipsiprymnus*	ARTIODACTYLA: HIPPOTRAGINAE
WILDEBEEST		
black	*Connochaetes gnou*	ARTIODACTYLA: HIPPOTRAGINAE
blue	*C. taurinus*	ARTIODACTYLA: HIPPOTRAGINAE
WOLF	*Canis lupus*	CARNIVORA: CANIDAE
WOMBAT		
hairy-nosed	*Lasiorhinus latifrons*	MARSUPIALIA: PHASCOLOMIDAE
ZEBRA		
Burchell's	*Equus burchelli*	PERISSODACTYLA: EQUIDAE
Grevy's	*E. grevyi*	PERISSODACTYLA: EQUIDAE

* The name *Phacochoerus aethiopicus* (Pallas) refers to the now extinct Cape Warthog. In common with many palaeontologists, I therefore prefer to use the name *P. africanus* (Gmelin) for the extant animal—although admittedly it is possible that the two differed only subspecifically, not specifically.

BIBLIOGRAPHY

ADAMSON, J. 1960. *Born free*. Collins & Harvill Press, London.

ALEXANDER, A. J. & EWER, R. F. 1959. Observations on the biology and behaviour of the smaller African polecat. *Afr. wild Life*, **13**, 313–320.

ALEXANDER, G. & WILLIAMS, D. 1966. Teat-seeking activity in lambs during the first hours of life. *Anim. Behav.*, **14**, 166–176.

ALLAN, P. F. 1946. Notes on *Dipodomys ordii richardsoni*. *J. Mammal.*, **27**, 271–272.

ALTMANN, M. 1963. Naturalistic studies of maternal care in moose and elk. *Maternal Behavior in Mammals* ed. H. L. Rheingold. John Wiley & Sons, New York, 233–253.

ALTMANN, S. A 1962. A field study of the sociobiology of rhesus monkeys, *Macaca mulatta*. *Ann. N.Y. Acad. Sci.*, **102**, 338–435.

ANAND, B. K. 1961. Nervous regulation of food intake. *Physiol. Rev.*, **41**, 677–708.

ANDERSSON, B., LARSSON, S. & PERSSON, N. 1960. Some characteristics of the hypothalamic "drinking centre" as shown by the use of permanent electrodes. *Acta physiol. scand.*, **50**, 140–152.

ANDERSSON, B., PERSSON, N. & STRÖM, L. 1960. Post-stimulatory electrical activity in the preoptic "heat loss centre" concomitant with persistent thermoregulatory response. *Ibid.*, **50**, 54–61.

ANDERSSON, B. & WYRWICKA, W. 1957. The elicitation of a drinking motor conditioned reaction by electrical stimulation of the hypothalamic "drinking area" in the goat. *Ibid.*, **41**, 194–198.

ANDREW, R. J. 1963. The origin and evolution of the calls and facial expressions of primates. *Behaviour*, **20**, 1–109.

ANTONIUS, O. 1937. Über Herdenbildung und Paarungseigentumlichkeiten der Einhufer. *Z. Tierpsychol.*, **1**, 259–289.

ANTONIUS, O. 1943. Nachtrag zu "Symbolhandlungen und Verwandtes bei Säugetieren". *Ibid*, **5**, 38–42.

ANTONIUS, O. 1955. Beobachtungen an Einhufern. *Ibid.*, **12**, 169–174.

ARMITAGE, K. B. 1962. Social behaviour of a colony of the yellow-bellied marmot (*Marmota flaviventris*). *Anim. Behav.*, **10**, 319–331.

ARVOLA, A., ILMEN, M. & KOPONEN, T. 1962. On the aggressive behaviour of the Norwegian lemming (*Lemmus lemmus*), with special reference to the sounds produced. *Arch. Soc. zool. bot. fenn. Vanamo*, **17**, 80–101.

BACKHAUS, D. 1960. Über das Kampverhalten beim Steppen-Zebra (*Equus quagga* H. Smith 1841). *Z. Tierpsychol.*, **17**, 345–350.

BAILEY, V. 1939. The solitary lives of two little pocket mice. *J. Mammal.*, **20**, 325–328.

BAILLIE, P. & MORRISON, S. D. 1963. The nature of the suppression of food intake by lateral hypothalamic lesions in rats. *J. Physiol.*, **165**, 227–245.

BAKKEN, A. 1959. Behavior of gray squirrels. *Symposium on gray squirrel*. Contr. 162, Maryland Dept. Res. Ed., 393–407.

BALLY, G. 1945. *Vom Ursprung und von den Grenzen der Freiheit, eine Deutung des Spieles bei Tier und Mensch*. Birkhäuser, Basel.

BALPH, D. E. & STOKES, A. W. 1963. On the ethology of a population of Uinta ground squirrels. *Amer. Midl. Nat.*, **69**, 106–126.

BANNIKOV, A. G. 1964. Biologie du chien viverrin en U.R.S.S. *Mammalia*, **28**, 1–39.

BARCROFT, J. 1946. *Researches on prenatal life*. Blackwell, Oxford.

BARNETT, S. A. 1955. Competition among wild rats. *Nature, Lond.*, **175**, 126–127.

BARNETT, S. A. 1963. *A study in behaviour*. Methuen & Co., London.

BARNETT, S. A. & BURN, J. 1967. Early stimulation and maternal behaviour. *Nature, Lond.*, **213**, 150–152.

BARRASS, R. 1959. The courtship behaviour of *Mormoniella vitripennis* Walk. (Hymenoptera, Pteromalidae). *Behaviour*, **15**, 185–209.

BARRASS, R. 1961. A quantitative study of the behaviour of the male *Mormoniella vitripennis* (Walker) towards two constant stimulus situations. *Ibid.*, **18**, 288–312.

BARTLETT, D. & BARTLETT, J. 1961. Observations while filming African game. *S. Afr. J. Sci.*, **57**, 313–320.

BEACH, F. A. 1939. Maternal behavior of the pouchless marsupial *Marmosa cinerea*. *J. Mammal.*, **20**, 315–322.

BEACH, F. A. 1942. Comparison of copulatory behavior in male rats reared in isolation, cohabitation and segregation. *J. genet. Psychol.*, **60**, 212–236.

BEACH, F. A. 1945. Current concepts of play in animals. *Amer. Nat.*, **79**, 523–541.

BEATTY, H. 1951. A note on the behavior of the chimpanzee. *J. Mammal.* **32**, 118.

BENNET, A. L., DIAMOND, M. C., ROSENZWEIG, M. R. & KRECH, D. 1964. Chemical and anatomical plasticity of brain. *Science*, **146**, 610–619.

BERGAMINI, D. 1965. *The land and wildlife of Australasia*. Life Nature Library. Time Life Books, Amsterdam.

BINDRA, D. 1948. What makes a rat hoard? *J. comp. physiol. Psychol.*, **41**, 397–402.

BOLWIG, N. 1959. A study of the behaviour of the chacma baboon, *Papio ursinus*. *Behaviour*, **14**, 136–163.

BOOTH, A. H. 1957. Observations on the natural history of the olive colobus monkey, *Procolobus verus* (van Beneden). *Proc. zool. Soc. Lond.*, **129**, 421–430.

BOURLIÈRE, F. 1955. *The natural history of mammals*. Harrap, London.

BROWN, J. L. & HUNSPERGER, R. W. 1963. Neuroethology and the motivation of agonistic behaviour. *Anim. Behav.*, **11**, 438–448.

BROWNLOW, A. L'E. 1940. Crab-eating mongoose (*Herpestes urva* (Hodgs.)) in captivity. *J. Bombay nat. Hist. Soc.*, **41**, 893–894.

BRUELL, J. H. 1962. Dominance and segregation in the inheritance of quantitative behavior in mice. *Roots of behavior*, ed. E. L. Bliss, Harper & Row, New York, 48–67.

BUBENIK, A. B. 1965. Beitrag zur Geburtskunde und zu den Mutter-Kind-Beziehungen des Reh- (*Capreolus capreolus L.*) und Rotwildes (*Cervus elaphus L.*). *Z. Säugetierk.* **30**, 65–128.

BUECHNER, H. K. 1961. Territorial behaviour in Uganda kob. *Science*, **133**, 698–699.

BUECHNER, H. K. & SCHLOETH, R. 1965. Ceremonial mating behaviour in Uganda kob (*Adenota kob thomasi* Neuman). *Z. Tierpsychol.*, **22**, 209–225.

BUNN, D. S. 1966. Fighting and moult in shrews. *J. Zool.*, **148**, 580–582.

BURCKHARDT, D. 1958. Kindliches Verhalten als Ausdrucksbewegung im Fortpflanzungszeremoniell einiger Wiederkauer. *Rev. suisse Zool.*, **65**, 311–316.

BURRELL, H. 1927. *The platypus*. Angus & Robertson Ltd., Sydney.

BURTON, M. 1959. *Phoenix reborn*. Hutchinson, London.

BUSNEL, R. G. 1963. On certain aspects of animal acoustic signals. *Acoustic behaviour of animals*, ed. R. G. Busnel. Elsevier, New York, 69–111.

BUSS, I. O. & SMITH, N. S. 1966. Observations on reproduction and breeding behavior of the African elephant. *J. Wildl. Mgmt.*, **30**, 375–388.

CALABY, J. H. 1960. Observations on the banded ant-eater *Myrmecobius f. fasciatus* Waterhouse (Marsupialia), with particular reference to its food habits. *Proc. zool. Soc. Lond.*, **135**, 183–206.

CARLIER, C. & NOIROT, E. 1965. Effects of previous experience on maternal retrieving by rats. *Anim. Behav.*, **13**, 423–426.

CARPENTER, C. R. 1934. A field study of the behavior and social relations of the howling monkeys (*Alouatta palliata*). *Comp. Psychol. Monogr.*, **10**, 1–168

CARPENTER, C. R. 1965. The howlers of Barro Colorado Island. *Primate behavior—field studies of monkeys and apes*. ed. I. DeVore, Holt, Rinehart & Winston, New York. 250–291.

CARR, N. 1962. *Return to the wild*. 1965 Edition, Fontana Books, London.

CARRINGTON, R. 1958. *Elephants*. Penguin Books, Harmondsworth.

CLARKE, J. R. 1956. The aggressive behaviour of the vole. *Behaviour*, **9**, 1–23.

COHEN, H. B. & DEMENT, W. C. 1966. Sleep: suppression of rapid eye movement phase in the cat after electroconvulsive shock. *Science*, **154**, 396–398.

CRISLER, L. 1959. *Arctic wild*. Secker & Warburg, London.

CROWCROFT, P. 1957. *The life of the shrew*. Max Reinhardt, London.

CROWCROFT, P. 1966. *Mice all over*. G. T. Foulis & Co., London.

CULBERTSON, A. E. 1946. Observations on the natural history of the Fresno kangaroo rat. *J. Mammal.*, **27**, 189–203.

CURIO, E. 1955. Der Jungentransport einer Gelbhalsmaus (*Apodemus f. flavicollis* Melch.). *Z. Tierpsychol.*, **12**, 459–462.

DARCHEN, R. 1964. Notes éthologiques sur le rat musque, *Ondatra zibetica* L., et en particulier sur la construction de la hutte d'hiver. *Mammalia*, **28**, 137–168.

DARLING, F. FRASER, 1937. *A herd of red deer*. Oxford University Press.

DARWIN, C. 1904. *The expression of the emotions in man and animals*. Popular edition, John Murray, London.

DATHE, H. 1963. Vom Harnspritzen des Ursons (*Erethizon dorsatus*) *Z. Säugetierk.*, **28**, 369–375.

DAVIS, J. A. 1966. Notes on 'M'Tundu'. A banded mongoose in the Bronx zoo. *Anim. Kingd.*, **69**, 58–59.

DAVIS, R. B., HERREID, C. F. & SHORT, H. L. 1962. Mexican free-tailed bats in Texas. *Ecol. Monogr.*, **32**, 311–346.

DELGADO, J. M. R. 1966. Aggressive behavior evoked by radio stimulation in monkey colonies. *Amer. Zool.*, **6**, 669–681.

DELGADO, J. M. R. & ANAND, B. K. 1953. Increase of food intake induced by electrical stimulation of the lateral hypothalamus. *Amer. J. Physiol.*, **172**, 162–168.

DETHIER, V. G. 1962. *To know a fly*. Holden-Day Inc., San Francisco.

DEVORE, I. 1963. Mother-infant relations in free-ranging baboons. *Maternal behavior in mammals*, ed. H. L. Rheingold. John Wiley & Sons, New York, 305–335.

DE VOS, A. 1967. Rubbing of conifers by white-tailed deer in successive years. *J. Mammal.*, **48**, 146–147.

DIETERLEN, F. 1959. Das Verhalten des syrischen Goldhamsters (*Mesocricetus auratus* Waterhouse): Untersuchungen zur Frage seiner Entwicklung und seiner angeborener Anteil durch geruschsisolierte Aufzuchten. *Z. Tierpsychol.*, **16**, 47–103.

DIETERLEN, F. 1960. Bemerkungen zu Zucht und Verhalten der Zwergmaus (*Micromys minutus soricinus* Hermann). *Ibid.*, **17**, 552–554.

DIETERLEN, F. 1962. Geburt und Geburtshilfe bei der Stachelmaus, *Acomys cahirinus*. *Ibid.*, **19**, 191–222.

DIETERLEN, F. 1963. Vergleichende Untersuchungen zur Ontogenese von Stachelmaus (*Acomys*) und Wanderratte (*Rattus norvegicus*). Beiträge zum Nesthocker-Nestflüchter-Problem bei Nagetiere. *Z. Säugetierk.*, **28**, 193–227.

DIMELOW, E. J. 1963. The behaviour of the hedgehog (*Erinaceus europaeus* L.) in the routine of life in captivity. *Proc. zool. Soc. Lond.*, **141**, 281–289.

DOBRORUKA, L. J. 1960. Einige Beobachtungen an Ameisenigeln *Echidna aculeata* Shaw (1792). *Z. Tierpsychol.*, **17**, 178–181.

DOBRORUKA, L. J. 1961. Eine Verhaltensstudie des Przewalski-Urwildpferdes (*Equus przewalskii* Poliakov 1881) in dem zoologischen Garten Prag. *Proc. 1st internat. Symp. Przewalski Horse*, 89–104.

DONALD, H. P. 1937. Suckling and suckling preference in pigs. *Emp. J. exp. Agric.*, **5**, 361–368.

DÜCKER, G. 1957. Farb- und Helligkeitssehen und Instinkte bei Viverriden und Feliden. *Zool. Beitr., Berl.*, **3**, 25–99.

DÜCKER. G. 1965. Das Verhalten der Viverriden. *Handb. Zool., Berl.*, **8** (10) No. 38, 1–48.

ECCLES, J. C. 1957. *The physiology of nerve cells.* Oxford University Press.

EDWARDS, R. L. 1946. Some notes on the life history of the Mexican ground squirrel in Texas. *J. Mammal.*, **27**, 105–115.

EIBL-EIBESFELDT, I. 1950. Über die Jugendentwicklung des Verhaltens eines männlichen Dachses (*Meles meles* L.) unter besonderer Berücksichtigung des Spieles. *Z. Tierpsychol.*, **7**, 327–355.

EIBL-EIBESFELDT, I. 1951a. Gefangenschaftsbeobachtungen an der persischen Wüstenmaus (*Meriones persicus persicus* Blanford): ein Beitrag zur vergleichenden Ethologie der Nager. *Ibid.*, **8**, 400–423.

EIBL-EIBESFELDT, I. 1951b. Beobachtungen zur Fortpflanzungsbiologie und Jugendentwicklung des Eichhörnchens. *Ibid.*, **8**, 370–400.

EIBL-EIBESFELDT, I. 1953a. Zur Ethologie des Hamsters (*Cricetus cricetus* L.). *Ibid.*, **10**, 204–254.

EIBL-EIBESFELDT, I. 1953b. Eine besondere Form des Duftmarkierens beim Riesengalago, *Galago crassicaudatus* E. Geoffroy. *Säugetierkundl. Mitt.*, **1**, 171–173.

EIBL-EIBESFELDT, I. 1955. Ethologischen Studien am Galapagos-Seelöwen *Zalophus wollebaeki* Sivertsen. *Z. Tierpsychol.*, **12**, 286–303.

EIBL-EIBESFELDT, I. 1957. Ausdrucksformen der Säugetiere. *Handb. Zool., Berl.*, **8**, (10) No. 2, 1–26.

EIBL-EIBESFELDT, I. 1958. Das Verhalten der Nagetiere. *Ibid.*, **8** (10) No. 6, 1–88.

EIBL-EIBESFELDT, I. 1963. Angeborenes und Erworbenes im Verhalten einiger Säuger. *Z. Tierpsychol.*, **20**, 705–754.

EIBL-EIBESFELDT, I. 1965. Das Duftmarkierung des Igeltanrec (*Echinops telfairi* Martin). *Ibid.*, **22**, 810–812.

ELOFF, F. C. 1964. On the predatory habits of lions and hyaenas. *Koedoe*, No 7, 1964, 105–112.

ESPMARK, Y. 1964. Studies in dominance-subordination relationships in a group of semi-domestic reindeer (*Rangifer tarandus* L.). *Anim. Behav.*, **12**, 420–426.

ESTES, R. D. 1966. Behaviour and life history of the wildebeest (*Connochaetes taurinus* Burchell). *Nature, Lond.*, **212**, 999–1000.

EWBANK, R. 1964. Observations on the suckling habits of twin lambs. *Anim. Behav.*, **12**, 34–37.

EWBANK, R. 1967. Nursing and suckling behaviour amongst Clun Forest ewes and lambs. *Ibid.*, **15**, 251–258.

EWEN, D. 1956. My grysbok family. *Afr. wild. Life*, **10**, 249–252.

EWER, R. F. 1959. Suckling behaviour in kittens. *Behaviour*, **15**, 146–162.

EWER, R. F. 1960. Natural selection and neoteny. *Acta biotheoret.*, Leiden, **13**, 161–184.

EWER, R. F. 1961. Further observations on suckling behaviour in kittens, together with some general considerations of the interrelations of innate and acquired responses. *Behaviour*, **18**, 247–260.

EWER, R. F. 1963. The behaviour of the meerkat, *Suricata suricatta* (Schreber). *Z. Tierpsychol.*, **20**, 570–607.

EWER, R. F. 1965. Food burying in the African ground squirrel, *Xerus erythropus* (E. Geoff.). *Ibid.*, **22**, 321–327.

EWER, R. F. 1966. Juvenile behaviour in the African ground squirrel, *Xerus erythropus* (E. Geoff.). *Ibid.*, **23**, 190–216.

EWER, R. F. 1967. The behaviour of the African giant rat (*Cricetomys gambianus* Waterhouse). *Ibid.*, **24**, 6–79.

EWER, R. F. (1968). A preliminary survey of the behaviour in captivity of the dasyurid marsupial, *Sminthopsis crassicaudata* (Gould). *Ibid.*, **25**, 319–365.

EYRE, M. 1961. A tame otter. *Afr. wild. Life*, **15** 49–53.

FENTRESS, J. C. 1966. Observations on the behavioral development of a hand reared male timber wolf. *Amer. Zool.*, **6**, 333.

FISHER, A. E. 1964. Chemical stimulation of the brain. *Sci. Amer.*, **210**, (6), 60.

FISHER, E. M. 1939. Habits of the southern sea-otter. *J. Mammal.*, **20**, 21–36.

FITZGERALD, A. 1935. Rearing marmosets in captivity. *Ibid.*, **16**, 181–188.

FLEAY, D. 1965. Breeding the mulgara. *Wildlife in Australia*, **3** **(1)** 2–5.

FORMAZOV, A. N. 1966. Adaptive modification of behaviour in mammals of the Eurasian steppes. *J. Mammal.*, **47**, 208–223.

FRÄDRICH, H. 1965. Zur Biologie und Ethologie des Warzenschweines (*Phacochoerus aethiopicus* Pallas), unter Berücksichtigung des Verhalten anderer Suiden. *Z. Tierpsychol.*, **22**, 328–393.

FRANK, F. 1956. Das Duftmarkieren der grossen Wühlmaus, *Arvicola terrestris* (L.). *Z. Säugetierk.*, **21**, 172–175.

FRECHKOP, S. 1955. *Traité de Zoologie* **17** (1), ed. P. P. Grassé, Masson et Cie, Paris.

GANDER, F. F. 1929. Experiences with wood rats, *Neotoma fuscipes macrotis*. *J. Mammal.*, **10**, 52–58.

GASHWILER, J. S., ROBINETTE, W. L. & MORRIS, O. W. 1961. Breeding habits of bobcats in Utah. *Ibid.*, **42**, 76–84.

GAUTHIER-PILTERS, H. 1959. Einige Beobachtungen zum Droh-Angriffs- und Kampverhalten des Dromedarhengstes, sowie über Geburt und Verhaltensentwickling des Jungtiers, in der nordwestlichen Sahara. *Z. Tierpsychol.*, **16**, 593–604.

GAUTHIER-PILTERS, H. 1962. Beobachtungen an Feneks (*Fennecus zerda* Zimm.). *Ibid.*, **19**, 440–464.

GEIST, V. 1964. On the rutting behavior of the mountain goat. *J. Mammal.*, **45**, 551–568.

GEIST, V. 1966. The evolution of horn-like organs. *Behaviour*, **27**, 175–214.

VAN GELDER, R. G. 1953. The egg-opening technique of a spotted skunk. *J. Mammal.*, **34**, 255–256.

GERALL, A. A. 1963. An exploratory study of the effects of social isolation on the sexual behaviour of guinea pigs. *Anim. Behav.*, **11**, 274–282.

GERALL, H. D., WARD, I. L. & GERALL, A. A. 1967. Disruption of the male rat's sexual behaviour induced by social isolation. *Ibid.*, **15**, 54–58.

GETZ, L. L. 1962. Aggressive behavior of the meadow and prairie voles. *J. Mammal.*, **43**, 351–358.

GODDARD, J. 1966. Mating and courtship of the black rhinoceros (*Diceros bicornis* L.). *East Afr. Wildl. J.*, **4**, 69–75.

GODFREY, G. & CROWCROFT, P. 1960. *The life of the mole.* Museum Press, London.

GOETHE, F. 1938. Beobachtungen über das Absetzen von Witterungsmarken beim Baummarder. *Dtsch. Jäger*, **13**.

GOETHE, F. 1940. Beiträge zur Biologie des Iltis. *Z. Säugetierk.*, **15**, 180–223.

GOODALL, J. 1963. Feeding behaviour of wild chimpanzees. *Symp. zool. Soc. Lond.*, No. 10, 39–47.

GOODALL, J. 1964. Tool-using and aimed throwing in a community of free-living chimpanzees. *Nature, Lond.*, **201**, 1264–1266.

GOODALL, J. 1965. Chimpanzees of the Gombe Stream Reserve. *Primate behavior—field studies of monkeys and apes*, ed. I. DeVore. Holt, Rinehart & Winston, New York, 425–473.

GOY, R. W. & JAKWAY, J. S. 1959. The inheritance of patterns of sexual behaviour in female guinea pigs. *Anim. Behav.*, **7**, 142–149.

GRAF, W. 1956. Territorialism in deer. *J. Mammal.*, **37**, 165–170.

GRANT, E. C. & MACKINTOSH, J. H. 1963. A comparison of the social postures of some common laboratory rodents. *Behaviour*, **21**, 246–259.

GRASSÉ, P. P. 1955. *Traité de Zoologie*, **17** (2). Masson et Cie, Paris.

GRIFFIN, D. R., WEBSTER, F. A. & MICHAEL, C. R. 1960. The echolocation of flying insects by bats. *Anim. Behav.*, **8**, 141–154.

GRIFFIN, D. R., FRIEND, J. H. & WEBSTER, F. A. 1965. Target discrimination by the echolocation of bats. *J. exp. Zool.*, **158**, 155–168.

GRIFFITHS, M. 1965. Rate of growth and intake of milk in a suckling echidna. *J. comp. Biochem. Physiol.*, **16**, 383–392.

GRIMPE, G. 1923. Neues über die Geschlechtsverhaltnisse der gefleckten Hyäne (*Crocotta crocuta* Erxl.). *Verhl. dtsch. zool. Ges.*, **28**, 77–78.

GROSSMAN, M. I. 1955. Integration of current views on the regulation of hunger and appetite. *Ann. N.Y. Acad. Sci.*, **63**, 76–91.

GUGGISBERG, C. A. W. 1960. *Simba.* Verlag Hallwag, Bern.

HAFEZ, E. S. E., SUMPTION, L. J. & JAKWAY, J. S. 1962. The behaviour of swine. *The behaviour of domestic animals*, ed. E. S. E. Hafez. Baillière, Tindall & Cox, London, 334–369.

HAGA, R. 1960. Observations on the ecology of the Japanese pika. *J. Mammal.*, **41**, 200–212.

HALL, C. S. 1951. The genetics of behavior. *Handbook of experimental psychology*, ed. S. S. Stevens. John Wiley & Sons, New York, 304–329.

HALL, K. R. L. 1962a. Numerical data, maintenance activities and locomotion of the wild chacma baboon, *Papio ursinus*. *Proc. zool. Soc. Lond.*, **139**, 181–220.

HALL, K. R. L. 1962b. The sexual, agonistic and derived social behaviour patterns of the wild chacma baboon, *Papio ursinus*. *Ibid.*, **139**, 283–327.

HALL, K. R. L. 1965. Aggression in monkey and ape societies. *Symp. Inst. Biol.*, No. 13, 51–64.

HALL, K. R. L. 1966. Behaviour and ecology of the wild patas monkey, *Erythrocebus patas*, in Uganda. *J. Zool.*, **148**, 15–87.

HALL, K. R. L. & DEVORE, I. 1965. Baboon social behavior. *Primate behavior—field studies of monkeys and apes*, ed. I. DeVore. Holt, Rhinehart & Winston, New York, 53–110.

HALLIDAY, M. S. 1966a. Effects of previous exploratory activity on the exploration of a simple maze. *Nature, Lond.*, **209**, 432–433.

HALLIDAY, M. S. 1966b. Exploration and fear in the rat. *Symp. zool. Soc. Lond.* No. 18. 45–59.

HAMILTON, W. J. 1962. Reproductive adaptations of the red tree mouse. *J. Mammal.*, **43**, 486–504.

HAMMOND, J. & MARSHALL, F. H. A. 1930. Oestrus and pseudopregnancy in the ferret. *Proc. roy. Soc. B.*, **105**, 607–630.

HARDER, W. 1949. Zur Morphologie und Physiologie des Blinddarmes der Nagetiere. *Verh. dtsch. zool. Ges.*, **1949**, 95–109.

HARLOW, H. F. 1962. Development of affection in primates. *Roots of behavior*, ed. E. L. Bliss. Harper & Row, New York, 157–166.

HARLOW, H. F., & HARLOW, M. K. 1965. The affectional systems. *Behavior of nonhuman primates*, ed. A. M. Schrier, H. F. Harlow & F. Stollnitz. Academic Press, New York, 287–334.

HARLOW, H. F., HARLOW, M. K. & HANSEN, E. W. 1963. The maternal affective system of rhesus monkeys. *Maternal behavior in mammals*, ed. H. L. Rheingold. John Wiley & Sons, New York, 254–281.

HARTMAN, C. G. 1952. *Possums*. University of Texas Press, Austin.

HEDIGER, H. 1951. Observations sur la psychologie animale dans les parcs nationaux du Congo Belge. *Explorations des parcs nationaux du Congo Belge*. Inst. P.N.C.B., Brussels, 1–194.

HEDIGER, H. & KUMMER, H. 1956. Das Verhalten der Schnabeligel (Tachyglossidae). *Handb. Zool. Berl.*, **8** (10) No. 8, 1–8.

HEINROTH-BERGER, K. 1959. Beobachtungen an handaufgezogenen Mantelpavianen (*Papio hamadryas* L.). *Z. Tierpsychol.*, **16**, 706–732.

HEINROTH-BERGER, K. 1965. Über Geburt und Aufzucht eines männ-. lichen Schimpansen im Zoologischen Garten Berlin. *Ibid.*, **22**, 15–35.

HEPPES, J. B. 1958. The white rhinoceros in Uganda. *Afr. wild Life*, **12**, 273–280.

HEPTNER, V. G. 1939. The Turkestan desert shrew, its biology and adaptive peculiarities. *J. Mammal.* **20**, 139–149.

HERTER, K. 1957. Das Verhalten der Insektivoren. *Handb. Zool. Berl.*, **8** (10) No. 10, 1–50.

HILL, W. C. O. 1944. An undescribed feature in the Drill (*Mandrillus leucophaeus*). *Nature, Lond.*, **153**, 199.

HILL, W. C. O. 1955. *Primates: comparative anatomy and taxonomy. II Haplorhini: Tarsoidea.* Edinburgh University Press.

HILL, W. C. O. 1957. *Primates: comparative anatomy and taxonomy. III Pithecoidea: Platyrhini.* Edinburgh University Press.

HINDE, R. A. 1966. Ritualisation and social communication in rhesus monkeys. *Philos. Trans. B.*, **251**, 285–294.

HINDE, R. A. & ROWELL, T. E. 1962. Communication by postures and facial expressions in the rhesus monkey (*Macaca mulatta*). *Proc. zool. Soc. Lond.*, **138**, 1–22.

HINDE, R. A. & SPENCER-BOOTH, Y. 1967. The behaviour of socially living rhesus monkeys in their first two and a half years. *Anim. Behav.*, **15**, 169–196.

HIRSCHER, L. RICHMOND, J. B. & MOORE, A. L. 1963. Maternal behavior in sheep and goats. *Maternal behavior in mammals*, ed. H. L. Rheingold. John Wiley & Sons, New York, 203–232.

VON HOLST, E. & VON SAINT PAUL, U. 1963. On the functional organisation of drives. *Anim. Behav.*, **11**, 1–20.

HORNER, B. E., TAYLOR, J. M. & PADYKULA, H. A. 1964. Food habits and gastric morphology of the grasshopper mouse. *J. Mammal*, **45**, 513–535.

HOWELL, A. H. 1920. The Florida spotted skunk as an acrobat. *Ibid.*, **1**, 88.

ILSE, D. 1955. Olfactory marking of territory in two young male loris, *Loris tardigradus lydekkerianus*, kept in captivity in Poona. *Brit. J. anim. Behav.*, **3**, 118–120.

INGLES, L. G. 1941. Natural history observations on the Audubon cottontail. *J. Mammal.*, **22**, 227–250.

INGRAM, W. M. 1942. Snail associates of *Blarina brevicauda talpoides* (Say). *Ibid.*, **23**, 255–258.

INHELDER, E. 1955. Zur Psychologie einiger Verhaltensweisen—besonders des Spiels—von Zootieren. *Z. Tierpsychol.*, **12**, 88–144.

INNIS, A. C. 1958. The behaviour of the giraffe, *Giraffa camelopardalis*, in the eastern Transvaal. *Proc. zool. Soc. Lond.*, **131**, 245–278.

JAKWAY, J. S. 1959. Inheritance of patterns of mating behaviour in the male guinea pig. *Anim. Behav.*, **7**, 150–162.

JAY, P. 1963. Mother-infant relations in langurs. *Maternal behavior in mammals*, ed. H. L. Rheingold. John Wiley & Sons, New York, 282–304.

JAY, P. 1965. The common langurs of north India. *Primate behavior—field studies of monkeys and apes*, ed. I. DeVore. Holt, Rinehart & Winston, New York, 197–249.

JEWELL, P. A. 1966. The concept of home range in mammals. *Symp. zool. Soc. Lond.*, No. 18, 85–109.

JOHNSON, C. E. 1921. The "hand-stand" habit of the spotted skunk. *J. Mammal.*, **2**, 87–89.

JOLLY, A. 1966. *Lemur behavior: a Madagascar field study.* University of Chicago Press.

JORDAN, P. A., SHELTON, P. C. & ALLEN, D. L. 1967. Numbers, turnover, and social structure of the Isle Royale wolf population. *Amer. Zool.*, **7**, 233–252.

JOUVET, M. 1967. Neurophysiology of the states of sleep. *Physiol. Rev.* **47**, 117–177.

KAGAN, J. & BEACH, F. A. 1953. Effects of early experience on mating behavior in male rats. *J. Comp. physiol. Psychol.*, **46**, 204–208.

KAUFMANN, J. H. 1962. Ecology and social behavior of the coati, *Nasua narica*, on Barro Colorado Island, Panama. *Univ. Calif. Publ. Zool.*, **60**, 95–222.

KAUFMANN, J. H. & KAUFMANN, A. 1965. Observations on the behavior of tayras and grisons. *Z. Säugetierk.*, **30**, 146–155.

KEAN, R. I. 1967. Behaviour and territorialism in *Trichosurus vulpecula* (Marsupialia). *Proc. N.Z. ecol. Soc.*, **14**, 71–78.

KENNEDY, J. S. 1954. Is modern ethology objective? *Brit. J. anim. Behav.*, **2**, 12–19.

KENNEDY, J. S. 1966a. The balance between antagonistic induction and depression of flight activity in *Aphis fabae* Scopoli. *J. exp. Biol.*, **45**, 215–228.

KENNEDY, J. S. 1966b. Some outstanding questions in insect behaviour. *Insect behavior*, ed. P. T. Haskell. Royal Entomological Society, London, 97–112.

VON KETELHODT, H. F. 1966. Der Erdwolf, *Proteles cristatus* (Sparrman, 1783). *Z. Säugetierk.*, **31**, 300–308.

KILEY-WORTHINGTON, M. 1965. The waterbuck (*Kobus defassa* Ruppell 1835 & *K. ellipsiprimnus* Ogilby 1833) in East Africa: spatial distribution. A study of the sexual behaviour. *Mammalia*, **29**, 199–204.

KILHAM, L. 1954. Territorial behavior of red squirrel. *J. Mammal.*, **35**, 252–253.

KING, J. A. 1955. Social behavior, social organisation and population dynamics in a black-tailed prairiedog town in the Black Hills of south Dakota. *Contr. Lab. Verteb. Biol. Univ. Mich.*, No. 67, 1–123.

KING, J. A. 1963. Maternal behavior in *Peromyscus*. *Maternal behavior in mammals*, ed. H. L. Rheingold. John Wiley & Sons, New York, 58–93.

KINLOCH, B. 1964. *Sauce for the mongoose.* Harvill Press, London.

KIRSCH, J. A. W. & POOLE, W. E. 1967. Serological evidence for speciation in the grey kangaroo, *Macropus giganteus* Shaw 1790 (Marsupialia: Macropodidae). *Nature Lond.*, **215**, 1097–1098.

KLEIMAN, D. G. 1966a. Scent marking in the Canidae. *Symp. zool. Soc. Lond.*, No. 18, 167–177.

KLEIMAN, D. G. 1966b. The comparative social behavior of the Canidae. *Amer. Zool.*, **6**, 335.

KLEIMAN, D. G. 1967. Some aspects of social behavior in the Canidae. *Ibid.*, **7**, 365–372.

KLINGEL, H. 1967. Soziale Organisation und Verhalten freilebender Steppenzebras. *Z. Tierpsychol.*, **24**, 580–624.

KLOPFER, P. H., ADAMS, D. K. & KLOPFER, M. S. 1964. Maternal 'imprinting' in goats. *Proc. nat. Acad. Sci., Wash.*, **52**, 911–914.

KNAPPE, H. 1964. Zur Funktion des Jacobsonschen Organs (*Organon vomeronasale Jacobsoni*). *Zool. Gart., Lpz.*, **28**, 188–194.

KOCH, W. 1951. Psychogene Beeinflussung des Geburtsterminus bei Pferden. *Z. Tierpsychol.*, **8**, 441–443.

KOENIG, L. 1957. Beobachtungen über Reviermarkierung sowie Droh-, Kampf- und Abwehrverhalten des Murmeltieres (*Marmota marmota* L.). *Ibid.*, **14**, 510–521.

KOENIG, L. 1960. Das Aktionssystem des Siebenschlafers (*Glis glis* L.). *Ibid.*, **17**, 427–505.

KOFORD, C. B. 1957. The vicuña and the puna. *Ecol. Monogr.*, **27**, 153–219.

KOVACH, J. K. & KLING, A. 1967. Mechanisms of neonate sucking behaviour in the kitten. *Anim. Behav.*, **15**, 91–101.

KRIEG, H. & RAHM, U. 1956. Das Verhalten der Ameisenbären (Myrmecophagidae). *Handb. Zool., Berl.*, **8** (10) No 13, 2–19.

KROTT, P. 1961. Der gefährliche Braunbär (*Ursus arctos* L. 1758). *Z. Tierpsychol.*, **18**, 245–256.

KROTT, P. & KROTT, G. 1963. Zum Verhalten des Braunbären (*Ursus arctos* L. 1758) in den Alpen. *Ibid.*, **20**, 160–206.

KRUIJT, J. P. 1964. Ontogeny of social behaviour in Burmese red jungle fowl (*Gallus gallus spadiceus*). *Behaviour, Suppl.* **12**, 1–201.

KRUUK, H. 1966. Clan-system and feeding habits of spotted hyaenas (*Crocuta crocuta* Erxleben). *Nature Lond.*, **209**, 1257–1258.

KRUUK, H. & TURNER, M. 1967. Comparative notes on predation by lion, leopard, cheetah and wild dog in the Serengeti area, East Africa. *Mammalia*, **31**, 1–27.

KÜHME, W. 1961. Beobachtungen am afrikanischen Elefanten (*Loxodonta africana* Blumenbach 1797) in Gefangenschaft. *Z. Tierpsychol.* **18**, 285–296.

KÜHME, W. 1963. Ergänzende Beobachtungen am afrikanischen Elefanten (*Loxodonta africana* Blumenbach 1797) im Freigehege. *Ibid.* **20**, 66–79.

KÜHME, W. 1964. Die Ernährungsgemeinschaft der Hyänenhundes (*Lycaon pictus lupinus* Thomas 1902). *Naturwissenschaften*, **20**, 495.

KÜHME, W. 1965a. Freilandstudien zur Soziologie des Hyänenhundes (*Lycaon pictus lupinus* Thomas 1902). *Z. Tierpsychol.*, **22**, 495–541.

KÜHME, W. 1965b. Über die soziale Bindung innerhalb eines Hyänenhund–Rudels. *Naturwissenschaften*, **23**, 567–568.

KÜHME, W. 1965c. Communal food distribution and division of labour in African hunting dogs. *Nature, Lond.*, **205**, 443–444.

KÜHME, W. 1966. Beobachtungen zur Soziologie des Löwens in der Serengeti–Steppe Ostafrikas. *Z. Säugetierk.*, **31**, 205–213.

KUMMER, H. & KURT, F. 1963. Social units of a free-living population of hamadryas baboons. *Folia Primatol.*, **1**, 4–19.

KUNKEL, P. & KUNKEL, I. 1964. Beiträge zur Ethologie des Hausmeerschweinchens, *Cavia aperea* f. *porcellus* (L.). *Z. Tierpsychol.*, **21**, 603–641.

LACK, D. 1943. *The life of the robin.* H. F. & G. Witherby Ltd., London.

LANG, H. 1923. Caudal and pectoral glands of African elephant shrews (*Elephantulus*). *J. Mammal.*, **4**, 261–263.

LANG, H. 1925. How squirrels and other rodents carry their young. *Ibid.*, **6**, 18–24.

LANYON, W. E. 1960. The ontogeny of vocalisations in birds. *Animal sounds and communications*, ed. W. E. Lanyon & W. N. Tavolga. *Amer. Inst. biol. Sci.* Washington, 321–347.

LARSSON, S. 1954. On the hypothalamic organisation of the nervous mechanism regulating food intake. *Acta physiol. scand., Suppl.*, **115**, 7–63.

LAWS, R. M. & CLOUGH, G. 1966. Observations on reproduction in the hippopotamus, *Hippopotamus amphibius* Linn. *Symp. zool. Soc. Lond.*, No. 15, 117–140.

LEHRMAN, D. S. 1953. A critique of Konrad Lorenz's theory of instinctive behavior. *Quart. Rev. Biol.*, **28**, 337–363.

LEHRMAN, D. S. 1958. Induction of broodiness by participation in courtship and nestbuilding in the ring dove (*Streptopelia risoria*). *Behaviour*, **7**, 241–286.

LEHRMAN, D. S. 1959 Hormonal responses to external stimuli in birds. *Ibis*, **101**, 478–496.

LEHRMAN, D. S. 1961. Hormonal regulation of parental behaviour in birds and infrahuman mammals. *Sex and internal secretions*, ed. W. C. Young, vol 2. Williams and Wilkins Co., Baltimore, 1268–1382.

LEUTHOLD, W. 1966. Variations in territorial behaviour of Uganda kob, *Adenota kob thomasi* (Neumann, 1896). *Behaviour*, **27**, 215–258.

LEVINE, S. & MULLINS, R. F. 1966. Hormonal influences on brain organisation in infant rats. *Science*, **152**, 1585–1592.

LEYHAUSEN, P. 1956. Verhaltensstudien an Katzen. *Z. Tierpsychol.*, *Beiheft* **2**, 1–120.

LEYHAUSEN, P. 1962. *Felis nigripes*—Katzenzwerg aus Südwestafrika. *Umschau*, **1962**, 768–770.

LEYHAUSEN, P. 1964. The communal organisation of solitary animals. *Symp. zool. Soc. Lond.*, No. 14, 249–263.

LEYHAUSEN, P. 1965. Über die Funktion der relativen Stimmungshierarchie (dargestellt am Beispiel der phylogenetischen und ontogenetischen Entwicklung des Beutefangs von Raubtieren). *Z. Tierpsychol.*, **22**, 412–494.

LEYHAUSEN, P. & WOLFF, R. 1959 Das Revier einer Hauskatze. *Ibid.*, **16**, 666–670.

LIERS, E. E. 1951. Notes on the river otter (*Lutra canadensis*). *J. Mammal.*, **32**, 1–9.

LINDEMANN, W. 1951. Zur Psychologie des Igels. *Z. Tierpsychol.*, **8**, 224–251.

LINDEMANN, W. 1955. Über die Jugendentwicklung beim Luchs (*Lynx l. lynx* Kerr) und bei der Wildkatze (*Felis s. silvestris* Schreb.). *Behaviour*, **8**, 1–46.

LINDEMANN, W. & RIEK, W. 1953. Beobachtungen bei der Aufzucht von Wildkatzen. *Z. Tierpsychol.*, **10**, 92–119.

LOCKIE, J. D. 1966. Territory in small carnivores. *Symp. zool. Soc. Lond.*, No. 18, 143–165.

LORENZ, K. 1935. Der Kumpan in der Umwelt des Vogels. *J. Ornithol.*, **83**, 137–213, 289–413. (Abbreviated English version 1957—"Companionship in bird life". *Instinctive behavior—the development of a modern concept*, ed. C. H. Schiller. Methuen, London, 83–128.

LORENZ, K. 1950. The comparative method of studying innate behaviour patterns. *Symp. Soc. exp. Biol.*, **4**, 229–269.

LORENZ, K. 1954. *Man meets dog*. Methuen, London.

LORENZ, K. 1961. Phylogenetische Anpassung und adaptive Modifikation des Verhaltens. *Z. Tierpsychol.*, **18**, 139–187. (English version 1966 *Evolution and modification of behavior*. Methuen, London.)

LORENZ, K. 1963. *Das sogennante Böse*. G. Borotha-Schoeler, Vienna. (English version, 1966 *On aggression*. Methuen, London).

LORENZ, K. 1964. Ritualised fighting. *Symp. Inst. Biol.* No. 13, 39–50.

LORENZ, K. 1966a. Stammes- und Kulturgeschichtliche Ritenbildung. *Mitt. Max-Planck Ges.*, **1**, 3–30.

LORENZ, 1966b. Evolution of ritualisation in the biological and cultural spheres. *Philos. Trans. B*, **251**, 273–284.

LUCAS, N. S., HUME, E. M. & HENDERSON, H. 1927. On the breeding of the common marmoset (*Hapale jaccus* L.) in captivity when irradiated with ultra-violet rays. *Proc. zool. Soc. Lond.*, *1927*, 447–451.

LUDWIG, J. 1965. Beobachtungen über das Spiel bei Boxern. *Z. Tierpsychol.*, **22**, 813–838.

LYALL-WATSON, M. 1963. A critical re-examination of food "washing" behaviour in the raccoon (*Procyon lotor* Linn.). *lroc. zool. Soc. Lond.*, **141**, 371–393.

MCBRIDE, G. 1963. The 'teat order' and communication in young pigs *Anim. Behav.*, **11**, 53–56.

MCCLEARN, G. E. 1959. The genetics of mouse behavior in novel situations. *J. Comp. physiol. Psychol.*, **52**, 62–67.

MCCRADY, E. 1938. The embryology of the opossum. *Amer. anat. Mem. Philadelphia*, No. 16, 1–234.

MANN, G. 1961. *Bulbus olfactorius accessorius* in Chiroptera. *J. comp. Neurol.*, **116**, 135–144.

MARLER, P. R. & HAMILTON, W. J. 1966. *Mechanisms of animal behavior*. John Wiley & Sons, New York.

MARLOW, B. J. 1961. Reproductive behaviour of the marsupial mouse, *Antechinus flavipes* (Waterhouse) (Marsupialia) and the development of the young. *Aust. J. Zool.*, **9**, 203–218.

MARSDEN, H. M. & HOLLER, N. R. 1964. Social behavior in confined populations of the cottontail and the swamp rabbit. *Wildl. Monogr. Chestertown*, No. 13.

MARTAN, J. 1962. Effect of castration and androgen replacement on the supracaudal gland of the male guinea pig. *J. Morph.*, **110**, 285–298.

MARTIN, R. D. 1966. Tree shrews: unique reproductive mechanism of systematic importance. *Science*, **152**, 1402–1404.

MARTINS, T. 1949. Disgorging of food to the puppies by the lactating dog. *Physiol. Zool.*, **22**, 169–172.

MASON, W. A. 1960. The effects of social restriction on the behavior of rhesus monkeys. I. Free social behavior. *J. comp. physiol. Psychol.* **53**, 582–589.

MEESTER, J. 1960. Shrews in captivity. *Afr. wild Life*, **14**, 57–63.

MEIER, G. W. 1965a. Maternal behaviour of feral- and laboratory-reared monkeys following the surgical delivery of their infants. *Nature Lond.*, **206**, 492–493.

MEIER, G. W. 1965b. Other data on the effects of social isolation during rearing upon adult reproductive behaviour in the rhesus monkey (*Macaca mulatta*). *Anim. Behav.*, **13**, 228–231.

MERCHANT J. C. & SHARMAN, G. B. 1966. Observations on the attachment of marsupial pouch young to the teats and on the rearing of pouch young by foster mothers of the same or different species. *Aust. J. Zool.*, **14**, 593–609.

VAN DER MERWE, N. J. 1953. *The jackal.* Fauna & Flora, Publ. Transvaal prov. Admin. No. 4, Pretoria.

MEYER-HOZLAPFEL, M. 1956. Das Speil bei Säugetieren. *Handb. Zool., Berl.*, **8** (10) No. 2, 1–36.

MEYER-HOZLAPFEL, M. 1957. Das Verhalten der Bären (Ursidae). *Ibid.*, **8** (10) No. 17, 1–28.

MINCHIN, A. K. 1937. Notes on the weaning of a young koala. *Rec. S. Aust. Mus.*, **6**, 1–3.

MOHR, E. 1956. Das Verhalten der Pinnipedier. *Handb. Zool., Berl.*, **8** (10) No. 22, 1–20.

MORRIS, B. 1962. A denizen of the evergreen forest. *Afr. wild Life*, **16**, 117–121.

MORRIS, B. 1963. Notes on the giant rat (*Cricetomys gambianus*) in Nyasaland. *Ibid.*, **17**, 102–107.

MORRIS, D. 1957. "Typical intensity" and its relation to the problem of ritualisation. *Behaviour*, **11**, 1–12.

MORRIS, D. 1962. The behaviour of the green acouchi (*Myoprocta pratti*) with special reference to scatter hoarding. *Proc. zool. Soc. Lond.*, **139**, 701–732.

MORRIS, D. 1967. (Ed.) *Primate ethology*. Weidenfeld & Nicolson, London.

MORRIS, D. 1967. *The naked ape*. Jonathan Cape, London.

MORRIS, R. & MORRIS, D. 1966. *Men and apes*. Hutchinson, London.

MOYNIHAN, M. 1966. Communication in the titi monkey, *Callicebus. J. Zool.*, **150**, 77–127.

Müller-Using, D. 1956. Zum Verhalten des Murmeltieres (*Marmota marmota* (L.)). *Z. Tierpsychol.*, **13**, 135–142.

Mundy, K. R. D. & Flook, D. R. 1964. Notes on the mating activity of grizzly and black bears. *J. Mammal.*, **45**, 637–638.

Munn, N. L. 1964. Discrimination-reversal learning in kangaroos. *Aust. J. Psychol.*, **16**, 62–86.

Myer, J. S. 1964. Stimulus control of mouse-killing rats. *J. comp. physiol. Psychol.*, **58**, 112–117.

Myer, J. S. 1967. Prior killing experience and the suppressive effects of punishment on the killing of mice by rats. *Anim. Behav.*, **15**, 59–61.

Myer, J. S. & White, R. T. 1965. Aggressive motivation in the rat. *Ibid.*, **13**, 430–433.

Myers, K. & Poole, W. E. 1959. A study of the biology of the wild rabbit, *Oryctolagus cuniculus* (L.) in confined populations. I The effects of density on home range and the formation of breeding groups. *C.S.I.R.O. Wildl. Res.*, **4**, 14–26.

Myers, K. & Poole, W. E. 1961. *Op. cit.* II The effects of season and population increase on behaviour. *Ibid.*, **6**, 1–41.

Mykytowycz, R. 1958. Social behaviour of an experimental colony of wild rabbits, *Oryctolagus cuniculus* (L.) I Establishment of the colony. *Ibid.*, **3**, 7–25.

Mykytowycz, R. 1959. *Op. cit.* II First breeding season. *Ibid.*, **4**, 1–13.

Mykytowycz, R. 1960. *Op. cit.* III Second breeding season. *Ibid.*, **5**, 1–20.

Mykytowycz, R. 1962. Territorial function of chin gland secretion in the rabbit, *Oryctolagus cuniculus* (L.). *Nature Lond.*, **193**, 799.

Mykytowycz, R. 1964. Territoriality in rabbit populations. *Aust. nat. Hist.*, **14**, 326–329.

Mykytowycz, R. 1965. Further observations on the territorial function and histology of the submandibular cutaneous (chin) glands in the rabbit, *Oryctolagus cuniculus* (L.). *Anim. Behav.*, **13**, 400–412.

Naaktgeboren, C. & Van Wagtendonk, A. W. 1966. Wahre Knoten in der Nabelschnuur nebst Bemerkungen über Plazentophagie bei Menschaffen. *Z. Säugetierk.*, **31**, 376–382.

Neumann, G. H. 1961. Die visuelle Lernfähigkeit primitiver Säugetiere. *Z. Tierpsychol.*, **18**, 71–83.

Noirot, E. 1964. Changes in responsiveness to young in the adult mouse. I The problematical effect of hormones. *Anim. Behav.*, **12**,52–58.

Nolte, A. 1958. Beobachtungen über das Instinktverhalten von Kapuzineraffen (*Cebus apella* L.) in der Gefangenschaft. *Behaviour*, **12**, 183–207.

Orr, R. T. 1967. The Galapagos sea lion. *J. Mammal.*, **48**, 62–69.

Packard, R. L. 1955. Release, dispersal and reproduction of fallow deer in Nebraska. *Ibid.*, **36**, 471–473.

Paintal, A. S. 1953. Impulses in vagal afferent fibres from stretch receptors in the stomach and their role in the peripheral mechanism of hunger. *Nature, Lond.* **172**, 1194–1195.

PAULIAN, P. 1964. Contribution à l'étude de l'otarie de l'île Amsterdam. *Mammalia*, **28**, Suppl. 1, 1–146.

PETERSEN, E. 1965. Biologische Beobachtungen über Verhaltensweisen einiger einheimischer Nager beim Offnen von Nüssen und Kernen. *Z. Säugetierk.*, **30**, 156–162.

PETTER, F., 1964. Affinités du genre *Cricetomys*. Une nouvelle sous-famille de rongeurs Cricetidae les Cricetomyinae. *C.R. Acad. Sci. Paris*, **258**, 6516–6518.

PETTER, J. J. 1965. The lemurs of Madagascar. *Primate behavior—field studies of monkeys and apes*, ed. I. DeVore. Holt, Rinehart & Winston, New York, 292–319.

PFEFFER, P. 1967. Le mouflon de Corse (*Ovis ammon musimon* Schreber 1782). *Mammalia*, **31**, Suppl., 1–262.

PILTERS, H. 1956. Das Verhalten der Tylopoden. *Handb. Zool., Berl.*, **8** (10) No. 27, 1–24.

PLOOG, D. W. & MACLEAN, P. D. 1963. Display of penile erection in the squirrel monkey (*Saimiri sciureus*). *Anim. Behav.*, **11**, 32–39.

POGLAYEN-NEUWALL, I. & POGLAYEN-NEUWALL, I. 1966. Gefangenschaftsbeobachtungen an Makibären (*Bassaricyon* Allen 1876). *Z. Säugetierk.*, **30**, 321–366.

POLLARD, J. S. & LYSONS, A. M. 1967. Possums in the closed field test. *Anim. Behav.*, **15**, 129–133.

PRUITT, W. O. 1954. Rutting behavior of the whitetail deer (*Odocoileus virginianus*). *J. Mammal.*, **35**, 129–130.

PUROHIT, K. G., KAMETKAR, L. R. & PRAKASH, I. 1966. Reproduction biology and post-natal development in the northern palm squirrel, *Funambulus pennanti* Wroughton. *Mammalia*, **30**, 538–546.

QUAY, W. B. 1962. Apocrine sweat glands in the *angulus oris* of microtine rodents. *J. Mammal*, **43**, 303–310.

QUAY, W. B. 1965. Comparative survey of the sebaceous and sudoriferous glands of the oral lips and angle in rodents. *Ibid.*, **46**, 23–37.

QUAY, W. B. & TOMICH, P. Q. 1963. A specialised midventral sebaceous glandular area in *Rattus exulans*. *Ibid.*, **44**, 537–542.

QUICK, H. 1965. Ecology of the African elephant. *BSCS Pamphlet No. 21*, D. C. Heath & Co., Boston.

RAHM, U. 1961. Beobachtungen an der ersten in Gefangenschaft gehalten *Mesopotamogale ruwenzorii* (Mammalia—Insectivora). *Rev. suisse Zool.*, **68**, 73–90.

RANKIN, J. 1965. Notes on the ecology, capture and behaviour in captivity of the elephant shrew, *Nasilio brachyrhynchus brachyrhynchus* (A. Smith). *Zool. afr.*, **1**, 73–80.

REED, C. A. 1946. The copulatory behaviour of small mammals. *J. comp. Psychol.*, **39**, 185–206.

RENSCH, B. 1956. Increase in learning capacity with increase in brain size. *Amer. Nat.*, **90**, 81–96.

RENSCH, B. & DÜCKER, G. 1959. Die Spiele von *Mungo* und *Ichneumon*. *Behaviour*, **14**, 185–213.

Reynolds, H. 1952. Studies on reproduction in the opossum (*Didelphis virginiana virginiana*). *Univ. Calif. Publ. Zool.*, **53**, 133–275.

Reynolds, V. & Reynolds, F. 1965. Chimpanzees of the Budongo Forest. *Primate behavior—field studies of monkeys and apes*, ed. I. DeVore. Holt, Rinehart & Winston, New York, 368–424.

Rheingold, H. L. 1963. Maternal behavior in the dog. *Maternal behavior in mammals*, ed. H. L. Rheingold. John Wiley & Sons, New York, 169–202.

Richard, P. B. 1964. Les materiaux de construction du castor (*Castor fiber*), leur signification pour ce rongeur. *Z. Tierpsychol.* **21**, 592–601.

Richards, M. P. M. 1966a. Maternal behaviour in virgin female golden hamsters (*Mesocricetus auratus* Waterhouse): the role of the age of the test pup. *Anim. Behav.*, **14**, 303–309.

Richards, M. P. M. 1966b. Maternal behaviour in the golden hamster: responsiveness to young in virgin, pregnant and lactating females. *Ibid.*, **14**, 310–313.

Richardson, W. B. 1943. Wood rats (*Neotoma albigula*): their growth and development. *J. Mammal.*, **24**, 130–143.

Riess, P. E. 1954. The effect of altered environment and of age on the mother-young relationship among animals. *Ann. N.Y. Acad. Sci.*, **57**, 606–610.

Roberts, A. 1951. *The mammals of South Africa.* Central News Agency, Cape Town.

Roberts, W. W. & Kiess, H. O. 1964. Motivational properties of hypothalamic aggression in cats. *J. comp. physiol. Psychol.* **58**, 187–193.

Roberts, W. W. & Carey, R. J. 1965. Rewarding effect of performance of gnawing aroused by hypothalamic stimulation in the rat. *Ibid.*, **59**, 317–324.

Rongstad, O. J. 1966. Biology of penned cottontail rabbits. *J. Wildl. Mgmt.*, **30**, 312–319.

Rosenblatt, J. S. & Lehrman, D. S. 1963. Maternal behavior of the laboratory rat. *Maternal behavior in mammals*, ed. H. L. Rheingold. John Wiley & Sons, New York, 8–57.

Ross, S. Sawin, P. B., Zarrow, M. X. & Denenberg, V. H. 1963. Maternal behavior in the rabbit. *Ibid.*, 94–121.

Roth, H. H. 1964. Note on the early growth development of *Hystrix africaeaustralis*. *Z. Säugetierk.*, **29**, 313–316.

Roth-Kolar, H. 1957. Beiträge zu einem Aktionssystem des Augti (*Dasyprocta aguti aguti* L.). *Z. Tierpsychol.*, **14**, 362–375.

Rowell, T. E. 1961. Maternal behaviour in non-maternal golden hamsters (*Mesocricetus auratus*). *Anim Behav.*, **9**, 11–15.

Rowell, T. E. 1966. Forest living baboons in Uganda. *J. Zool.*, **149**, 344–364.

Rowell, T. E. 1967. Female reproductive cycles and the behavior of baboons and rhesus macaques. *Social communication among primates*, ed. S. A. Altmann, University of Chicago Press, 15–32.

Rudd, R. L. 1966. The midventral gland in Malaysian murid rodents. *J. Mammal.*, **47**, 331–332.

Ruffer, D. G. 1966. Sexual behaviour of the northern grasshopper mouse (*Onychomys leucogaster*). *Anim. Behav.*, **13**, 447–452.

Rundquist, E. A. 1933. Inheritance of spontaneous activity in rats. *J. comp. Psychol.*, **16**, 415–438.

Russek, M. & Morgane, P. J. 1963. Anorexic effect of intraperitoneal glucose in the hypothalamic hyperphagic cat. *Nature, Lond.*, **199**, 1004–1005.

Sale, J. B. 1965. Observations on parturition and related phenomena in the hyrax (Procaviidae). *Acta trop.*, **22**, 37–54.

Sälzle, K. 1936. Untersuchungen über das Farbsehvermögen von Opossum, Waldmäusen, Rötelmäusen und Eichhörnchen. *Z. Säugetierk.*, **11**, 106–148.

Sauer, E. G. F. 1967. Mother–infant relationship in galagos and the oral child-transport among primates. *Folia primatol.* **7**, 127–149.

Schaller, G. B. 1963. *The mountain gorilla.* University of Chicago Press.

Schaller, G. B. 1965a. My year with the tigers. *Life* (*internat.*), **39** (1), 58–64.

Schaller, G. B. 1965b. Behavioral comparisons of the apes. *Primate behavior—field studies of monkeys and apes*, ed. I. DeVore. Holt, Rinehart & Winston, New York, 471–481.

Schaller, G. B. 1965c. The behavior of the mountain gorilla. *Ibid.*, 324–367.

Schaller, G. B. 1967. *The deer and the tiger.* University of Chicago Press.

Schenkel, R. 1947. Ausdrucks-studien an Wölfen. *Behaviour*, **1**, 81–129.

Schenkel, R. 1966a. Play, exploration and territoriality in the wild lion. *Symp. zool. soc. Lond.*, No. 18, 11–22.

Schenkel, R. 1966b. On sociology and behaviour in impala (*Aepyceros melampus suara* Matschie). *Z. Säugetierk.*, **31**, 177–205.

Schenkel, R. 1967. Submission: its features and functions in the wolf and dog. *Amer. Zool.* **7**, 319–329.

Schiller, P. H. 1957. Innate motor action as a basis of learning. *Instinctive behavior: the development of a modern concept*, ed. C. H. Schiller. Methuen & Co., London, 264–287.

Schloeth, R. 1961. Das Socialleben des Camargue-Rindes. *Z. Tierpsychol.*, **18**, 574–627.

Schnierla, T. C., Rosenblatt, J. S. & Tobach, E. 1963. Maternal behavior in the cat. *Maternal behavior in mammals*, ed. H. L. Rheingold. John Wiley & Sons, New York, 122–168.

Schnitzler, H. U. 1967. Kompensation von Döpplereffekten bei Hufeisen-Fledermäusen. *Naturwissenschaften*, **19**, 523.

Schönberner, D. 1965. Beobachtungen zur Fortpflanzungsbiologie des Wolfes, *Canis lupus*. *Z. Säugetierk.*, **30**, 171–178.

Schrier, A. M., Harlow, H. F. & Stollnitz, F. 1965. *Behavior of nonhuman primates* (2 vols). Academic Press, New York & London.

SCHULTZ, A. H. 1921. The occurrence of a sternal gland in orang-utan *J. Mammal.*, **2**, 194–196.

SCHULTZE-WESTRUM, T. 1965. Innerartliche Verständigung durch Düfte beim Gleitbeutler *Petaurus breviceps papuanus* Thomas (Marsupialia, Phalangeridae). *Z. vergl. Physiol.*, **50**, 151–220.

SCOTT, J. P. 1962. Critical periods in behavioral development. *Science*, **138**, 949–958.

SCOTT, J. P. 1966. Agonistic behavior of mice and rats: a review. *Amer. Zool.*, **6**, 683–701.

SCOTT, J. P. & FULLER, J. L. 1965. *Genetics and the social behavior of the dog*. University of Chicago Press.

SEARLE, L. V. 1949. The organisation of hereditary maze-brightness and maze-dullness. *Genet. Psychol. Monogr.*, **39**, 279–325.

SEITZ, A. 1955. Untersuchungen über angeborene Verhaltensweisen bei Caniden III Beobachtungen an Marderhunden (*Nyctereutes procyonoides* Gray). *Z. Tierpsychol.*, **12**, 463–489.

SEITZ, A. 1959. Beobachtungen an handaufgezogenen Goldschakalen (*Canis aureus algirensis* Wagner 1843). *Ibid.*, **16**, 747–771.

SETON, E. T. 1920. Acrobatic skunks. *J. Mammal.*, **1**, 140.

SHADLE, A. R. 1946. Copulation in the porcupine. *J. Wildl Mgmt.* **10**, 159–162.

SHADLE, A. R. 1950. Feeding, care and handling of captive porcupines (*Erethizon*). *J. Mammal.*, **31**, 411–416.

SHADLE, A. R., SMELZER, M. & METZ, M. 1946. The sex reactions of porcupines (*Erethizon d. dorsatum*) before and after copulation. *Ibid.*, **27**, 116–121.

SHARMA, K. N., ANAND, B. K., DUA, S. & SINGH, B. 1961. Role of stomach in regulation of activities of hypothalamic feeding centers. *Amer. J. Physiol.*, **201**, 593–598.

SHARMAN, G. B. 1965. Marsupials and the evolution of viviparity. *Viewpoints in Biology*, No. 4. Butterworths, London.

SHARMAN, G. B. & CALABY, J. H. 1964. Reproductive behaviour in the red kangaroo, *Megaleia rufa*, in captivity. *C.S.I.R.O. Wildl. Res.*, **9**, 58–85.

SHARMAN, G. B. & PILTON, P. E. 1964. The life history and reproduction of the red kangaroo (*Megaleia rufa*). *Proc. zool. Soc. Lond.*, **142**, 29–48.

SHAW, W. T. 1934. The ability of the giant kangaroo rat as a harvester and storer of seeds. *J. Mammal.*, **15**, 275–286.

SHAW, W. T. 1936. Moisture and its relation to the cone-storing habit of the western pine squirrel. *Ibid.*, **17**, 337–349.

SHEPPE, W. 1966. Exploration by the deer mouse, *Peromyscus leucopus*. *Amer. midl. Nat.*, **76**, 257–276.

SHERRINGTON, C. S. 1906. *The integrative action of the nervous system.* Yale University Press, New Haven.

SHILLITO, E. E. 1963. Exploratory behaviour in the short-tailed vole, *Microtus agrestis*. *Behaviour*, **21**, 145–154.

SHILLITO, J. F. 1963. Field observations on the growth, reproduction and activity of a woodland population of the common shrew, *Sorex araneus* L. *Proc. zool. Soc. Lond.*, **140**, 99–114.
SHORTRIDGE, G. C. 1934. *The mammals of South West Africa.* Heinemann, London.
SIKES, S. 1962. Pangolins. *Afr. wild Life*, **16**, 275–281.
SIKES, S. 1964. The ratel or honey badger. *Ibid.* **18**, 29–37.
SIMONDS, P. E. 1965. The bonnet macaque in south India. *Primate behavior—field studies of monkeys and apes*, ed. I. DeVore. Holt, Rinehart & Winston, New York, 175–196.
SKINNER, B. F. 1966. The phylogeny and ontogeny of behavior. *Science*, **153**, 1205–1213.
SMITH, F. V., VAN-TOLLER, C. & BOYES, T. 1966. The 'critical period' in the attachment of lambs and ewes. *Anim. Behav.*, **14**, 120–125.
SNOW, C. J. 1967. Some observations on the behavioral and morphological development of coyote pups. *Amer. Zool.*, **7**, 353–355.
SOOTER, C. A. 1946. Mule deer attacks young Canada geese. *J. Mammal.*, **27**, 86.
SOUTHERN, H. N. 1948. Sexual and aggressive behaviour in the wild rabbit. *Behaviour*, **1**, 173–195.
SOUTHERN, H. N. & THOMPSON, H. V. 1964. *Handbook of British mammals*, ed. H. N. Southern. Blackwell, Oxford.
STEINBACHER, G. 1951. Nüsse offnender Sumpfichneumon. *Zool. Gart. Lpz.*, **18**, 58.
STEINIGER, F. 1950. Beiträge zur Soziologie und sonstigen Biologie der Wanderratte. *Z. Tierpsychol.*, **7**, 356–379.
STEVENSON-HAMILTON, J. 1947. *Wild life in South Africa.* Cassell & Co., London.
STODART, E. 1966. Management and behaviour of breeding groups of the marsupial *Perameles nasuta* Geoffroy in captivity. *Aust. J. Zool.*, **14**, 611–623.
SUTHERS, R. A. 1965. Acoustic orientation by fish-catching bats. *J. exp. Zool.*, **158**, 319–348.
TALBOT, L. M. & TALBOT, M. H. 1963. The wildebeest in western Masailand. *Wildl. Monogr., Chestertown*, No. 12, 1–88.
TEMBROCK, G. 1958. Spielverhalten beim Rotfuchs. *Zool. Beitr., Berl.*, **3**, 423–496.
TEMBROCK, G. 1959. Beobachtungen zur Fuchsranz unter besonderer Berücksichtigung der Lautgebung. *Z. Tierpsychol.*, **16**, 351–368.
TEMBROCK, G. 1963a. Acoustic behaviour of mammals. *Acoustic behaviour of animals*, ed. R. G. Busnel. Elsevier, New York, 751–786.
TEMBROCK, G. 1963b. Mischlaute beim Rotfuchs (*Vulpes vulpes* L.). *Z. Tierpsychol.*, **20**, 616–623.
TERKEL, J. & ROSENBLATT, J. S. (in press) *J. comp. physiol. Psychol.*
THOMSON, J. A. & PEARS, F. N. 1962. The functions of the anal glands of the brushtail possum. *Victoria Nat.*, **78**, 306–308.
THORPE, W. H. 1956. *Learning and instinct in animals.* Methuen, London.

TINBERGEN, N. 1952. "Derived" activities; their causation, biological significance, origin and emancipation during evolution. *Quart. Rev. Biol.*, **27**, 1–32.

TINBERGEN, N. 1962. The evolution of animal communication—a critical examination of methods. *Symp. zool. Soc. Lond.*, No. **8**, 1–6.

TINBERGEN, N. 1963. On aims and methods of ethology. *Z. Tierpsychol.*, **20**, 410–433.

TINBERGEN, N. 1964. The evolution of signalling devices. *Social behavior and organisation among vertebrates*, ed. W. Etkin. University of Chicago Press.

TINBERGEN, N. 1965a. Von der Vorratskammern des Rotfuchses (*Vulpes vulpes* L.). *Z. Tierpsychol.*, **22**, 119–149.

TINBERGEN, N. 1965b. *Animal behaviour*. Life Nature Library. Time Life Books, Amsterdam.

TREVIS, L. 1950. Summer behavior of a family of beavers in New York State. *J. Mammal.*, **31**, 40–65.

TRUMLER, E. 1959. Das "Rossigkeitsgesicht" und ähnliches Ausdrucksverhalten bei Einhufern. *Z. Tierpsychol.*, **16**, 478–488.

TRYON, R. C. 1942. Individual differences. *Comparative psychology*, ed. F. A. Moss. Prentice Hall, New York, 330–365.

VARLEY, M. & SYMMES, D. 1966. The hierarchy of dominance in a group of macaques. *Behaviour*, **27**, 54–75.

VENGE, O. 1963. The influence of nursing behaviour and milk production on early growth in rabbits. *Anim. Behav.*, **11**, 500–506.

VERHEYEN, R. 1954. Monographie éthologique de l'hippopotame (*Hippopotamus amphibius* Linné). *Exploration du Parc National Albert*, Inst. parcs nat. Congo Belge, Brussels, 1–91.

VERHEYEN, R. 1955. Contribution à l'éthologie du waterbuck *Kobus defassa ugandae* Neumann et de l'antilope harnachée *Tragelaphus scriptus* (Pallas). *Mammalia*, **19**, 309–319.

VOSSELER, J. 1929. Beitrag zur Kenntnis der Fossa (*Cryptoprocta ferox* Benn.) und ihrer Fortpflanzung. *Zool. Gart. Lpz.*, **2**, 1–9.

WADDINGTON, C. H. 1953. Genetic assimilation of an acquired character. *Evolution*, **7**, 118–126.

WALTHER, F. 1958. Zum Kampf- und Paarungsverhalten einiger Antilopen. *Z. Tierpsychol.*, **15**, 340–382.

WALTHER, F. 1960. "Antilopenhafte" Verhaltensweisen im Paarungszeremoniell des Okapi (*Okapia johnstoni* Sclater 1901). *Ibid.*, **17**, 188–210.

WALTHER, F. 1964a. Einige Verhaltungsbeobachtungen an Thomsongazellen (*Gazella thomsoni* Günther, 1884) im Ngorongoro-Krater. *Ibid.*, **21**, 871–890.

WALTHER, F. 1964b. Verhaltensstudien an der Gattung *Tragelaphus* De Blainville, 1816 in Gefangenschaft, unter besonderer Berücksichtigung des Socialverhaltens. *Ibid.*, **21**, 393–467 & 642–646.

WALTHER, F. 1965. Verhaltensstudien und der Grantgazelle (*Gazella granti* Brooke, 1872) im Ngorongoro-Krater. *Ibid.*, **22**, 167–208.

WALTHER. F. 1966. Zum Liegeverhalten des Weissschwanzgnus (*Connochaetes gnou* Zimmerman, 1780). *Z. Säugetierk.*, **31**, 1–16.

WANDELER, I. & PILLERI, G. 1965. Weitere Beobachtungen zum Verhalten von *Aplodontia rufa* Rafinesque (Rodentia, Aplodontoidea) in Gefangenschaft. *Z. Tierpsychol.*, **22**, 570–583.

WEBSTER, F. A. & GRIFFIN, D. R. 1962. The role of the flight membranes in insect capture by bats. *Anim. Behav.*, **10**, 332–340.

WICKLER, W. 1963. Die biologische Bedeutung auffallend farbiger nackter Hautstellen und innerartliche Mimikry der Primaten. *Naturwissenschaften*, **50**, 481–482.

WICKLER, W. 1964. Vom Gruppenleben einiger Säugetiere Afrikas. *Mitt. Max-Planck Ges.*, **6**, 296–309.

WICKLER, W. 1967. Socio-sexual signals and their intra-specific imitation among primates. *Primate ethology* ed. D. Morris, Weidenfeld & Nicolson, London, 69–147.

WIEPKEMA, P. R. 1963. Hyperphagia, aphagia and finickiness. *Anim. Behav.*, **11**, 609–610.

WIESEL, T. N. & HUBEL, D. H. 1965. Comparison of the effects of unilateral and bilateral eye closure on cortical unit responces in kittens. *J. Neurophysiol.*, **28**, 1029–1072.

WIGHT, H. M. 1931. Reproduction in the eastern skunk (*Mephitis mephitis nigra*). *J. Mammal.*, **12**, 42–47.

WOOLLEY, M. P. 1966. Reproduction in *Antechinus* spp. and other dasyurid marsupials. *Symp. zool. Soc. Lond.*, No. 15, 281–294.

WÜNSCHMANN, A. 1966. Einige Gefangenschaftsbeobachtungen am Breitstirn-Wombats (*Lasiorhinus latifrons* Owen 1845). *Z. Tierpsychol.* **23**, 56–71.

WÜSTEHUBE, C. 1960. Beiträge zur Kenntnis besonders des Spiel- und Beuteverhaltens einheimischer Musteliden. *Ibid.*, **17**, 578–613.

YOUNG, W. C. 1961. The hormones and mating behavior. *Sex and internal secretions*, ed. W. C. Young. Williams & Wilkins, Baltimore, 1173–1239.

ZANNIER, F. 1965. Verhaltensuntersuchungen an der Zwergmanguste, *Helogale undulata rufula* im Zoologischen Garten Frankfurt am Main. *Z. Tierpsychol.*, **22**, 672–695.

ZARROW, M. X., DENENBERG, V. H. & ANDERSON, C. O. 1965. Rabbit: frequency of suckling in the pup. *Science*, **150**, 1835–1836.

ZEEB, K. 1959. Die "Unterlegenheitsgebärde" des noch nicht ausgewachsenen Pferds (*Equus caballus*). *Z. Tierpsychol.*, **16**, 489–496.

AUTHOR INDEX

SPECIES AND SUBJECT INDEX

GPSR Compliance

The European Union's (EU) General Product Safety Regulation (GPSR) is a set of rules that requires consumer products to be safe and our obligations to ensure this.

If you have any concerns about our products, you can contact us on ProductSafety@springernature.com

In case Publisher is established outside the EU, the EU authorized representative is:

Springer Nature Customer Service Center GmbH
Europaplatz 3
69115 Heidelberg, Germany

Batch number: 10620169

Printed by Printforce, the Netherlands